Lectures on Quantum Mechanics and Attractors

Lectures on Quantum Mechanics and Attractors

Alexander Komech

Institute for Information Transmission Problems
of Russian Academy of Sciences, Russia

NEW JERSEY · LONDON · SINGAPORE · BEIJING · SHANGHAI · HONG KONG · TAIPEI · CHENNAI · TOKYO

Published by

World Scientific Publishing Co. Pte. Ltd.

5 Toh Tuck Link, Singapore 596224

USA office: 27 Warren Street, Suite 401-402, Hackensack, NJ 07601

UK office: 57 Shelton Street, Covent Garden, London WC2H 9HE

Library of Congress Control Number: 2022006347

British Library Cataloguing-in-Publication Data
A catalogue record for this book is available from the British Library.

LECTURES ON QUANTUM MECHANICS AND ATTRACTORS

ISBN 978-981-124-889-4 (hardcover)
ISBN 978-981-124-890-0 (ebook for institutions)
ISBN 978-981-124-891-7 (ebook for individuals)

For any available supplementary material, please visit
https://www.worldscientific.com/worldscibooks/10.1142/12602#t=suppl

To the memory of Vera Komech and Alexey Komech

Preface

The goal of this book is to give an introduction to Quantum Mechanics for mathematically-minded readers. We present in detail the basic physical applications and the required mathematical tools. However, we do not touch methods of contemporary quantum field theory, and we also do not touch topics such as quantum computing and quantum entanglement.

Many students and professionals in mathematics, chemistry, and physics are interested in learning the basic concepts and applications of quantum theory. However, many existing sources are insufficient for various reasons:

• Many theoretical arguments rely on subtle mathematical tools, such as the Hamiltonian formalism, short-wave asymptotics, classical electrodynamics, spectral theory, scattering and diffraction theory, limiting amplitude and limiting absorption principles, representation of Lie groups and algebras, and so on. At the same time, a serious discussion of these methods in lecture courses and textbooks is impossible due to the insufficient and non-uniform mathematical level of the audience.

• A transparent exposition of quantum theory requires detailed motivations of all concepts and a careful separation of definitions from assertions. However, they are rarely found in available manuals, while in professional physical works such discussions are considered indecent and superfluous.

• Finally, the theory includes certain provisions, which are still not clarified in the framework of the Schrödinger theory: Bohr's postulates, de Broglie's wave-particle duality, Born's probabilistic interpretation, and so on. As a rule, no lecturer or textbook cares to discuss these issues, although they are of cardinal interest, especially for mathematicians.

That is why we decided to break this vicious circle. We make this possible by relying on basic methods and results of the scattering theory and on recent progress in the theory of attractors of nonlinear Hamiltonian PDEs.

In Chapters I–IV we present nonrelativistic and relativistic quantum mechanics. We restrict ourselves with basic principles and applications which clarify these principles. We carefully separate definitions from assertions.

We do not try to formalize the standard mathematical details (this can be easily done by the reader) so as not to overburden the exposition. In particular, we do not specify the domains of self adjoint operators and do not detail the assumptions on smoothness and decay of potentials and initial data.

We discuss some open questions in Chapter V on the basis of a new *conjecture* on global attractors of nonlinear Hamiltonian PDEs in the context of the coupled Maxwell–Schrödinger equations. In Chapter VI, we give a brief survey of rigorous results obtained during 1990–2020 which justify this conjecture for a number of model Hamiltonian PDEs.

We do not aim to present recent advances in quantum theory; our goal is to give a streamlined presentation of its physical and mathematical principles. We separate i) genesis of basic mathematical concepts and equations from empirical facts; ii) formulation of mathematical problems describing physical phenomena; iii) formal (although sometimes not totally rigorous) mathematical solution of arising problems.

The fundamental differences between our presentation and the existing ones are as follows:

i) we always give a deep and meaningful motivation for introduced concepts and constructions;

ii) we describe physical applications involving the straightforward calculations and full explanations of all mathematical methods used;

iii) we do not just talk about landmark experiments; instead, we present their detailed theory, thus making the interplay between physics and mathematics as clear as possible.

All calculations are explained in detail. Sometimes we give exact references, including pages and formulae, where certain technical details are fully provided (in particular, we frequently refer to [51] for routine calculations). This allows us not to overload the exposition.

The book contains numerous problems and exercises that clarify and complete the main exposition, and makes the book suitable for individual study.

The book is intended

i) for graduate and advanced undergraduate students in mathematics, physics, and chemistry, studying quantum mechanics;

ii) for lecturers in quantum mechanics;

iii) for engineers working with lasers and masers, klystrons and magnetrons, traveling-wave tubes, synchrotrons, electronic microscopes, superconductors, and semiconductors;

iv) for all researchers working in mathematical physics;

v) for any mathematically-minded reader who is interested in the subject.

Prerequisites

Calculus and linear algebra (exponent of a matrix); variational derivatives; distributions; Fourier transform, PDE with constant coefficients (d'Alembert formula, solution of the Poisson equation); Lagrangian and Hamiltonian formalism; Maxwell equations; definitions of Lie group and Lie algebra and of their representations.

Only Sections I.7.1, I.5, and VI.4 require some additional knowledge:

Hilbert space, the Hilbert–Schmidt spectral theorem for compact self adjoint operators, the weak compactness of the unit ball in Hilbert space, complex analysis, Sobolev spaces, Parseval theorem for the Fourier transform.

Nothing is required from the theory of attractors and dynamical systems.

Further reading

Our main goal is to give a concise explanation of the basic physical concepts and of the mathematical principles of quantum mechanics. More technical details and a systematic comparison with experimental data can be found in [5, 7, 8, 11, 14, 18, 31, 36, 41, 58, 60, 66, 67, 70, 74, 77, 82, 85, 92]. The books [48, 88] and [24, 69], respectively, present classical electrodynamics and the basic concepts of quantum mechanics. An introduction to the modern mathematical methods of quantum theory can be found in [6, 40, 45].

We develop the methods of quantum mechanics for the hydrogen atom, and we do not touch multi-electron problems of quantum chemistry [18, 21, 84, 89]. We also do not touch stability of matter [15, 62], quantum electrodynamics, and quantum field theory [9, 16, 23, 25, 30, 38, 64, 65, 75, 76, 80, 91].

Acknowledgments. The author thanks Herbert Spohn for long-term collaboration and Alexander Shnirelman for many useful discussions, and Andrew Comech for numerous improvements and corrections in the manuscript.

The author is indebted to Moscow State University, Munich Technical University, INRIA (Rocquencourt, France), Max-Planck Institute for Mathematics (Leipzig), Morelia Institute for Physics and Mathematics (University of Michoacan, Mexico), and Vienna University for hospitality and for support of this work during many years.

The author was supported by the Alexander von Humboldt Foundation, the Austrian Science Fund (FWF), the German Research Foundation (DFG), and the Russian Foundation for Basic Research (RFBR).

Alexander Komech
Moscow–Munich–Paris–Leipzig–Morelia–Vienna

Contents

Introduction

The goal of this book is to give an introduction to quantum mechanics for mathematically-minded readers and an exposition of the interplay of empirical facts with related mathematics. We present nonrelativistic and relativistic quantum mechanics relying on the Schrödinger, Pauli and Dirac equations. In this book, we do not prove any new mathematical results.

We carefully discuss the key points, such as

a) introduction of the Schrödinger, Pauli and Dirac equations;

b) introduction of quantum observables, including spin;

c) rotational covariance of the Pauli equation, relativistic covariance of the Dirac equation and its nonrelativistic approximations;

d) construction of spherical functions in the nonrelativistic and relativistic context, calculation of irreducible representations for the corresponding Lie algebras, and the proof of the Clebsch–Gordan theorem in a special case.

We calculate in detail

i) the spectrum of the hydrogen atom in the framework of the Schrödinger and Dirac equations;

ii) the scattering of light and of charged particles;

iii) the normal and anomalous Zeeman splitting of spectral lines in a magnetic field;

iv) the angular distribution of the photoelectric current;

v) the diffraction amplitude for scattering of electron beams.

The outline of this book is as follows.

In Chapter I, we present de Broglie's deduction of wavelength formula and justify introduction of the Schrödinger equation relying on short-wave WKB-asymptotics. We establish basic conservation laws and justify Bohr's postulates via perturbation theory. Further, we introduce the semiclassical coupled nonlinear Maxwell–Schrödinger equations [75, 129, 130, 131, 132]. Finally, we calculate the spectrum of the hydrogen atom in the framework of the Schrödinger equation.

In Chapter II, we calculate scattering of light and particles. We relate Einstein's rules for the photoelectric effect with the limiting amplitude principle in a continuous spectrum, and calculate the Kramers–Kronig formula for polarization. Further, we calculate the diffraction amplitude of the electron beam in the Kirchhoff approximation. We check that the amplitude agrees with recent experimental data of R. Bach *et al.* for double-slit diffraction [108].

Chapter III deals with the Pauli theory of spin. A simple proof of the rotational covariance of the Pauli equation with respect to the action of the $SO(3)$ group is given, as well as relevant conservation laws. The Pauli equation is applied to the calculation of the anomalous Zeeman effect.

In Chapter IV, we present Dirac's relativistic quantum mechanics, establish its Lorentz covariance and the corresponding conservation laws. We calculate the nonrelativistic limits and the spectrum of the hydrogen atom in this framework.

In Chapter V, we formulate a *conjecture on attractors* for generic G-invariant nonlinear Hamiltonian partial differential equations. We discuss possible application of this conjecture to dynamical interpretation of basic postulates of quantum theory in the framework of the coupled Maxwell–Schrödinger equations.

In Chapter VI, we survey the results obtained since 1990 that justify the conjecture on attractors for a list of model Hamiltonian nonlinear PDEs. Besides, we describe the results of numerical simulation of relativistic nonlinear PDEs obtained jointly with Arkadii Vinnichenko (1945–2009) which also justify the conjecture.

In Appendix A, we present an updated version of the Old Quantum Theory, which is necessary for understanding the emergence of the Schrödinger theory and for the introduction of the electron spin. Finally, in Appendices B and C, we present the Noether theory of invariants and the perturbation formula for eigenvalues, respectively.

All derivatives throughout the book are understood in the sense of distributions.

Notation

The components of the coordinate vectors are denoted by superscripts: $\mathbf{x} = (x^1, x^2, x^3) \in \mathbb{R}^3$. The components of the "dual" vectors, the wave vector and the momentum, are denoted by subscripts:

$$\mathbf{k} = (k_1, k_2, k_3) \in \mathbb{R}^3, \qquad \mathbf{p} = (p_1, p_2, p_3) \in \mathbb{R}^3.$$

For three-component vectors $\mathbf{u} = (u^1, u^2, u^3)$ and $\mathbf{v} = (v^1, v^2, v^3)$, the inner product and cross product are defined respectively as

$$\mathbf{u} \cdot \mathbf{v} = \sum_{k=1}^{3} u^k v^k, \quad \mathbf{u} \times \mathbf{v} = \left(\det \begin{pmatrix} u^2 & u^3 \\ v^2 & v^3 \end{pmatrix}, -\det \begin{pmatrix} u^1 & u^3 \\ v^1 & v^3 \end{pmatrix}, \det \begin{pmatrix} u^1 & u^2 \\ v^1 & v^2 \end{pmatrix}\right),$$

including the cases when the components are complex-valued, matrix-valued, and operator-valued.

For a matrix $a \in \mathbb{C}^{m \times n}$, $a^* \in \mathbb{C}^{n \times m}$ denotes its Hermitian conjugate. In particular, for

$$\psi = \begin{pmatrix} \psi_1 \\ \psi_2 \\ \psi_3 \\ \psi_4 \end{pmatrix} \in \mathbb{C}^4, \qquad \varphi = \begin{pmatrix} \varphi_1 \\ \varphi_2 \\ \varphi_3 \\ \varphi_4 \end{pmatrix} \in \mathbb{C}^4,$$

one has

$$\psi^* = (\overline{\psi}_1, \overline{\psi}_2, \overline{\psi}_3, \overline{\psi}_4), \qquad \psi^* \varphi = \overline{\psi}_1 \varphi_1 + \overline{\psi}_2 \varphi_2 + \overline{\psi}_3 \varphi_3 + \overline{\psi}_4 \varphi_4.$$

Following Wigner [94], we write $\boldsymbol{K}$ for the (antilinear) operator of componentwise complex conjugation,

$$\boldsymbol{K} : \mathbb{C}^{m \times n} \to \mathbb{C}^{m \times n}.$$

That is, for a matrix $a \in \mathbb{C}^{m \times n}$, we have $(\boldsymbol{K}a)_{ij} = \overline{a_{ij}}$.

For a function $\psi(x, t)$ of $(x, t) \in \mathbb{R} \times \mathbb{R}$,

$$\psi'(x, t) = \partial_x \psi(x, t), \qquad \dot{\psi}(x, t) = \partial_t \psi(x, t).$$

For a function $\mathcal{H}(\mathbf{x}, \mathbf{p})$ of $\mathbf{x}, \mathbf{p} \in \mathbb{R}^3$, we denote by $\mathcal{H}_{\mathbf{x}}(\mathbf{x}, \mathbf{p})$ and $\mathcal{H}_{\mathbf{p}}(\mathbf{x}, \mathbf{p})$ the corresponding derivatives. For a function $\psi(\mathbf{x})$ of $\mathbf{x} \in \mathbb{R}^3$, we denote

$$\nabla \psi(\mathbf{x}) = (\partial_{x^1} \psi(\mathbf{x}), \partial_{x^2} \psi(\mathbf{x}), \partial_{x^3} \psi(\mathbf{x})), \quad \partial_{x^k} \psi(\mathbf{x}) = \frac{\partial \psi(\mathbf{x})}{\partial x^k}, \quad 1 \le k \le 3.$$

For a functional $\mathcal{H}(\Psi)$ on a Hilbert space, $\mathcal{H}_{\Psi}(\Psi)$ denotes the corresponding variational derivative.

The Hermitian (sesquilinear) form $\langle , \rangle$ and the bilinear form $\langle\!\langle , \rangle\!\rangle$ on $L^2(\mathbb{R}^d)$ are defined by

$$\langle f, g \rangle = \int \overline{f(\mathbf{x})} g(\mathbf{x}) \, d\mathbf{x}, \qquad \langle\!\langle f, g \rangle\!\rangle = \int f(\mathbf{x}) g(\mathbf{x}) \, d\mathbf{x}.$$

We use the same notation for their different extensions (for example, f can be a distribution when g is a smooth function).

Chapter I

Nonrelativistic Quantum Mechanics

In this chapter, we introduce the Schrödinger equation. The introduction relies on de Broglie's wave-particle duality and is justified by quasiclassical asymptotics and the Hamilton–Jacobi equation. Introduction of quantum observables is justified by the correspondence principle.

Bohr's postulates are justified by perturbation procedure applied to the coupled nonlinear Maxwell–Schrödinger equations.

The calculation of the hydrogen spectrum relies on the construction of representations of the Lie algebra of the rotation group $SO(3)$.

I.1 Photons and Wave-Particle Duality

The emergence of Schrödinger theory and Pauli spin theory was prepared by the entire development of the *Old Quantum Theory* in 1890–1925; this is described in Appendix A. A decisive impetus for the appearance of Schrödinger's theory came from the Einstein theory of photoeffect and de Broglie's theory of wave-particle duality.

I.1.1 Planck's law and Einstein's photons

The key role in the emergence of quantum theory was played by the investigation of the *spectral intensity* of radiation which is in thermodynamic equilibrium with matter at some fixed temperature $T > 0$. The study of this spectral intensity was one of the main directions of physics in XIX century since R. Bunsen and G. Kirchhoff's experiments during 1859–1862; see details in Appendix A.

In 1900, M. Planck stated his famous formula for this intensity that earned him the 1918 Nobel Prize. Now this formula is known as the *Kirchhoff–Planck law* or the *black-body radiation law*: in the Gaussian units,

$$I_T(\omega) = \frac{\hbar\omega^3}{\pi^2 c^3} \frac{e^{-\frac{\hbar\omega}{kT}}}{1 - e^{-\frac{\hbar\omega}{kT}}}, \tag{I.1.1}$$

where $\hbar$ is the famous *Planck constant*; see details in [51, Section 1.3.4]. The introduction of this constant by M. Planck symbolizes the emergence of quantum theory. In the *unrationalized* Gaussian units (also called the Heaviside–Lorentz units [93, p. 221], [48, p. 781], [49]), the physical constants — the electron charge and mass, the speed of light in vacuum, the Boltzmann constant, and the Planck constant — have the following approximate values:

$$e = -4.8 \times 10^{-10} \text{esu}, \quad m = 9.1 \times 10^{-28} \text{g}, \quad c = 3.0 \times 10^{10} \text{cm/s}$$

$$k = 1.38 \times 10^{-16} \text{erg/K}, \qquad \hbar = 1.1 \times 10^{-27} \text{erg} \cdot \text{s} \tag{I.1.2}$$

Planck's derivation of the formula (I.1.1) relied on rather delicate thermodynamic arguments [83, Section 20]. In 1905, A. Einstein proposed a new derivation of the formula introducing the discretization with the step

$$\varepsilon = \hbar\omega > 0 \tag{I.1.3}$$

for allowed energies of waves with frequency ω; see [51, Section 1.3.4] for details. At the same time, Einstein proposed to identify this portion of energy with the energy of a *photon* (hypothetical "particle of light") to explain the *photoelectric effect* (this earned him the 1921 Nobel Prize). Indeed, the experimental observations showed that light of frequency ω, when falling onto a metallic surface, knocks electrons out of the metal, and the *maximal kinetic energy* of the "photoelectrons" is

$$K_{\max} = \hbar\omega - W, \tag{I.1.4}$$

where W is the *work function* of the metal [51, Section 8.4]. To explain these *empirical facts*, Einstein suggested identifying the light wave with a beam of particles:

light wave $\psi(\mathbf{x}, t) = Ce^{i(\mathbf{k} \cdot \mathbf{x} - \omega t)} \Leftrightarrow$ *beam of photons with energy* $E = \hbar\omega$, (I.1.5)

and he treated the formula (I.1.4) as the energy balance in the collision of the photons with an electron in the metal:

the electron leaves the metal losing the energy W
to overcome the attraction to the metal.

Remark I.1.1. The Kirchhoff–Planck law (I.1.1) is of great practical significance. In particular, it is the basis for design and tuning of the electronic devices which measure body temperature by its spectrum (for example, in night-vision devices).

I.1.2 De Broglie's wave-particle duality

The energy E and momentum $\mathbf{p}$ of free *nonrelativistic particles* (with velocities $|\mathbf{v}| \ll c$) and *relativistic particles* (with velocities $|\mathbf{v}| \sim c$) are related by the formulae

$$E = \frac{\mathbf{p}^2}{2\mathrm{m}} \qquad \text{and} \qquad \frac{E^2}{c^2} = \mathbf{p}^2 + \mathrm{m}^2 c^2, \tag{I.1.6}$$

respectively, where c is the speed of light in vacuum, and m is the rest mass of the particle (for photons, m = 0). The former formula is classical. The latter formula follows directly from Einstein's Special Relativity Principle; see [35], [48, Chapter 11] and [51, Sections 12.2 and 12.3] for details. Indeed, this principle prescribes that the Lagrangian action for the free particle is

$$S = A \int \sqrt{(c\,dt)^2 - (d\mathbf{x})^2} = A \int \sqrt{c^2 - \dot{\mathbf{x}}^2}\,dt, \tag{I.1.7}$$

with an appropriate constant A. This action is *Lorentz-invariant*, since it depends only on the Lorentz interval

$$(c\,dt)^2 - (d\mathbf{x})^2. \tag{I.1.8}$$

The corresponding Lagrangian is

$$\Lambda = A\sqrt{c^2 - \dot{\mathbf{x}}^2}.$$

Then the conjugate momentum is

$$\mathbf{p} := \Lambda_{\dot{\mathbf{x}}} = -A\dot{\mathbf{x}}/\sqrt{c^2 - \dot{\mathbf{x}}^2},$$

and the Hamiltonian (energy) is given by the Legendre transformation

$$E = \mathbf{p} \cdot \dot{\mathbf{x}} - \Lambda = -\frac{Ac}{\sqrt{1 - \dot{\mathbf{x}}^2/c^2}} = -Ac \left[1 + \frac{\dot{\mathbf{x}}^2}{2c^2} + \mathcal{O}\left(\frac{|\dot{\mathbf{x}}|^4}{c^4} \right) \right], \quad |\dot{\mathbf{x}}| \ll c. \quad (I.1.9)$$

Finally, the constant is

$$A = -mc, \qquad (I.1.10)$$

since the energy for small velocities $\dot{\mathbf{x}}$ must coincide with the one given by the first formula (I.1.6) up to an additive constant. This gives

$$E = \frac{mc^2}{\sqrt{1 - \dot{\mathbf{x}}^2/c^2}}, \qquad \mathbf{p} = \frac{m\dot{\mathbf{x}}}{\sqrt{1 - \dot{\mathbf{x}}^2/c^2}}, \qquad (I.1.11)$$

which implies the second relation of (I.1.6).

Exercise I.1.2. Check (I.1.10).

Exercise I.1.3. Check that (I.1.11) implies the second relation from (I.1.6).

In particular, the second formula (I.1.6) gives the energy of a relativistic particle at rest, i.e. when $\dot{\mathbf{x}} = 0$ and $\mathbf{p} = 0$, as follows:

$$E_0 = mc^2. \qquad (I.1.12)$$

This is the famous Einstein's *mass-energy equivalence* which predicted the enormous possibilities of nuclear energy.

In 1923, L. de Broglie suggested a wave theory of matter as an *antithesis* of Einstein's corpuscular theory of light (I.1.5). The main postulate of de Broglie's PhD Thesis was the following one:

$$\left. \begin{array}{l} a \ beam \ of \ free \ \text{relativistic particles} \ with \ momentum \ \mathbf{p} \ and \ energy \\ E \ can \ be \ described \ by \ the \ wave \ function \ \psi(\mathbf{x}, t) = Ce^{i(\mathbf{k} \cdot \mathbf{x} - \omega t)} \end{array} \right|, \quad (I.1.13)$$

where the 4-vector $(\mathbf{k}, \omega)$ is a certain function of the vector $(\mathbf{p}, E)$. The key idea of de Broglie was that

i) this correspondence must be relativistically covariant, and

ii) $\mathbf{k} \cdot \mathbf{x} - \omega t$ is a Lorentz-invariant scalar product.

The basic postulate of Einstein's Special Relativity Principle is that the vectors

$$(\mathbf{x}, \tau) := (\mathbf{x}, ct)$$

are transformed by Lorentz matrices, which are 4×4-matrices preserving the quadratic form $\tau^2 - \mathbf{x}^2$ (the Lorentz interval).

De Broglie postulated that the phase function $\mathbf{k} \cdot \mathbf{x} - \omega t$ of the wave (I.1.13) must be Lorentz-invariant, and this suggests that the vector $(\mathbf{k}, \omega/c)$ will also be transformed by the same Lorentz matrix. On the other hand, the second relation (I.1.6) suggests that the vector $(\mathbf{p}, E/c)$ also transforms by the Lorentz matrix, since its Lorentz interval $E^2/c^2 - \mathbf{p}^2 = m^2 c^2$ is identical in all frames.

These arguments suggest that the four-dimensional vectors, $(\mathbf{p}, E/c)$ and $(\mathbf{k}, \omega/c)$, must be proportional:

$$(\mathbf{p}, E/c) = \mathrm{const}\,(\mathbf{k}, \omega/c). \qquad (I.1.14)$$

Finally, the Einstein identification (I.1.5) gives $E = \hbar\omega$, so (I.1.14) becomes

$$(\mathbf{p}, E) = \hbar(\mathbf{k}, \omega). \qquad (I.1.15)$$

This relation was experimentally confirmed by C. Davisson and L. Germer in 1924–1927 for the diffraction of electrons (see Section II.7 below). In particular, *de Broglie's wavelength* is given by

$$\lambda = \frac{2\pi}{|\mathbf{k}|} = \frac{2\pi\hbar}{|\mathbf{p}|}. \qquad (I.1.16)$$

This great discovery of de Broglie plays a key role in nuclear physics for calculating the energy and momentum of neutrons and other elementary particles in terms of the wavelength, which is measured via their diffraction.

Problem I.1.4. Deduce (I.1.14) from the Lorentz-invariance of the inner product $\mathbf{p} \cdot \mathbf{k} - E\omega/c^2$. **Hint:** construct the Lorentz transformation Λ such that $\Lambda(\mathbf{p}, E/c) = (0, 0, 0, m)$ (see [51, pp. 36–37]).

Now (I.1.15) and the second relation in (I.1.6) give

$$\frac{\hbar^2 \omega^2}{c^2} = \mathbf{p}^2 + \mathrm{m}^2 c^2 = \hbar^2 \mathbf{k}^2 + \mathrm{m}^2 c^2. \qquad (I.1.17)$$

In the case $\mathrm{m} = 0$, we obtain $|\mathbf{k}| = \omega/c$, and hence formula (I.1.16) becomes

$$\lambda = 2\pi c/\omega, \qquad (I.1.18)$$

which is well known for light. Hence, the particles of light ("photons") have zero rest mass. In the nonrelativistic limit $\dot{\mathbf{x}} \to 0$, (I.1.17) reads similarly to (I.1.9):

$$\hbar\omega = \mathrm{m}c^2 + \frac{\mathbf{p}^2}{2\mathrm{m}} + \mathcal{O}\left(\frac{|\mathbf{p}|^4}{c^4}\right), \qquad |\mathbf{p}| \ll \mathrm{m}c. \qquad (I.1.19)$$

In other words, the difference $E - \mathrm{m}c^2$ approximately reads as the *nonrelativistic energy*:

$$E - \mathrm{m}c^2 \approx E_* = \frac{\mathbf{p}^2}{2\mathrm{m}}, \qquad |\mathbf{p}| \ll \mathrm{m}c. \qquad (I.1.20)$$

Exercise I.1.5. Check (I.1.19).

Exercise I.1.6. Calculate de Broglie's wavelength of the electron beam in a television tube with voltage $U \approx 10\,\mathrm{KV}$. **Hints:** i) By (I.1.12) and (I.1.2), the electron energy at rest is

$$E_0 := \mathrm{m}c^2 \approx 0.51\,\mathrm{MeV}.$$

Hence, the electron energy after acceleration is $E = E_0 + eU \approx 0.52\,\mathrm{MeV}$, provided that the electron was initially at rest. Now the second formula of (I.1.6) gives

$$\mathbf{p}^2 c^2 = E^2 - E_0^2 \approx (52^2 - 51^2) \times 10^8 (\mathrm{e} \cdot \mathrm{V})^2 \approx 10^{10} (\mathrm{e} \cdot \mathrm{V})^2. \qquad (\mathrm{I.1.21})$$

ii) Using $\mathrm{e} \cdot \mathrm{V} \approx 1.6 \times 10^{-12}\,\mathrm{erg}$, one has

$$|\mathbf{p}|\,c \approx 10^5 \cdot 1.6 \times 10^{-12}\mathrm{erg} = 1.6 \times 10^{-7}\mathrm{erg}. \qquad (\mathrm{I.1.22})$$

Taking $c \approx 3 \times 10^{10}\,\mathrm{cm/s}$, we see that that $|\mathbf{p}| \approx 0.5 \times 10^{-17}\,\mathrm{g} \cdot \mathrm{cm/s}$. Now, (I.1.16) together with the value for $\hbar$ from (I.1.2) gives

$$\lambda \approx \frac{6 \cdot 10^{-27}}{0.5 \cdot 10^{-17}}\,\mathrm{cm} \approx 10^{-9}\mathrm{cm} = 0.1\,\text{Å}, \qquad (\mathrm{I.1.23})$$

where $1\,\text{Å} = 10^{-8}\,\mathrm{cm}$ is one Ångström, which is about the size of the hydrogen atom.

For comparison, the wavelengths of visible light, which was measured by Young around 1802, when he observed the interference of light in thin films, are as follows:

$$\begin{array}{ll}
\lambda = 3800\,\text{Å} & \text{for violet light;} \\
\lambda = 4200\,\text{Å} & \text{for indigo light;} \\
\lambda = 4400\,\text{Å} & \text{for blue light;} \\
\lambda = 5000\,\text{Å} & \text{for cyan light;} \\
\lambda = 5200\,\text{Å} & \text{for green light;} \\
\lambda = 5650\,\text{Å} & \text{for yellow light;} \\
\lambda = 5900\,\text{Å} & \text{for orange light;} \\
\lambda = 6250\,\text{Å} & \text{for red light.}
\end{array}$$

These wavelengths explain why the scattering of light by droplets of fog (of size $10^{-3}\,\mathrm{mm} = 10^4\,\text{Å}$) or by dust in the atmosphere (of size $10^{-2}\,\mathrm{mm} = 10^5\,\text{Å}$) or by bacteria (of size $10^{-2}\,\mathrm{mm} = 10^5\,\text{Å}$) is easily observable.

On the other hand, for much smaller particles, like viruses (of size $10^2\,\text{Å}$), the scattering of visible light is negligible. Such smaller particles are visualized by using electron beams with wavelength $\lambda \ll 10^2\,\text{Å}$; see, e.g., (I.1.23). This explains the high efficiency of electron microscopes.

I.2 The Schrödinger Equation

De Broglie's wave-particle theory of 1923 assigns a wave function to free particles. The next step towards the new quantum mechanics was made by E. Schrödinger in 1925–1926: he introduced a new wave equation for electrons in the presence of external Maxwell field. In this section, we present the Schrödinger equation on the basis of canonical quantization suggested by de Broglie's wave-particle duality, and also justify this equation by short-wave asymptotics.

I.2.1 Canonical quantization

Free particles. In the case of free relativistic particles, de Broglie's relation (I.1.15) implies that the wave function $\psi(\mathbf{x}, t) = Ce^{i(\mathbf{k}\cdot\mathbf{x} - \omega t)}$ satisfies the identities

$$i\hbar\partial_t\psi(\mathbf{x}, t) = E\psi(\mathbf{x}, t), \qquad -i\hbar\nabla\psi(\mathbf{x}, t) = \mathbf{p}\psi(\mathbf{x}, t). \qquad (I.2.1)$$

Hence, the second formula (I.1.17) implies the free Klein–Gordon equation

$$\frac{1}{c^2}[i\hbar\partial_t]^2\psi(\mathbf{x}, t) = [(-i\hbar\nabla)^2 + m^2c^2]\psi(\mathbf{x}, t). \qquad (I.2.2)$$

In the nonrelativistic limit (I.1.20), the relations (I.2.1) lead to the free Schrödinger equation

$$i\hbar\partial_t\psi(\mathbf{x}, t) = -\frac{\hbar^2}{2m}\Delta\psi(\mathbf{x}, t) + mc^2\psi(\mathbf{x}, t). \qquad (I.2.3)$$

Here the last term can be eliminated by the *gauge transform*

$$\psi(\mathbf{x}, t) \mapsto \psi_*(\mathbf{x}, t) := e^{\frac{i}{\hbar}mc^2t}\psi(\mathbf{x}, t). \qquad (I.2.4)$$

This replacement results in the standard form of the free Schrödinger equation

$$i\hbar\partial_t\psi_*(\mathbf{x}, t) = -\frac{\hbar^2}{2m}\Delta\psi_*(\mathbf{x}, t). \qquad (I.2.5)$$

Mnemonically, equations (I.2.5) and (I.2.2) are derived from the energy-momentum relations (I.1.6) by the replacements

$$E \mapsto \hat{E} := i\hbar\partial_t, \qquad \mathbf{p} \mapsto \hat{\mathbf{p}} := -i\hbar\nabla, \qquad (I.2.6)$$

known as *canonical quantization*. The relation (I.1.20) implies that for de Broglie's wave function $\psi(\mathbf{x}, t) = Ce^{i(\mathbf{k}\cdot\mathbf{x} - \omega t)}$, the corresponding *nonrelativistic wave function* (I.2.4) reads as

$$\psi_*(\mathbf{x}, t) = Ce^{i(\mathbf{k}\cdot\mathbf{x} - \omega_* t)}, \qquad \omega_* = \omega - \frac{mc^2}{\hbar} \approx E_*/\hbar, \qquad \hbar|\mathbf{k}| \ll mc, \quad (I.2.7)$$

where $E_* = \dfrac{\mathbf{p}^2}{2\mathrm{m}}$ is the nonrelativistic energy. This means that the wave vector $(\mathbf{k}, \omega_*)$ of the wave function $\psi_*(\mathbf{x}, t)$ approximately satisfies the relation of type (I.1.15):

$$(\mathbf{p}, E_*) \approx \hbar(\mathbf{k}, \omega_*), \qquad |\mathbf{p}| \ll \mathrm{m}c. \tag{I.2.8}$$

This relation extends de Broglie's duality (I.1.13) to *nonrelativistic particles*.

Bound particles. Now let us consider *electrons* with charge e in the presence of an *external Maxwell field* with magnetic vector potential

$$\mathbf{A}^{\mathrm{ext}}(\mathbf{x}, t) = \big(A_1^{\mathrm{ext}}(\mathbf{x}, t), A_2^{\mathrm{ext}}(\mathbf{x}, t), A_3^{\mathrm{ext}}(\mathbf{x}, t)\big)$$

and a scalar potential $A_0^{\mathrm{ext}}(\mathbf{x}, t)$. The corresponding electric and magnetic fields are expressed in these Maxwell potentials as follows:

$$\mathbf{E}^{\mathrm{ext}}(\mathbf{x}, t) = -\frac{1}{c}\dot{\mathbf{A}}^{\mathrm{ext}}(\mathbf{x}, t) - \nabla A_0^{\mathrm{ext}}(\mathbf{x}, t), \quad \mathbf{B}^{\mathrm{ext}}(\mathbf{x}, t) = \mathrm{curl}\ \mathbf{A}^{\mathrm{ext}}(\mathbf{x}, t), \tag{I.2.9}$$

as explained in (I.4.15) below. We will assume the *Coulomb gauge*

$$\mathrm{div}\,\mathbf{A}^{\mathrm{ext}}(\mathbf{x}, t) \equiv 0. \tag{I.2.10}$$

The motion of a classical charged particle with charge e in the case of small velocities $|\dot{\mathbf{x}}(t)| \ll c$ in the Maxwell field is governed by the nonrelativistic Lorentz equation. In the Gaussian and Heaviside–Lorentz units [48, p. 781], [49], the equation reads as

$$\mathrm{m}\ddot{\mathbf{x}}(t) = e\left[\mathbf{E}^{\mathrm{ext}}(\mathbf{x}(t), t) + \frac{1}{c}\dot{\mathbf{x}}(t) \times \mathbf{B}^{\mathrm{ext}}(\mathbf{x}(t), t)\right], \tag{I.2.11}$$

where $\mathrm{m} > 0$ is the mass of the electron. The validity of this equation for electrons was confirmed by the fundamental experiments of J.J. Thomson in 1893–1897. This equation is equivalent to the Hamiltonian system

$$\left\{ \begin{array}{rl} \dot{\mathbf{x}}(t) &= \mathcal{H}_{\mathbf{p}}(\mathbf{x}, \mathbf{p}, t) = \dfrac{1}{\mathrm{m}}\left[\mathbf{p} - \dfrac{e}{c}\mathbf{A}^{\mathrm{ext}}(\mathbf{x}, t)\right] \\[2ex] \dot{\mathbf{p}}(t) &= -\mathcal{H}_{\mathbf{x}}(\mathbf{x}, \mathbf{p}, t) \\[2ex] &= -e\nabla A_0^{\mathrm{ext}}(\mathbf{x}, t) + \dfrac{e}{\mathrm{m}c}\left[\mathbf{p} - \dfrac{e}{c}\mathbf{A}^{\mathrm{ext}}(\mathbf{x}, t)\right] \cdot \nabla \mathbf{A}^{\mathrm{ext}}(\mathbf{x}, t) \end{array} \right. \tag{I.2.12}$$

Here the Hamiltonian function (the energy) is given by the formula (8.35) of [35] (see also [51, (12.90)]). In the Gaussian and Heaviside–Lorentz units [48, p. 781], [49], this Hamiltonian reads as

$$E = \mathcal{H}(\mathbf{x}, \mathbf{p}, t) = \frac{1}{2\mathrm{m}}\left[\mathbf{p} - \frac{e}{c}\mathbf{A}^{\mathrm{ext}}(\mathbf{x}, t)\right]^2 + eA_0^{\mathrm{ext}}(\mathbf{x}, t). \tag{I.2.13}$$

In this case, the canonical quantization (I.2.6) leads to the Schrödinger equation

$$i\hbar\dot\psi(\mathbf{x}, t) = H(t)\psi(t)$$

$$:= \frac{1}{2m}\left[-i\hbar\nabla - \frac{e}{c}\mathbf{A}^{\text{ext}}(\mathbf{x}, t)\right]^2 \psi(\mathbf{x}, t) + eA_0^{\text{ext}}(\mathbf{x}, t)\psi(\mathbf{x}, t). \quad (\text{I.2.14})$$

For relativistic electrons, the momentum is given by (I.1.11) and the Lorentz equation now reads

$$\dot{\mathbf{p}}(t) = e\left[\mathbf{E}^{\text{ext}}(\mathbf{x}(t), t) + \frac{1}{c}\dot{\mathbf{x}}(t) \times \mathbf{B}^{\text{ext}}(\mathbf{x}(t), t)\right]. \quad (\text{I.2.15})$$

This equation can also be written in the Hamiltonian form

$$\dot{\mathbf{x}}(t) = \mathcal{H}_{\mathbf{p}}(\mathbf{x}, \mathbf{p}, t), \qquad \dot{\mathbf{p}}(t) = -\mathcal{H}_{\mathbf{x}}(\mathbf{x}, \mathbf{p}, t), \quad (\text{I.2.16})$$

with the Hamiltonian function $\mathcal{H}$ defined by the formula (8.56) of [35] (see also [51, (12.93)]). In the Gaussian and Heaviside–Lorentz units, this Hamiltonian reads as

$$\left[\frac{E}{c} - \frac{e}{c}A_0^{\text{ext}}(\mathbf{x}, t)\right]^2 = \left[\mathbf{p} - \frac{e}{c}\mathbf{A}^{\text{ext}}(\mathbf{x}, t)\right]^2 + \mathrm{m}^2 c^2. \quad (\text{I.2.17})$$

Now the canonical quantization (I.2.6) gives the Klein–Gordon equation

$$\left[\frac{i\hbar\partial_t}{c} - \frac{e}{c}A_0^{\text{ext}}(\mathbf{x}, t)\right]^2 \psi(\mathbf{x}, t) = \left[-i\hbar\nabla - \frac{e}{c}\mathbf{A}^{\text{ext}}(\mathbf{x}, t)\right]^2 \psi(\mathbf{x}, t) + \mathrm{m}^2 c^2 \psi(\mathbf{x}, t).$$
$$(\text{I.2.18})$$

Problem I.2.1. Prove the equivalence of the Hamiltonian systems (I.2.12) and (I.2.16) with the Lorentz equations (I.2.11) and (I.2.15), respectively. **Hint:** see Section 12.6 of [51].

I.2.2 Quasiclassical asymptotics and geometrical optics

The derivation of the wave equations (I.2.14) and (I.2.18) relies on de Broglie's relativistic arguments and on formulae (I.2.13) and (I.2.17) from classical electrodynamics. However, a deeper basis for these equations lies in *short-wave asymptotics* of solutions of these equations. These asymptotics reconcile the equations with the classical Lorentz equation and were in the center of Schrödinger's first papers on quantum mechanics [78, II]. Indeed, the key argument for the Schrödinger equation (I.2.14) is the short-wave asymptotics of type

$$\psi(\mathbf{x}, t) = a(\mathbf{x}, t)e^{iS(\mathbf{x}, t)/\hbar}, \qquad \hbar \ll 1 \qquad (\text{I.2.19})$$

for solutions with short-wave initial conditions

$$\psi(\mathbf{x}, 0) = a_0(\mathbf{x})e^{iS_0(\mathbf{x})/\hbar}, \qquad \text{where} \quad a_0(\mathbf{x}) = 0 \quad \text{for} \quad |\mathbf{x} - \mathbf{x}_0| \geq \varepsilon \quad (\text{I.2.20})$$

with small $\varepsilon > 0$. The key fact is that for any $N > 0$, we have

$$a(\mathbf{x}, t) = a_N(\mathbf{x}, t) + \mathcal{O}(\hbar^N), \qquad \hbar \to 0, \qquad (\text{I.2.21})$$

and for sufficiently small $|t - t_0|$,

> $a_N(\mathbf{x}, t) = 0$ *outside of a thin tubular neighborhood of the*
> *trajectory of the Hamiltonian system (I.2.12) with initial data*   (I.2.22)
> $$\mathbf{x}(t_0) = \mathbf{x}_0, \quad \mathbf{p}(t_0) = \nabla S_0(\mathbf{x}_0).$$

In other words,

> *short-wave solutions of the Schrödinger equation propagate along*
> *trajectories of the classical nonrelativistic Lorentz equation.*   (I.2.23)

This short-wave asymptotics is called *quasiclassical asymptotics*, or *geometrical optics*. The proof of the asymptotics (I.2.21), (I.2.22) relies on the following two fundamental facts [3, 35]:

I. The phase function $S(\mathbf{x}, t)$ satisfies the Hamilton–Jacobi equation

$$-\partial_t S(\mathbf{x}, t) = \mathcal{H}(\mathbf{x}, \nabla S(\mathbf{x}, t), t), \qquad (\text{I.2.24})$$

where $\mathcal{H}(\mathbf{x}, \mathbf{p}, t)$ is the Hamiltonian (I.2.13);

II. The relation

$$\nabla S(\mathbf{x}(t), t) = \mathbf{p}(t) \qquad (\text{I.2.25})$$

holds for any solution $(\mathbf{x}(t), \mathbf{p}(t))$ of the system (I.2.12).

These relations immediately imply the remarkable ODE

$$\frac{d}{dt}S(\mathbf{x}(t), t) = \mathbf{p}(t)\dot{\mathbf{x}}(t) - \mathcal{H}(\mathbf{x}(t), \mathbf{p}(t), t) = \Lambda(\mathbf{x}(t), \dot{\mathbf{x}}(t), t), \qquad (\text{I.2.26})$$

where $\Lambda(\mathbf{x}(t), \dot{\mathbf{x}}(t), t)$ is the corresponding Lagrangian. The ODE implies the well-known formula for the phase function as the *Lagrangian action* along the Hamiltonian trajectories:

$$S(\mathbf{x}(t), t) = S_0(\mathbf{x}(t_0)) + \int_{t_0}^{t} \Lambda(\mathbf{x}(s), \dot{\mathbf{x}}(s), s)\, ds. \qquad (\text{I.2.27})$$

Remarks I.2.2. i) The Hamilton–Jacobi equation (I.2.24) and the relation (I.2.25) uniquely dictate the rules of canonical quantization (I.2.6).

ii) The asymptotics (I.2.19) sheds new light on Thomson's 1893–1897 experiments with electrons. It appears that he observed exactly the short-wave asymptotics of cathode rays, as shown on Fig. I.1.

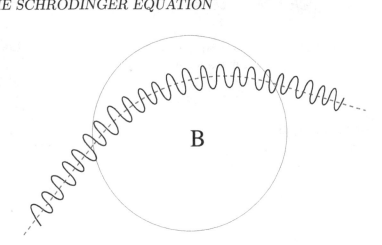

Figure I.1: Deflection of electron beam in a magnetic field.

Remark I.2.3. The extension of geometrical optics to the Klein–Gordon equation is also possible by the same Hamilton–Jacobi theory. Geometrical optics was extended to the Dirac equation by W. Pauli [72] and to general hyperbolic systems by P. Lax [153]. Further developments culminated in the Maslov–Hörmander theory of Fourier integral operators; see the monograph [139].

Exercise I.2.4. Obtain the Hamilton–Jacobi equation (I.2.24). **Hint:** Substitute the asymptotics (I.2.19) into the Schrödinger equation (I.2.14) and then set $\hbar = 0$.

Problem I.2.5. Prove (I.2.25). Hint: i) differentiate the difference $\mathbf{D}(t) := \nabla S(\mathbf{x}(t), t) - \mathbf{p}(t)$ and check that $\dot{\mathbf{D}}(t) \equiv 0$; ii) use the fact that $\mathbf{D}(t_0) = 0$ by the initial data (I.2.22). **Solution:** using (I.2.24) and (I.2.12), we obtain for the components of the vector $\mathbf{D}(t) = (D_1(t), D_2(t), D_3(t))$,

$$\dot{D}_k(t) = \partial_{x^k} \frac{d}{dt} S(\mathbf{x}(t), t) - \dot{p}_k(t)$$

$$= \partial_{x^k} [\nabla S(\mathbf{x}(t), t) \cdot \dot{\mathbf{x}}(t) - \mathcal{H}(\mathbf{x}, \nabla S(\mathbf{x}, t), t)] + \partial_{x^k} \mathcal{H}$$

$$= \partial_{x^k} \nabla S(\mathbf{x}(t), t) \cdot \dot{\mathbf{x}}(t) - \partial_{x^k} \mathcal{H} - \mathcal{H}_{\mathbf{p}} \cdot \partial_{x^k} \nabla S(\mathbf{x}(t), t) + \partial_{x^k} \mathcal{H}$$

$$= \partial_{x^k} \nabla S(\mathbf{x}(t), t) \cdot \dot{\mathbf{x}}(t) - \mathcal{H}_{\mathbf{p}} \cdot \partial_{x^k} \nabla S(\mathbf{x}(t), t) = 0, \quad k = 1, 2, 3. \quad (I.2.28)$$

Problem I.2.6. Prove (I.2.22). **Hint:** i) substitute (I.2.19) into the Schrödinger equation (I.2.14) and obtain the *transport ODEs* for $a(\mathbf{x}(t), t)$, using the first equation of (I.2.12) and (I.2.25); ii) substitute $a(\mathbf{x}, t) = \sum_{k \geq 0} \hbar^k b_k(\mathbf{x}, t)$

into these transport equations and equate the terms with identical factors $\hbar^k$; iii) use (I.2.20) to prove that, for sufficiently small $|t - t_0|$,

$$b_k(\mathbf{x}, t) = 0, \qquad (\mathbf{x}, t) \notin \mathcal{T}. \qquad (\text{I.2.29})$$

Here $\mathcal{T}$ is the union of all trajectories $(\mathbf{x}(t), t)$, where $\mathbf{x}(t)$ are solutions of the Lorentz equation (I.2.11) with $\mathbf{x}(t_0) \in \operatorname{supp} a_0$ and $\mathbf{p}(t_0) = \nabla S_0(\mathbf{x}(t_0))$. For details, see [51, Sections 3.2 and 13.2]).

I.3 Quantum Observables

Quantum observables are defined as invariants of the Schrödinger dynamics. This identification goes back to Ehrenfest's adiabatic hypothesis that only adiabatic invariants are quantized [96, p. 110], [68].

The invariance is due to the symmetry of external potentials: the total energy of solutions is conserved if the external potentials are independent of time, the total charge is conserved for any real potentials, the projection of the momentum on a certain direction is conserved if the potentials are invariant with respect to shifts in this direction, and the projection of the angular momentum onto a certain direction is conserved if the potentials are invariant (in a suitable sense) with respect to rotations about this direction.

The *correspondence principle* in particular states that the quantum observables and the corresponding classical observables asymptotically coincide as $\hbar \to 0$; see (I.3.20).

I.3.1 Hamiltonian structure

The linear Schrödinger equation (I.2.14) can be written in the Hamiltonian form (B.0.1) as

$$i\hbar\dot{\psi}(t) = \frac{1}{2}D_\psi\mathcal{H}(\psi(t), t) = H(t)\psi(t), \qquad t \in \mathbb{R}, \qquad (\text{I.3.1})$$

where D_ψ is the variational derivative, and the Hamiltonian functional $\mathcal{H}$ (which is the *quantum energy* $E(t)$) reads as

$$E(t) := \mathcal{H}(\psi(t), t) := \langle \psi(t), H(t)\psi(t) \rangle$$

$$= \int \left[\frac{1}{2m} |[-i\hbar\nabla - \frac{e}{c}\mathbf{A}^{\text{ext}}(\mathbf{x}, t)]\psi(\mathbf{x}, t)|^2 + eA_0^{\text{ext}}(\mathbf{x}, t)|\psi(\mathbf{x}, t)|^2 \right] d\mathbf{x}. \quad (\text{I.3.2})$$

Here the brackets $\langle \cdot, \cdot \rangle$ denote the Hermitian inner product in the complex Hilbert space $\mathcal{E} := L^2(\mathbb{R}^3) \otimes \mathbb{C}$:

$$\langle f, g \rangle := \int_{\mathbb{R}^3} \overline{f(\mathbf{x})}g(\mathbf{x}) \, d\mathbf{x}. \qquad (\text{I.3.3})$$

Exercise I.3.1. Check that equation (I.3.1) can be written in the Hamiltonian form (B.0.1). **Hints:** The space $\mathcal{E}$ should be identified with the real Hilbert space $L^2(\mathbb{R}^3) \otimes \mathbb{R}^2$; then $J = \frac{1}{2\hbar}i^{-1}$, where i is identified with the skew-symmetric matrix:

$$i = \begin{pmatrix} 0 & -1 \\ 1 & 0 \end{pmatrix}. \qquad (\text{I.3.4})$$

The Hamiltonian structure implies that the quantum energy (I.3.2) is conserved if the external Maxwell potentials are time-independent; see (I.3.21) and (I.3.27) (and also (B.0.2)).

I.3.2   Charge and current densities

The comparison of (I.3.2) with the energy in electrostatics shows that the *electric charge density* should be defined as

$$\rho(\mathbf{x}, t) = e|\psi(\mathbf{x}, t)|^2. \tag{I.3.5}$$

Now equation (I.3.1) implies *conservation of total charge*

$$Q(t) := \int \rho(\mathbf{x}, t) \, d\mathbf{x} \equiv \text{const.} \tag{I.3.6}$$

This follows from the Schrödinger equation $\dot{\psi}(t) = \frac{1}{i\hbar}H(t)\psi(t)$:

$$\begin{aligned}
\dot{Q}(t) &:= e\langle \dot{\psi}(t), \psi(t)\rangle + e\langle \psi(t), \dot{\psi}(t)\rangle \\
&= -\frac{e}{i\hbar}\langle H(t)\psi(t), \psi(t)\rangle + \frac{e}{i\hbar}\langle \psi(t), H(t)\psi(t)\rangle = 0
\end{aligned} \tag{I.3.7}$$

since the Schrödinger operator $H(t)$ is *symmetric* with respect to the Hermitian inner product (I.3.3).

In particular, for one electron, we have $\int \rho(\mathbf{x}, t) \, d\mathbf{x} = e$, hence according to (I.3.5) and (I.3.6) the *normalization condition* holds:

$$\int |\psi(\mathbf{x}, t)|^2 \, d\mathbf{x} = 1, \qquad t \in \mathbb{R}. \tag{I.3.8}$$

Further, the *electric current density* should be defined according to [77, (7.3)], [31, (21.12)]). In the Gaussian and Heaviside–Lorentz units ([48, p. 781], [49]), this formula reads as

$$\mathbf{j}(\mathbf{x}, t) = \frac{e}{m} \operatorname{Re}\left(\overline{\psi}(\mathbf{x}, t)\left[-i\hbar\nabla - \frac{e}{c}\mathbf{A}^{\text{ext}}(\mathbf{x}, t)\right]\psi(\mathbf{x}, t)\right). \tag{I.3.9}$$

This expression is justified by the following two arguments:

i) The agreement with the asymptotic behavior of the corresponding classical current for *quasiclassical solutions* (I.2.19): substituting the quasiclassical asymptotics (I.2.19) with $\hbar \ll 1$ into (I.3.9), for any trajectory $(\mathbf{x}(t), \mathbf{p}(t))$ of the Hamiltonian system (I.2.12), we obtain:

$$\begin{aligned}
\mathbf{j}(\mathbf{x}(t), t) &\sim \operatorname{Re}\left(\frac{e}{m}\bar{\psi}(\mathbf{x}(t), t)\left[\nabla S(\mathbf{x}(t), t) - \frac{e}{c}\mathbf{A}^{\text{ext}}(\mathbf{x}(t), t)\right]\psi(\mathbf{x}(t), t)\right) \\
&= \frac{e}{m}\operatorname{Re}\left(\bar{\psi}(\mathbf{x}(t), t)\left[\mathbf{p}(t) - \frac{e}{c}\mathbf{A}^{\text{ext}}(\mathbf{x}(t), t)\right]\psi(\mathbf{x}(t), t)\right) \\
&= e\operatorname{Re}\left\{\bar{\psi}(\mathbf{x}(t), t)\dot{\mathbf{x}}(t)\psi(\mathbf{x}(t), t)\right\} \\
&= e|\psi(\mathbf{x}(t), t)|^2\dot{\mathbf{x}}(t) = \rho(\mathbf{x}(t), t)\dot{\mathbf{x}}(t), \tag{I.3.10}
\end{aligned}$$

where we used (I.2.25) and the first equation from (I.2.12).

ii) This current density and the charge density (I.3.5) satisfy the *continuity equation*

$$\dot{\rho}(\mathbf{x}, t) + \operatorname{div} \mathbf{j}(\mathbf{x}, t) \equiv 0. \tag{I.3.11}$$

The last identity follows by direct differentiation: equation (I.2.14) implies that

$$
\begin{aligned}
\dot{\rho} &= 2e \operatorname{Re}\left(\overline{\dot{\psi}}\psi\right) \\
&= 2e \operatorname{Re}\left(\overline{\psi}\frac{1}{i\hbar}\left[\frac{1}{2\mathrm{m}}\left[-i\hbar\nabla - \frac{e}{c}\mathbf{A}^{\mathrm{ext}}\right]^2 \psi + eA_0^{\mathrm{ext}}\psi\right]\right) \\
&= 2e \operatorname{Re}\left(\overline{\psi}\frac{1}{i\hbar}\left[-\frac{i\hbar}{2\mathrm{m}}\nabla\cdot\left[-i\hbar\nabla\psi - \frac{e}{c}\mathbf{A}^{\mathrm{ext}}\psi\right] + \frac{ie\hbar}{2\mathrm{m}c}\mathbf{A}^{\mathrm{ext}}\cdot\nabla\psi\right]\right),
\end{aligned} \tag{I.3.12}
$$

where all the functions are evaluated at $(\mathbf{x}, t)$. At the same time, calculating the divergence of the current (I.3.9), we obtain

$$
\begin{aligned}
\nabla\cdot\mathbf{j} = \frac{2e}{\mathrm{m}}\operatorname{Re}\Big(&\overline{\psi}\nabla\cdot\left[-i\hbar\nabla\psi - \frac{e}{c}\mathbf{A}^{\mathrm{ext}}\psi\right] \\
&+ \nabla\overline{\psi}\cdot\left[-i\hbar\nabla\psi - \frac{e}{c}\mathbf{A}^{\mathrm{ext}}\psi\right]\Big).
\end{aligned} \tag{I.3.13}
$$

Taking into account that

$$\operatorname{Re}\left(\overline{\psi}\mathbf{A}^{\mathrm{ext}}\cdot\nabla\psi\right) = \operatorname{Re}\left(\nabla\overline{\psi}\cdot\mathbf{A}^{\mathrm{ext}}\psi\right),$$

and discarding purely imaginary terms, we see that (I.3.12) and (I.3.13) add up to zero. Thus the charge continuity equation (I.3.11) is proved.

I.3.3 Quantum momentum and angular momentum

The *quantum momentum* and the *quantum angular momentum* in the Schrödinger theory are defined for any state $\psi(t)$ as the following *mean values*:

$$\mathbf{p}(t) := \langle\psi(t), \hat{\mathbf{p}}\psi(t)\rangle, \qquad \mathbf{L}(t) := \langle\psi(t), \hat{\mathbf{L}}\psi(t)\rangle, \tag{I.3.14}$$

where

$$\hat{\mathbf{p}} := -i\hbar\nabla, \qquad \hat{\mathbf{L}} := \hat{\mathbf{x}}\times\hat{\mathbf{p}} \tag{I.3.15}$$

are the *selfadjoint operators* of quantum momentum and of quantum angular momentum, and $\hat{\mathbf{x}}$ is the operator of multiplication by $\mathbf{x}$. The fundamental commutation relations hold for the components of the vector-operators $\hat{\mathbf{x}} = (\hat{x}^1, \hat{x}^2, \hat{x}^3)$ and of $\hat{\mathbf{p}} = (\hat{p}_1, \hat{p}_2, \hat{p}_3)$:

$$[\hat{p}_k, \hat{x}^j] = -i\hbar\delta_{kj}, \qquad k, j = 1, 2, 3. \tag{I.3.16}$$

They imply highly important cyclic *commutation relations* for the components of the vector-operator $\hat{\mathbf{L}} = (\hat{L}_1, \hat{L}_2, \hat{L}_3)$:

$$[\hat{L}_1, \hat{L}_2] = i\hbar\hat{L}_3, \qquad [\hat{L}_2, \hat{L}_3] = i\hbar\hat{L}_1, \qquad [\hat{L}_3, \hat{L}_1] = i\hbar\hat{L}_2. \qquad (\text{I.3.17})$$

Let us note that

$$\hat{L}_k = -i\hbar\partial_{\theta_k} \qquad (\text{I.3.18})$$

in the cylindrical coordinate system with the longitudinal axis x^k and with the azimuthal angle denoted by θ_k.

Exercise I.3.2. Check (I.3.16).

Exercise I.3.3. Check (I.3.17) and (I.3.18).

Exercise I.3.4. Prove that the identities (I.3.16) imply more general relations

$$[\hat{p}_k, f(\hat{\mathbf{x}})] = -i\hbar\partial_{x^k} f(\hat{\mathbf{x}}), \qquad [f(\hat{\mathbf{p}}), \hat{x}^j] = -i\hbar\partial_{p_j} f(\hat{\mathbf{p}}) \qquad (\text{I.3.19})$$

for an appropriate class of functions f. **Hint:** consider polynomials f.

I.3.4   Correspondence principle

The names of the quantum observables *energy*, *momentum*, and *angular momentum* are justified by the *asymptotic coincidence* with the corresponding classical observables for *quasiclassical solutions* (I.2.19): using the quasiclassical asymptotics (I.2.19) with small $\hbar \ll 1$ and $\varepsilon \ll 1$, one can prove that for sufficiently small $|t - t_0|$,

$$\left. \begin{aligned} E(t) &\approx \tfrac{1}{2\mathrm{m}}[\mathbf{p}(t) - \tfrac{e}{c}\mathbf{A}^{\text{ext}}(\mathbf{x}(t), t)]^2 + eA_0^{\text{ext}}(\mathbf{x}(t), t) \\[2mm] \mathbf{p}(t) &\approx \mathrm{m}\dot{\mathbf{x}}(t) + \tfrac{e}{c}\mathbf{A}^{\text{ext}}(\mathbf{x}(t), t), \qquad \mathbf{L}(t) \approx \mathbf{x}(t) \times \mathbf{p}(t) \end{aligned} \right| \qquad (\text{I.3.20})$$

if the *one-electron normalization* (I.3.8) holds.

Problem I.3.5. Check (I.3.20). **Hint:** See [51, Section 3.3.3].

I.3.5   Conservation laws

The main motivation for the definitions of quantum observables for the Schrödinger equation (I.2.14) consists in the *conservation laws* which will be proved in the next section: for solutions to this equation,

i) the quantum energy $E(t)$ does not depend on time in the case of *static Maxwell potentials*, i.e., when

$$\mathbf{A}^{\text{ext}}(\mathbf{x}, t) \equiv \mathbf{A}^{\text{ext}}(\mathbf{x}), \qquad A_0^{\text{ext}}(\mathbf{x}, t) \equiv A_0^{\text{ext}}(\mathbf{x}). \qquad (\text{I.3.21})$$

In this case the Schrödinger operator (I.2.14) reads as

$$H(t) \equiv H = \mathcal{H}(\hat{\mathbf{x}}, \hat{\mathbf{p}}) = \frac{1}{2m} \left[\hat{\mathbf{p}} - \frac{e}{c} \mathbf{A}^{\text{ext}}(\hat{\mathbf{x}}) \right]^2 + e A_0^{\text{ext}}(\hat{\mathbf{x}}); \qquad (\text{I.3.22})$$

ii) the quantum momentum $p_k(t)$ does not depend on time if the Maxwell potentials $\mathbf{A}^{\text{ext}}(\mathbf{x}, t)$ and $A_0^{\text{ext}}(\mathbf{x}, t)$ are independent of x^k;

iii) the quantum angular momentum $L_k(t)$ does not depend on time if the external Maxwell potentials are invariant with respect to rotations around the x^k-axis: i.e.,

$$\mathbf{A}^{\text{ext}}(R_k(\theta_k)\mathbf{x}, t) \equiv R_k(\theta_k)\mathbf{A}^{\text{ext}}(\mathbf{x}, t), \quad A_0^{\text{ext}}(R_k(\theta_k)\mathbf{x}, t) \equiv A_0^{\text{ext}}(\mathbf{x}, t), \quad (\text{I.3.23})$$

where $\theta_k \in [0, 2\pi]$ and $R_k(\theta_k)$ denotes the rotation of the space $\mathbb{R}^3$ by the angle θ_k about the x^k-axis. Let us reformulate this rotational invariance for the case $k = 3$ in the cylindrical coordinates

$$(x^3, \quad r := |(x^1, x^2)|, \quad \varphi = \theta_3).$$

The second identity in (I.3.23) means that the scalar potential does not depend on the angle φ:

$$A_0^{\text{ext}}(\mathbf{x}, t) = a(x^3, r, t). \qquad (\text{I.3.24})$$

On the other hand, the first identity of (I.3.23) is equivalent to the representation

$$\mathbf{A}^{\text{ext}}(\mathbf{x}, t) = a_3(x^3, r, t)\mathbf{e}_3 + a_r(x^3, r, t)\mathbf{e}_r + a_\varphi(x^3, r, t)\mathbf{e}_\varphi, \qquad (\text{I.3.25})$$

where

$$\mathbf{e}_3 := (0, 0, 1), \qquad \mathbf{e}_r := (x^1, x^2, 0)/r, \qquad \mathbf{e}_\varphi := (-x^2, x^1, 0)/r,$$

and all the coefficients do not depend on the angle φ. In particular, $A_3^{\text{ext}}(\mathbf{x}, t) = a_3(x^3, r, t)$ does not depend on φ.

Example I.3.6. For a uniform magnetic field $\mathbf{B} = (0, 0, B)$, the magnetic potential (A.7.3) can be written as

$$\mathbf{A}^{\text{ext}}(\mathbf{x}) = \frac{1}{2}B(-x^2, x^1, 0) = \frac{1}{2}Br\mathbf{e}_\varphi, \qquad (\text{I.3.26})$$

which satisfies condition (I.3.23) with $k = 3$.

Exercise I.3.7. Prove the equivalence of (I.3.25) with the first identity of (I.3.23).

I.3.6   Proof of conservation laws

i) The conservation of energy (that is, the value of the Hamiltonian) holds for general *autonomous Hamiltonian systems*; see (B.0.2). In particular, for the autonomous Schrödinger equation (I.3.1) with $H(t) \equiv H$, we have

$$\dot{\psi}(t) = \frac{1}{i\hbar} H\psi(t).$$

Hence, differentiating (I.3.2), and using (I.3.1), we obtain

$$\dot{E}(t) = \langle \dot{\psi}(t), H\psi(t)\rangle + \langle \psi(t), H\dot{\psi}(t)\rangle = \left\langle \frac{1}{i\hbar}H\psi(t), H\psi(t)\right\rangle + \left\langle \psi(t), H\frac{1}{i\hbar}H\psi(t)\right\rangle$$

$$= -\frac{1}{i\hbar}\langle H\psi(t), H\psi(t)\rangle + \frac{1}{i\hbar}\langle H\psi(t), H\psi(t)\rangle = 0 \qquad (I.3.27)$$

since the Schrödinger operator H is *symmetric* with respect to Hermitian inner product (I.3.3).

ii) The conservation of momentum and angular momentum is deduced from the corresponding *commutation relations*. For instance, if the external Maxwell potentials are independent of x^k, then the commutator $[H(t), \hat{p}_k]$ vanishes. Hence, differentiating $p_k(t) := \langle \psi(t), \hat{p}_k\psi(t)\rangle$ and using (I.3.1), we obtain:

$$\begin{aligned}\dot{p}_k(t) &= \langle \dot{\psi}(t), \hat{p}_k\psi(t)\rangle + \langle \psi(t), \hat{p}_k\dot{\psi}(t)\rangle \\ &= -\frac{1}{i\hbar}\langle H(t)\psi(t), \hat{p}_k\psi(t)\rangle + \frac{1}{i\hbar}\langle \psi(t), \hat{p}_k H(t)\psi(t)\rangle \\ &= \frac{1}{i\hbar}\langle \psi(t), [\hat{p}_k, H(t)]\psi(t)\rangle = 0.\end{aligned} \qquad (I.3.28)$$

iii) Let us consider the case $k = 3$ and prove that the Schrödinger operator (I.2.14) admits the following expression:

$$H(t) = \sum_{j,k=1}^{3} a_{j,k}(x^3, r, t)\frac{\partial^2}{\partial z^j \, \partial z^k} + \sum_{j=1}^{3} a_j(x^3, r, t)\frac{\partial}{\partial z^j} + a(x^3, r, t), \quad (I.3.29)$$

where $z^1 := x^3$, $z^2 := r$, and $z^3 := \varphi$. Using the above expression and (I.3.18), we obtain the commutation

$$[\hat{L}_3, H(t)] = 0. \qquad (I.3.30)$$

Hence, differentiating $L_3(t) := \langle \psi(t), \hat{L}_3\psi(t)\rangle$ and using (I.3.1), we obtain:

$$\dot{L}_3(t) = \langle \dot{\psi}(t), \hat{L}_3\psi(t)\rangle + \langle \psi(t), \hat{L}_3\dot{\psi}(t)\rangle = \frac{1}{i\hbar}\langle \psi(t), [\hat{L}_3, H(t)]\psi(t)\rangle = 0. \quad (I.3.31)$$

It remains to justify the expression (I.3.29). For the scalar potential $A^0(\mathbf{x}, t)$, such an expression (without the terms with derivatives) holds due to the second identity from (I.3.23). It is left to prove that the operator

$$\left[-i\hbar\nabla - \frac{e}{c}\mathbf{A}^{\text{ext}}(\mathbf{x}, t)\right]^2 = -\Delta + \frac{2i\hbar e}{c}\mathbf{A}^{\text{ext}}(\mathbf{x}, t) \cdot \nabla + \frac{e^2}{c^2}|\mathbf{A}^{\text{ext}}(\mathbf{x}, t)|^2 \quad (\text{I.3.32})$$

can also be expressed in the form (I.3.29). For the first and last terms in the right-hand side, the expression (I.3.29) obviously holds (this is the **Exercise!**). The operator $\mathbf{A}^{\text{ext}}(\mathbf{x}, t) \cdot \nabla$ is the derivative in the direction of the vector field $\mathbf{A}^{\text{ext}}(\mathbf{x}, t)$, and hence (I.3.25) implies that (this is also the **Exercise!**)

$$\mathbf{A}^{\text{ext}}(\mathbf{x}, t) \cdot \nabla = a_3(x^3, r, t)\partial_{x^3} + a_r(x^3, r, t)\partial_r + a_\varphi(x^3, r, t)\partial_\varphi. \quad (\text{I.3.33})$$

Exercise I.3.8. Check the identity (I.3.32). **Hint:** use the Coulomb gauge (I.2.10).

Definition I.3.9. *A quantum observable is a selfadjoint operator $\hat{K}$ in the Hilbert space $\mathcal{E} = L^2(\mathbb{R}^3)$. Its quadratic form*

$$K(\psi) := \langle \psi, \hat{K}\psi \rangle \quad (\text{I.3.34})$$

is called the mean value of the quantum observable $\hat{K}$ in the state ψ.

As above, the mean value $K(\psi(t))$ does not depend on time for solutions of the Schrödinger equation (I.3.1) if

$$[H(t), \hat{K}] = 0, \qquad t \in \mathbb{R}. \quad (\text{I.3.35})$$

Example I.3.10. In the autonomous case (I.3.22), the energy (I.3.2) is the *mean value* of the Schrödinger operator in the state ψ:

$$\mathcal{H}(\psi) = \langle \psi, H\psi \rangle.$$

Remark I.3.11. i) In Appendix B, we present a general Noether theory of conservation laws for Hamiltonian systems with Lie symmetry group. The momentum and angular momentum conservations are consequences of this theory in the case of the symmetry groups of translations and rotations, respectively.

ii) The names *momentum* and *angular momentum* of these quantum observables are justified in particular by the fact that they are conserved for translation-invariant and rotation-invariant systems, respectively, like the same observables of finite-dimensional Hamiltonian systems.

I.3.7   The Heisenberg picture

In 1925, W. Heisenberg introduced his *matrix mechanics*; then, in 1926, Schrödinger introduced his *wave mechanics* and showed the equivalence of the two theories.

In the Heisenberg picture, the state of the system, *by definition*, does not depend on time, i.e., $\psi(t) \equiv \psi(0)$. On the other hand, the operators corresponding to quantum observables evolve *by definition* as follows:

$$\hat{K}(t) := U^*(t)\hat{K}U(t), \tag{I.3.36}$$

where $U(t) : \psi(0) \mapsto \psi(t)$ is the dynamical map corresponding to the Schrödinger equation (I.3.1) and $U^*(t)$ is the corresponding *adjoint operator*. By (I.3.6), both $U(t)$ and $U^*(t)$ are *unitary operators*. The Schrödinger equation (I.3.1) implies that

$$i\hbar\dot{U}(t) = H(t)U(t), \qquad i\hbar\dot{U}^*(t) = -U^*(t)H(t). \tag{I.3.37}$$

In this notation, the mean value (I.3.34) in the state $\psi(t) = U(t)\psi(0)$ can be written as

$$K(\psi(t)) = \langle U(t)\psi(0), \hat{K}U(t)\psi(0)\rangle = \langle \psi(0), \hat{K}(t)\psi(0)\rangle. \tag{I.3.38}$$

In the *autonomous case* corresponding to the Schrödinger operators (I.3.22), when $H(t)$ does not depend on time, the dynamics of all observables $\hat{K}(t)$ is governed by the *Heisenberg equation*

$$i\hbar\dot{\hat{K}}(t) = [\hat{K}(t), H], \qquad t \in \mathbb{R}, \tag{I.3.39}$$

which follows by differentiating (I.3.36). In particular, for the quantum observables $\hat{K} = \hat{\mathbf{x}}$ and $\hat{K} = \hat{\mathbf{p}}$, the Heisenberg equations read as

$$\dot{\hat{\mathbf{x}}}(t) = \frac{1}{i\hbar}[\hat{\mathbf{x}}(t), H(t)], \qquad \dot{\hat{\mathbf{p}}}(t) = \frac{1}{i\hbar}[\hat{\mathbf{p}}(t), H(t)]. \tag{I.3.40}$$

It is remarkable that these equations can be written as the "Hamiltonian system"

$$\dot{\hat{\mathbf{x}}}(t) = \mathcal{H}_{\mathbf{p}}(\hat{\mathbf{x}}(t), \hat{\mathbf{p}}(t)), \qquad \dot{\hat{\mathbf{p}}}(t) = -\mathcal{H}_{\mathbf{x}}(\hat{\mathbf{x}}(t), \hat{\mathbf{p}}(t)), \tag{I.3.41}$$

since (I.3.19) implies that

$$\frac{1}{i\hbar}[\hat{\mathbf{x}}(t), H(t)] = \mathcal{H}_{\mathbf{p}}(\hat{\mathbf{x}}(t), \hat{\mathbf{p}}(t)), \qquad \frac{1}{i\hbar}[\hat{\mathbf{p}}(t), H(t)] = -\mathcal{H}_{\mathbf{x}}(\hat{\mathbf{x}}(t), \hat{\mathbf{p}}(t)). \tag{I.3.42}$$

The system (I.3.40) for the observables $\hat{\mathbf{x}}(t)$ and $\hat{\mathbf{p}}(t)$ *formally* coincides with the Hamiltonian system (I.2.12) after the substitutions $\mathbf{x} \mapsto \hat{\mathbf{x}}$ and $\mathbf{p} \mapsto \hat{\mathbf{p}}$. This coincidence was the key idea of the Heisenberg approach to quantum theory. The Heisenberg picture is very efficient in quantum field theory, but this topic is outside the scope of our book; see [64, 75].

Exercise I.3.12. Check (I.3.39). **Hints:** differentiate (I.3.36) and use (I.3.37) and the commutation of $H(t) \equiv H$ with $U(t)$ and $U^*(t)$.

Exercise I.3.13. Check (I.3.42). **Hint:** use (I.3.19) and (I.3.22).

Remark I.3.14. The commutation (I.3.35) implies the conservation of the mean value (I.3.38), since $K(t) \equiv K$ satisfies (I.3.39).

I.3.8 Plane waves as electron beams

The wave-particle duality (I.1.13) was one of the main sources of the Schrö-dinger theory. Now we will dwell upon the duality in the framework of the Schrödinger formalism, calculating the corresponding "density of electrons" and of currents in plane waves. However, it is impossible to assign any positions to these "particles", because there are the restrictions due to the Heisenberg *uncertainty principle* (see the next section). Namely, consider the plane wave

$$\psi(\mathbf{x}, t) = A e^{i(\mathbf{k} \cdot \mathbf{x} - \omega t)} \tag{I.3.43}$$

satisfying the free Schrödinger equation without an external Maxwell field (I.2.5). Then

$$\hbar \omega = \frac{\hbar^2 \mathbf{k}^2}{2m}. \tag{I.3.44}$$

By (I.3.5) and (I.3.9), the charge and electric current densities are given by

$$\rho = e|\psi(\mathbf{x}, t)|^2 = e|A|^2, \quad \mathbf{j} = \frac{e}{m} \operatorname{Re} \left(\overline{\psi}(\mathbf{x}, t)[-i\hbar \nabla \psi(\mathbf{x}, t)] \right) = \frac{e\hbar \mathbf{k}}{m} |A|^2. \tag{I.3.45}$$

Respectively, the *hypothetical* density, velocity and momentum of electrons should be as follows:

$$n := \rho/e = |A|^2, \quad \mathbf{v} = \mathbf{j}/\rho = \frac{\hbar \mathbf{k}}{m}, \quad \mathbf{p} = m\mathbf{v} = \hbar \mathbf{k}. \tag{I.3.46}$$

In these terms, the energy density (i.e., the integrand in (I.3.2)) reads as

$$\mathcal{E} := \frac{1}{2m} |-i\hbar \nabla \psi(\mathbf{x}, t)|^2 = \frac{\hbar^2}{2m} \mathbf{k}^2 |A|^2 = \frac{m\mathbf{v}^2}{2} n. \tag{I.3.47}$$

Now, the *energy per one electron* is given by

$$E = \frac{\mathcal{E}}{n} = \frac{m\mathbf{v}^2}{2} = \frac{\hbar^2}{2m} \mathbf{k}^2 = \hbar \omega; \tag{I.3.48}$$

we took into account (I.3.44).

I.3.9 The Heisenberg uncertainty principle

Let us note that the interpretation of the plane wave (I.3.43) as a beam of free particles in (I.1.13) prescribes the exact value (I.3.46) to their momenta $\mathbf{p}$. On the other hand, this interpretation cannot be completed by an assignment of the coordinates to electrons, since the charge density (I.3.5) is constant. For the wave function $\psi(x)$ with sufficient decay at infinity, the mean value of the coordinate is defined by

$$\bar{\mathbf{x}} := \langle \psi, \hat{\mathbf{x}}\psi \rangle = \langle \psi, \mathbf{x}\psi \rangle. \qquad (\text{I.3.49})$$

The **uncertainty** of the mean value is measured by the *mean square error*,

$$\varDelta\mathbf{x} := |\langle \psi, (\hat{\mathbf{x}} - \bar{\mathbf{x}})^2\psi \rangle|^{1/2}/\|\psi\|, \qquad (\text{I.3.50})$$

where $\|\psi\|$ is the L^2-norm of ψ. Similarly, for the momentum, we have

$$\bar{\mathbf{p}} := \langle \psi, \hat{\mathbf{p}}\psi \rangle := \langle \psi, -i\hbar\nabla\psi \rangle, \qquad \varDelta\mathbf{p} := |\langle \psi, (\hat{\mathbf{p}} - \bar{\mathbf{p}})^2\psi \rangle|^{1/2}/\|\psi\|. \quad (\text{I.3.51})$$

It is easy to construct a wave functions ψ so that either $\varDelta\mathbf{x}(\psi)$ or $\varDelta\mathbf{p}(\psi)$ is arbitrarily small. However, the errors cannot be arbitrarily small simultaneously, and this is expressed in the exact *uncertainty principle*, as discovered by W. Heisenberg [41, 70]:

$$\varDelta x^k \, \varDelta p_{k'} \geq \frac{\hbar}{2}\delta_{kk'}, \qquad k, \, k' = 1, \, 2, \, 3. \qquad (\text{I.3.52})$$

Proof. First, note that the operators $\alpha := \hat{x}^k - \bar{x}^k$ and $\beta := \hat{p}_{k'} - \bar{\mathbf{p}}_{k'}$ are symmetric. Hence

$$\varDelta\mathbf{x} = \frac{\|\alpha\psi\|}{\|\psi\|}, \qquad \varDelta\mathbf{p} = \frac{\|\beta\psi\|}{\|\psi\|}. \qquad (\text{I.3.53})$$

Now the Cauchy–Schwarz inequality implies

$$\varDelta\mathbf{x} \, \varDelta\mathbf{p} \geq \frac{|\langle \alpha\psi, \beta\psi \rangle|}{\|\psi\|^2} = \frac{|\langle \psi, (\beta\alpha + \alpha\beta)\psi \rangle/2 + \langle \psi, (\beta\alpha - \alpha\beta)\psi \rangle/2|}{\|\psi\|^2}. \quad (\text{I.3.54})$$

It remains to observe that the operator $\beta\alpha + \alpha\beta$ is symmetric, while $\beta\alpha - \alpha\beta$ is skew-symmetric. Hence the first inner product in the numerator of the last term in the right-hand side is real, while the second is purely imaginary. Since $\beta\alpha - \alpha\beta = -i\hbar\delta_{kk'}$, the relation (I.3.52) follows. $\qquad \square$

Example I.3.15. Consider the "plane wave" $\psi_R(\mathbf{x}) = \zeta(\mathbf{x}/R)e^{i\mathbf{k}\cdot\mathbf{x}}$, with $\zeta \in C_0^\infty(\mathbb{R}^3)$, $\zeta(0) \neq 0$. Then $\varDelta p_k(\psi_R) \to 0$ as $R \to \infty$, while $\varDelta x^k(\psi_R) \to \infty$.

Remark I.3.16. For an eigenfunction ψ of the operator $\hat{\mathbf{x}}$ ($\hat{\mathbf{p}}$, respectively), the uncertainty of the mean vanishes: $\varDelta\hat{\mathbf{x}} = 0$ ($\varDelta\hat{\mathbf{p}} = 0$, respectively). Hence the uncertainty principle is related to the non-commutativity of the operators $\hat{\mathbf{p}}$ and $\hat{\mathbf{x}}$ which forbids the existence of their common eigenfunctions.

I.4 Bohr's postulates

Johann Balmer spent many years deciphering A.J. Ångström's numerical data on the hydrogen spectrum. His efforts culminated in 1885 in the astonishing discovery of the empirical formula for the wavelength of some hydrogen spectral lines, namely

$$\omega_{nn'} = \omega_{n'} - \omega_n, \qquad (I.4.1)$$

with the *spectral terms*

$$\omega_n = -B/n^2, \qquad n = 1, 2, \ldots, \qquad B \approx 2 \times 10^{16}\,\mathrm{s}^{-1}. \qquad (I.4.2)$$

In 1888, such a difference structure was proposed by J. Rydberg for all hydrogen spectral lines. In 1908, W. Ritz suggested extending this *Rydberg–Ritz combination principle* to all elements.

In 1913, N. Bohr formulated the following two fundamental postulates of the quantum theory of atoms:

I. *Each electron in atom lives in one of the quantum stationary orbits, and sometimes it jumps from one quantum stationary orbit to another: in the Dirac notation,*

$$|E_n\rangle \mapsto |E_{n'}\rangle. \qquad (I.4.3)$$

II. *On a stationary orbit, the electron does not radiate, while every jump is followed by a radiation of an electromagnetic wave with the frequency*

$$\omega_{nn'} = \frac{E_{n'} - E_n}{\hbar} = \omega_{n'} - \omega_n, \qquad \omega_n := E_n/\hbar. \qquad (I.4.4)$$

Both these postulates were inspired by stability of atoms, by Rydberg–Ritz combination principle, and by the Einstein relation (I.1.5) in the theory of the photoeffect. Indeed, the formula (I.4.4) can be written as

$$\hbar\omega_{nn'} = \hbar\omega_{n'} - \hbar\omega_n. \qquad (I.4.5)$$

The Einstein formula for the photon energy (I.1.5) implies that the left-hand side of (I.4.5) equals the energy E_p of the emitted photon. This suggests identifying the terms $\hbar\omega_n$ and $\hbar\omega_{n'}$ on the right-hand side with the quantum energies E_n and $E_{n'}$ of the quantum stationary orbits. Now (I.4.5) becomes

$$E_p = E_{n'} - E_n \qquad (I.4.6)$$

which was interpreted by N. Bohr as the energy balance in the transition (I.4.3).

With the discovery of the Schrödinger theory in 1926, the question arose about the implementation of the above Bohr axioms in the new theory.

I.4.1   Schrödinger's identification of stationary orbits

Besides equation (I.2.14) for the wave function, the Schrödinger theory contains a highly nontrivial definition of *quantum stationary orbits* (or *quantum stationary states*) in the case of static Maxwell external potentials (I.3.21). In this case, the Schrödinger equation (I.2.14) reads as

$$i\hbar\dot{\psi}(\mathbf{x},t) = H\psi(t) := \frac{1}{2\mathrm{m}} \left[-i\hbar\nabla - \frac{e}{c}\mathbf{A}^{\mathrm{ext}}(\mathbf{x}) \right]^2 \psi(\mathbf{x},t) + eA_0^{\mathrm{ext}}(\mathbf{x})\psi(\mathbf{x},t). \tag{I.4.7}$$

Definition I.4.1. *Quantum stationary orbits* (*or* quantum stationary states) *for the Schrödinger equation* (I.4.7) *are finite-energy solutions of the form*

$$\psi(\mathbf{x},t) \equiv \varphi(\mathbf{x})e^{-i\omega t}, \qquad \omega \in \mathbb{R}, \qquad \mathcal{H}(\varphi) < \infty. \tag{I.4.8}$$

Substitution into (I.4.7) leads to the famous Schrödinger *eigenvalue problem*

$$H\varphi = \hbar\omega\varphi. \tag{I.4.9}$$

The definition of quantum stationary orbits (I.4.8) is rather natural, since $|\psi(\mathbf{x},t)|$ does not depend on time. Most likely, this definition was suggested by de Broglie's wave function for *free particles* $\psi(\mathbf{x},t) = Ce^{i(\mathbf{k}\cdot\mathbf{x}-\omega t)}$, which factorizes as $Ce^{i\mathbf{k}\cdot\mathbf{x}}e^{-i\omega t}$. Indeed, in the case of particles *subject to an external force*, it is natural to change the spatial factor $Ce^{i k x}$, since the spatial properties have changed and ceased to be homogeneous. On the other hand, the homogeneous time factor $e^{-i\omega t}$ must be preserved, since the external Maxwell potentials (I.3.21) are independent of time. However, these "algebraic arguments" do not withdraw the question on the agreement of the Schrödinger definition with Bohr's postulate (I.4.3)!

Thus, the question arises of the mathematical interpretation of Bohr's postulate (I.4.3) in the Schrödinger theory. One of the *simplest interpretations* of the quantum jump (I.4.3) is the long-time asymptotics

$$\psi(\mathbf{x},t) \sim \varphi_{\pm}(\mathbf{x})e^{-i\omega_{\pm}t}, \qquad t \to \pm\infty, \tag{I.4.10}$$

for each finite energy solution, where $\omega_- = \omega_n$ and $\omega_+ = \omega_{n'}$. However, for the *linear Schrödinger equation* (I.4.7), such asymptotics are obviously wrong because of the *superposition principle:* for example, such asymptotics fail for solutions of the form $\psi(\mathbf{x},t) \equiv \varphi_1(\mathbf{x})e^{-i\omega_1 t} + \varphi_2(\mathbf{x})e^{-i\omega_2 t}$ with $\omega_1 \neq \omega_2$. It is exactly this contradiction which shows that the linear Schrödinger equation alone cannot serve as a basis for a theory compatible with Bohr's postulates.

Remark I.4.2. The Schrödinger definition (I.4.8) has been formalized by P. Dirac [24] and J. von Neumann [69]: they have identified quantum states with the rays in the Hilbert space, so the entire circular orbit (I.4.8) corresponds to one quantum state. Our main conjecture is that this formalization is indeed a *theorem*: namely, we suggest that the asymptotics (I.4.10) is an inherent property of the nonlinear Maxwell–Schrödinger equations (I.5.1). This conjecture is confirmed by perturbative arguments in the next section.

I.4.2 Perturbation theory

The remarkable success of the Schrödinger theory was in the explanation of Bohr's postulates via asymptotics (I.4.10) by means of perturbation theory in the case of *static external Maxwell potentials* (I.3.21). In this case, for "sufficiently good" external potentials and initial conditions, each finite energy solution of the Schrödinger equation (I.4.7) can be expanded into eigenfunctions:

$$\psi(\mathbf{x}, t) = \sum_n C_n \varphi_n(\mathbf{x}) e^{-i\omega_n t} + \psi_c(\mathbf{x}, t), \quad \psi_c(\mathbf{x}, t) = \int C(\omega) \varphi(x, \omega) e^{-i\omega t} \, d\omega, \quad (\text{I.4.11})$$

where integration is performed over the continuous spectrum of the Schrödinger operator H, and for any $R > 0$ we have

$$\int_{|\mathbf{x}| < R} |\psi_c(\mathbf{x}, t)|^2 \, d\mathbf{x} \to 0, \quad t \to \pm\infty; \quad (\text{I.4.12})$$

see, for example, [148, Theorem 21.1]. The substitution of the expansion (I.4.11) into the expression for currents (I.3.9) gives

$$\mathbf{j}(\mathbf{x}, t) = \sum_{nn'} \mathbf{j}_{nn'}(\mathbf{x}) e^{-i\omega_{nn'} t} + c.c. + \mathbf{j}_c(\mathbf{x}, t), \quad (\text{I.4.13})$$

where $\mathbf{j}_c(\mathbf{x}, t)$ contains the continuous frequency spectrum. This current enters the Maxwell equations. In the *unrationalized* Gaussian units (also called the Heaviside–Lorentz units [93, p. 221], [48, p. 781], [49]), the equations read as

$$\begin{cases} \operatorname{div} \mathbf{E}(\mathbf{x}, t) = \rho(\mathbf{x}, t), & \operatorname{curl} \mathbf{E}(\mathbf{x}, t) = -\frac{1}{c}\dot{\mathbf{B}}(\mathbf{x}, t) \\ \operatorname{div} \mathbf{B}(\mathbf{x}, t) = 0, & \operatorname{curl} \mathbf{B}(\mathbf{x}, t) = \frac{1}{c}\big(\mathbf{j}(\mathbf{x}, t) + \dot{\mathbf{E}}(\mathbf{x}, t)\big) \end{cases} . \quad (\text{I.4.14})$$

The second and third equations imply the representations of type (I.2.9):

$$\mathbf{B}(\mathbf{x}, t) = \operatorname{curl} \mathbf{A}(\mathbf{x}, t), \qquad \mathbf{E}(\mathbf{x}, t) = -\frac{1}{c}\dot{\mathbf{A}}(\mathbf{x}, t) - \nabla A_0(\mathbf{x}, t). \quad (\text{I.4.15})$$

We will assume the same Coulomb gauge as in (I.2.10):

$$\operatorname{div} \mathbf{A}(\mathbf{x}, t) \equiv 0. \quad (\text{I.4.16})$$

Then the Maxwell equations (I.4.14) are equivalent to the system

$$\frac{1}{c^2}\ddot{\mathbf{A}}(\mathbf{x}, t) = \Delta \mathbf{A}(\mathbf{x}, t) + \frac{1}{c}P\mathbf{j}(\mathbf{x}, t), \quad \Delta A_0(\mathbf{x}, t) = -\rho(\mathbf{x}, t), \quad \mathbf{x} \in \mathbb{R}^3, \quad (\text{I.4.17})$$

where P is the *orthogonal projection* in the real Hilbert space $L^2(\mathbb{R}^3) \otimes \mathbb{R}^3$ onto divergence-free vector fields.

Thus, the currents (I.4.13) on the right-hand side of the first of the two Maxwell equations (I.4.17) contain, besides the continuous spectrum, only the discrete frequencies $\omega_{nn'}$. Hence the discrete spectrum of the corresponding Maxwell radiated wave $\mathbf{A}(\mathbf{x}, t)$ also contains only these frequencies $\omega_{nn'}$. This justifies the Bohr rule (I.4.4) only in the first order of perturbation theory, since this calculation ignores the *back reaction* of radiation onto the atom.

The same arguments also suggest treating the jumps (I.4.3) as single-frequency asymptotics (I.4.10) (or rather, (V.1.1); see below) for solutions of the coupled Maxwell–Schrödinger equations (see (I.5.1) below) with static external Maxwell potentials (I.3.21). Indeed, the currents (I.4.13) on the right of the Maxwell equation from (I.5.1) produce radiation when nonzero frequencies $\omega_{nn'} \neq 0$ are present. This is due to the fact that $\mathbb{R} \setminus 0$ is the absolutely continuous spectrum of the Maxwell equations. However, this radiation cannot last forever, since it irrevocably carries the energy to infinity while the total energy is finite. Therefore, in the long-time limit only $\omega_{nn'} = 0$ survives, which means exactly that we have *single-frequency asymptotics* (I.4.10) in view of (I.4.12).

Remarks I.4.3. i) The perturbation arguments above cannot provide a rigorous justification of the long-time single-frequency asymptotics (I.4.10) for the coupled Maxwell–Schrödinger equations (I.5.1).

ii) The frequencies $\omega_{\pm}$ in these asymptotics correspond to some solutions of type (V.1.2) of the coupled Maxwell–Schrödinger equations. These frequencies are solutions of a *nonlinear eigenvalue problem* similar to (VI.4.12), and they are close to some eigenvalues of the linear Schrödinger operator (I.4.7).

iii) Similar asymptotics were justified in [212]–[220] for a list of model nonlinear Hamiltonian PDEs with the symmetry group $U(1)$; see Section VI.4. The strategy of the proofs relies on the radiation arguments above justified by subtle methods of harmonic analysis: the Titchmarsh convolution theorem, a nonlinear version of Kato's theorem on absence of embedded eigenvalues, and the theory of multipliers in the space of quasimeasures. However, for the coupled Maxwell–Schrödinger equations with static external Maxwell potentials, such a justification is still an open problem.

Exercise I.4.4. Prove the Maxwell representations (I.4.15). **Hint:** apply the Fourier transform to the second and third equations (I.4.14).

Exercise I.4.5. Prove the equivalence of the Maxwell equations (I.4.14) with the system (I.4.17) under the Coulomb gauge (I.4.16).

I.5 Coupled Maxwell–Schrödinger Equations

The Schrödinger equation (I.2.14) describes the evolution of the wave function in a given external Maxwell field with Maxwell potentials $\mathbf{A}^{\mathrm{ext}}(\mathbf{x}, t)$ and $A_0^{\mathrm{ext}}(\mathbf{x}, t)$. On the other hand, the charge and current densities (I.3.5), (I.3.9) generate their "own Maxwell field" with potentials satisfying the Maxwell equations (I.4.17). Hence, for a self-consistent description, these potentials $\mathbf{A}(\mathbf{x}, t)$ and $A_0(\mathbf{x}, t)$ should be added to the external Maxwell potentials in the Schrödinger equation (I.2.14). Thus, the coupled Maxwell–Schrödinger equations in the *unrationalized* Gaussian units (also called the Heaviside–Lorentz units [93, p. 221], [48, p. 781], [49]), read as (cf. [132])

$$\begin{cases} \frac{1}{c^2}\ddot{\mathbf{A}}(\mathbf{x}, t) = \Delta\mathbf{A}(\mathbf{x}, t) + \frac{1}{c}P\mathbf{j}(\mathbf{x}, t), \quad \Delta A_0(\mathbf{x}, t) = -\rho(\mathbf{x}, t) \\[2mm] i\hbar\dot{\psi}(\mathbf{x}, t) = \frac{1}{2\mathrm{m}}\left[-i\hbar\nabla - \frac{e}{c}(\mathbf{A}(\mathbf{x}, t) + \mathbf{A}^{\mathrm{ext}}(\mathbf{x}, t))\right]^2 \psi(\mathbf{x}, t) \\[2mm] \qquad\qquad + e(A_0(\mathbf{x}, t) + A_0^{\mathrm{ext}}(\mathbf{x}, t))\psi(\mathbf{x}, t) \end{cases}, \quad \mathbf{x} \in \mathbb{R}^3. \quad (\mathrm{I}.5.1)$$

The coupling is completed by expressions of the type (I.3.5) and (I.3.9) for the charge and current densities via the wave function: in the presence of the external Maxwell potentials,

$$\begin{cases} \rho(\mathbf{x}, t) = e|\psi(\mathbf{x}, t)|^2 \\[2mm] \mathbf{j}(\mathbf{x}, t) = \mathrm{Re}\left(\frac{e}{\mathrm{m}}\overline{\psi}(\mathbf{x}, t)[-i\hbar\nabla - \frac{e}{c}(\mathbf{A}(x, t) + \mathbf{A}^{\mathrm{ext}}(\mathbf{x}, t))]\psi(\mathbf{x}, t)\right) \end{cases}. \quad (\mathrm{I}.5.2)$$

These densities satisfy the charge continuity equation (I.3.11), which follows from the Schrödinger equation (I.5.1) by the same arguments as in (I.3.12)–(I.3.13).

Hamiltonian structure. The system (I.5.1) is formally Hamiltonian, with the Hamiltonian functional (which is the energy up to a factor)

$$\mathcal{H}(\mathbf{\Pi}, \mathbf{A}, \psi, t) = \frac{1}{2}\left[\frac{1}{c^2}\|\mathbf{\Pi}\|^2 + \|\mathrm{curl}\,\mathbf{A}\|^2\right] + \frac{1}{2}\langle\psi, H(\mathbf{A}, \psi, t)\psi\rangle, \quad (\mathrm{I}.5.3)$$

where $\|\cdot\|$ stands for the norm in the real Hilbert space $L^2(\mathbb{R}^3)$ and the brackets $\langle\cdot, \cdot\rangle$ stand for the Hermitian inner product in the complex Hilbert space $L^2(\mathbb{R}^3) \otimes \mathbb{C}$ (see (I.3.3)). The *Schrödinger operator* is defined by

$$H(\mathbf{A}, \psi, t) := \frac{1}{2\mathrm{m}}\left[-i\hbar\nabla - \frac{e}{c}(\mathbf{A}(\mathbf{x}) + \mathbf{A}^{\mathrm{ext}}(\mathbf{x}, t))\right]^2 + e\left[\frac{1}{2}A_0(\mathbf{x}) + A_0^{\mathrm{ext}}(\mathbf{x}, t)\right],$$

where $A_0(\mathbf{x}) := (-\Delta)^{-1}\rho$ with $\rho(\mathbf{x}) := e|\psi(\mathbf{x})|^2$. The system (I.5.1) can be written in the Hamiltonian form with variational derivatives (B.0.1) as

$$
\left\{
\begin{aligned}
&\frac{1}{c^2}\dot{\mathbf{A}}(t) = D_{\mathbf{\Pi}}\mathcal{H}(\mathbf{\Pi}(t), \mathbf{A}(t), \psi(t), t) \\[2mm]
&\frac{1}{c^2}\dot{\mathbf{\Pi}}(t) = -D_{\mathbf{A}}\mathcal{H}(\mathbf{\Pi}(t), \mathbf{A}(t), \psi(t), t) \\[2mm]
&i\hbar\dot{\psi}(t) = D_\psi\mathcal{H}(\mathbf{\Pi}(t), \mathbf{A}(t), \psi(t), t) = H(\mathbf{A}(t), \psi(t), t)\psi(t)
\end{aligned}
\right. , \qquad (\text{I.5.4})
$$

taking into account the fact that $(\psi, eA_0\psi) = (A_0, \rho) = ((-\Delta)^{-1}\rho, \rho)$ and hence

$$
D_\psi(\psi, eA_0\psi) = 4eA_0\psi.
$$

Therefore, the general *Noether theory of invariants* (see Section B) implies that the Hamiltonian (I.5.3) is conserved in the case of static external Maxwell potentials (I.3.21). For instance, in the case of the electron in an atom, $A_0^{\text{ext}}(\mathbf{x})$ is the Coulomb potential of the nucleus, while $\mathbf{A}^{\text{ext}}(\mathbf{x})$ is the magnetic potential of the magnetic field of the nucleus. On the other hand, the total charge

$$
Q(t) := \int \rho(\mathbf{x}, t)\, d\mathbf{x}
$$

is conserved for arbitrary time-dependent external potentials.

Symmetry group. The Hamiltonian (I.5.3) is invariant with respect to the action of the unitary group $U(1)$,

$$
T(e^{i\theta}) : (\mathbf{A}(\mathbf{x}), \mathbf{\Pi}(\mathbf{x}), \psi(\mathbf{x})) \mapsto (\mathbf{A}(\mathbf{x}), \mathbf{\Pi}(\mathbf{x}), \psi(\mathbf{x})e^{i\theta}), \qquad \theta \in [0, 2\pi]. \quad (\text{I.5.5})
$$

It is obvious that, given any solution $\mathbf{A}(\mathbf{x}, t)$, $\mathbf{\Pi}(\mathbf{x}, t)$, $\psi(\mathbf{x}, t)$ to the system (I.5.1), the functions $\mathbf{A}(\mathbf{x}, t)$, $\mathbf{\Pi}(\mathbf{x}, t)$, $\psi(\mathbf{x}, t)e^{i\theta}$ also form a solution.

Remark I.5.1. The existence of global solutions to the Cauchy problem with initial states of finite energy (I.5.3) for the system (I.5.1) in the entire space $\mathbb{R}^3$ without external potentials was proved in [131]. The uniqueness of solutions in the energy space was established in [129].

Remark I.5.2. The system (I.5.1) was essentially introduced by Schrödinger in his fundamental articles [78], and it underlies the entire theory of laser radiation [124]–[128].

Exercise I.5.3. Check that the Maxwell–Schrödinger system (I.5.4) admits the Hamiltonian representation (B.0.1) with the Hamiltonian (I.5.3) and calculate the corresponding operator J.

I.6 Hydrogen Spectrum

In this section, we calculate the spectrum of an atom with one electron and nucleus of charge $|e|Z$ (the hydrogen atom corresponds to $Z = 1$). In this case, the normalization condition (I.3.8) holds, which implies that

$$\psi \in \mathcal{E} := L^2(\mathbb{R}^3).$$

The Rutherford–Geiger–Marsden experiments of 1911 demonstrated that the positive charge $-e$ in atoms is concentrated in a relatively small region called the *nucleus*. Hence the corresponding electrostatic potential is Coulombic. The magnetic potential of the nucleus is assumed to be zero. In the *rationalized* Gaussian units [48, p. 781], [49], the potentials read as (see [49])

$$A_0^{\text{ext}}(\mathbf{x}) = -\frac{eZ}{|\mathbf{x}|}, \qquad \mathbf{A}^{\text{ext}}(\mathbf{x}) = 0. \tag{I.6.1}$$

Now the Schrödinger equation (I.4.7) becomes

$$i\hbar\dot{\psi}(t) = H\psi(t), \qquad H = -\frac{\hbar^2}{2\mathrm{m}}\Delta - \frac{e^2 Z}{|\mathbf{x}|}. \tag{I.6.2}$$

We must construct quantum stationary orbits $\psi(\mathbf{x}, t) = \varphi(\mathbf{x})e^{-i\omega t}$ as solutions of the eigenvalue problem (I.4.9):

$$H\varphi(\mathbf{x}) = E\varphi(\mathbf{x}), \qquad E = \hbar\omega. \tag{I.6.3}$$

The energy (I.3.2) of the quantum stationary orbit equals the eigenvalue E. This follows from the normalization condition (I.3.8),

$$\mathcal{H}(\varphi) := \langle \varphi, H\varphi \rangle = \hbar\omega\langle \varphi, \varphi \rangle = \hbar\omega. \tag{I.6.4}$$

The calculation of these eigenvalues yields the hydrogen spectrum in accordance with Bohr's rule for the frequency of radiation (I.4.4).

These eigenvalues were first calculated by Schrödinger [78, I] using the separation of variables in spherical coordinates. Later, these calculations were simplified by using irreducible representations of $\mathfrak{so}(3)$, the Lie algebra of the rotation group $SO(3)$; see [70]. These calculations confirmed the formula (A.6.9) of the Old Quantum Theory in Appendix A.

Theorem I.6.1. *The eigenvalues $E_n = \hbar\omega_n$ of the operator (I.6.2) are given by*

$$\omega_n = -\frac{B}{n^2}, \quad n = 1, 2, ..., \quad B = \frac{\mathrm{m}e^4}{2\hbar^3}Z^2 = 2\pi cRZ^2 \tag{I.6.5}$$

with the Rydberg constant

$$R = \frac{me^4}{4\pi\hbar^3 c}. \tag{I.6.6}$$

The currently recommended value of the Rydberg constant is

$$R = 109\ 737.315\ 681\ 60\ \text{cm}^{-1}; \tag{I.6.7}$$

see [13]. Formula (I.6.5) is also in a good agreement with the Balmer empirical terms (I.4.2), since $B = 2\pi cR \approx 18.8 \times 10^{15}\,\text{s}^{-1}$. In particular, the lowest energy level E_1 should be the (negative) ionization energy of the hydrogen atom, which is known experimentally to be about -13.6 eV. Using the data of (I.1.2) for $\hbar$ and c and (I.6.7) for R, we find that

$$E_1 = -2\pi\hbar cR \approx -21.79 \times 10^{-12}\,\text{erg} = -21.79 \times 10^{-19}\,\text{J}. \tag{I.6.8}$$

Dividing this by the electron charge $e = -1.602 \times 10^{-19}$ C (SI), we obtain $E_1/e = 13.60$ V.

We will show that the eigenfunctions and eigenvalues are numbered by the same quantum numbers n, l, m as in the formulae (A.6.9), (A.7.20) in Appendix A. However, the quantum number $l = n$ for the angular momentum is now forbidden, and formula (A.6.10) is slightly modified:

$$\begin{cases} H\varphi_{nlm} = E_n\varphi_{nlm}, \quad E_n = -\dfrac{me^4 Z^2}{2\hbar^2 n^2} \\[2mm] \hat{\mathbf{L}}^2\varphi_{nlm} = \hbar^2 l(l+1)\varphi_{nlm}, \ \hat{L}_3\varphi_{nlm} = \hbar m \varphi_{nlm} \end{cases} \left| \begin{array}{l} n = 1,\ 2,\ \ldots \\ l = 0,\ 1,\ \ldots,\ n-1 \\ m = -l,\ \ldots,\ l \end{array} \right| . \tag{I.6.9}$$

Here $\hat{\mathbf{L}}^2 := \hat{L}_1^2 + \hat{L}_2^2 + \hat{L}_3^2$. The indices n, l, m are called, respectively, the *principal*, *azimuthal*, and *magnetic quantum numbers*.

The eigenvalues for $\hat{\mathbf{L}}^2$ and $\hat{L}_3$ will be obtained from the general theorem on representations of the Lie algebra $\mathfrak{so}(3)$; see Lemma I.7.14. Let us note that the operators H, $\hat{\mathbf{L}}^2$, and $\hat{L}_3$ have common eigenfunctions due to the commutations

$$[H, \hat{\mathbf{L}}^2] = [H, \hat{L}_3] = [\hat{\mathbf{L}}^2, \hat{L}_3] = 0, \tag{I.6.10}$$

which hold by the spherical symmetry of the Coulomb potential of the nucleus. These commutations imply that the quantum observables $\mathbf{L}^2(\psi)$ and $L_3(\psi)$ are conserved along the solutions of the Schrödinger equation (I.6.2).

Exercise I.6.2. Check (I.6.10) for the Schrödinger operator (I.6.2).

I.6.1 Spherical symmetry and separation of variables

In this section, we obtain the eigenvalues (I.6.5) by solving the eigenvalue problem (I.6.3) for the Schrödinger operator (I.6.2).

The solution relies on the separation of variables, which is possible due to the spherical symmetry of the Schrödinger equation. The angular functions are chosen to be spherical functions that are the eigenfunctions of the spherical Laplacian. The key role in the calculation is played by decomposing the space $L^2(\mathbb{S}^2)$ into the sum of orthogonal eigenspaces of the *spherical Laplacian* which we will construct in the next section by analyzing the Lie algebra $\mathfrak{so}(3)$. The radial functions are obtained by solving the radial differential equation via the *Sommerfeld method of factorization.*

Rotational invariance

The basic argument in solving the eigenvalue problem (I.6.3) for the Schrödinger operator (I.6.2) is its spherical symmetry, which implies the conservation of the quantum angular momentum. Indeed, the Schrödinger operator H is invariant with respect to all rotations of the space $\mathbb{R}^3$. This means that we have the commutations

$$H\hat{R}_k(\theta_k) = \hat{R}_k(\theta_k)H, \qquad \theta_k \in \mathbb{R}, \qquad k = 1, 2, 3, \qquad (I.6.11)$$

where $R_k(\theta_k)$ is the space rotation by the angle θ_k (in radians) in the positive direction about the unit vector $\mathbf{e}_k$ (with $\mathbf{e}_1 = (1, 0, 0)$ and so on), and

$$(\hat{R}_k(\theta_k)f)(\mathbf{x}) := f(R_k(\theta_k)\mathbf{x}), \qquad \mathbf{x} \in \mathbb{R}^3.$$

The commutations (I.6.11) hold since the Laplacian Δ and the Coulomb potential are invariant under all rotations. Differentiating (I.6.11) in θ_k, we obtain:

$$[H, \partial_{\theta_k}] = 0, \qquad k = 1, 2, 3. \qquad (I.6.12)$$

Therefore, H also commutes with the operators

$$S_k := -i\partial_{\theta_k}, \qquad k = 1, 2, 3, \qquad (I.6.13)$$

and with the angular momentum operators $\hat{L}_k = \hbar S_k$ (see (I.3.18), (I.3.30)):

$$[H, S_k] = 0, \qquad [H, \hat{L}_k] = 0, \qquad k = 1, 2, 3. \qquad (I.6.14)$$

Remark I.6.3. The last commutation implies the conservation of quantum angular momenta $L_k(t) := \langle \psi(t), \hat{L}_k \psi(t) \rangle$ as was shown in Section I.3.6. Note that the conservation of the corresponding classical angular momentum played a crucial role in the determination of the hydrogen spectrum in the Bohr–Sommerfeld Old Quantum Theory (see Sections A.6 and A.7).

Exercise I.6.4. Check (I.6.11) for the Schrödinger operator (I.6.2).

Exercise I.6.5. Verify the commutation relations

$$[S_1, S_2] = iS_3, \qquad [S_2, S_3] = iS_1, \qquad [S_3, S_1] = iS_2. \qquad (I.6.15)$$

Hint: use (I.3.17).

Spherical functions

Let us now explain our general strategy in the proof of Theorem I.6.1. The commutations (I.6.12) suggest that we should solve the spectral problem (I.4.9) by the separation of variables using the following two general arguments:

I. The commutation relations (I.6.14) obviously imply that the operator

$$\mathbf{S}^2 := S_1^2 + S_2^2 + S_3^2 \qquad (I.6.16)$$

commutes with H:

$$[H, \mathbf{S}^2] = 0. \qquad (I.6.17)$$

Hence any eigenspace of the Schrödinger operator H is invariant with respect to $\mathbf{S}^2$ and any of the operators S_k. Moreover, $\mathbf{S}^2$ also commutes with each of S_k:

$$[\mathbf{S}^2, S_k] = 0, \qquad k = 1, 2, 3. \qquad (I.6.18)$$

Since the operators S_1, S_2, and S_3 do not commute, they cannot be diagonalized simultaneously. On the other hand, the operators H, $\mathbf{S}^2$, and S_k with a fixed value of k pairwise commute. Hence, we can expect that there is a basis of common eigenfunctions.

II. We will simultaneously diagonalize S_3 and $\mathbf{S}^2$. By (I.3.18), both of these operators act only on the angular variables in spherical coordinates. Hence, these operators act in the Hilbert space

$$\mathcal{E}_1 := L^2(\mathbb{S}^2), \qquad (I.6.19)$$

where $\mathbb{S}^2$ denotes the two-dimensional sphere $|\mathbf{x}| = 1$.

Theorem I.6.6. i) *For $\mathcal{E}_1$, there exists an orthonormal basis of spherical harmonics $Y_l^m(\theta, \varphi)$, which are common eigenfunctions of S_3 and $\mathbf{S}^2$:*

$$S_3 Y_l^m = m Y_l^m, \qquad \mathbf{S}^2 Y_l^m = l(l+1) Y_l^m, \qquad m = -l, -l+1, \ldots, l; \quad (I.6.20)$$

here $l = 0, 1, 2, \ldots$.
ii) *$Y_l^m(\theta, \varphi) = F_l^m(\theta) e^{im\varphi}$, where $F_l^m(\theta)$ are real-valued functions.*

We will prove this theorem in Section I.7. This theorem suggests constructing eigenfunctions of the Schrödinger operator H in the form

$$\psi(\mathbf{x}) = R(r) Y_l^m(\theta, \varphi). \qquad (I.6.21)$$

Each solution of the spectral problem (I.6.3) is a sum (or a series) of solutions of the particular form (I.6.21), since the spherical functions Y_l^m form the basis in $\mathcal{E}_1$. The solution of the spectral problem (I.6.20) relies on an investigation of commutation relations for the operators S_k, $k = 1, 2, 3$, i.e., on the Lie algebra that they generate.

Exercise I.6.7. Verify (I.6.18). **Hint:** use (I.6.15).

I.6.2   Spherical coordinates

To determine the radial functions in (I.6.21), let us express the Laplace operator Δ in the spherical coordinates r, θ, φ such that

$$x^1 = r \sin\theta \cos\varphi, \quad x^2 = r \sin\theta \sin\varphi, \quad x^3 = r \cos\theta. \tag{I.6.22}$$

The operator Δ is symmetric in the real Hilbert space $L^2(\mathbb{R}^3)$. Hence, it is defined uniquely by its quadratic form $\langle \psi, \Delta\psi \rangle$ with $\psi \in C_0^\infty(\mathbb{R}^3)$. In the spherical coordinates,

$$\langle \psi, \Delta\psi \rangle = -\langle \nabla\psi, \nabla\psi \rangle = -\int_0^\infty dr \int_0^\pi d\theta \int_0^{2\pi} d\varphi \, |\nabla\psi(r,\theta,\varphi)|^2 r^2 \sin\theta. \tag{I.6.23}$$

Geometrically, it is evident that

$$\nabla\psi(r,\theta,\varphi) = \mathbf{e}_r \partial_r \psi + \mathbf{e}_\theta \frac{\partial_\theta \psi}{r} + \mathbf{e}_\varphi \frac{\partial_\varphi \psi}{r \sin\theta}, \tag{I.6.24}$$

where $\mathbf{e}_r(r,\theta,\varphi)$, $\mathbf{e}_\theta(r,\theta,\varphi)$, and $\mathbf{e}_\varphi(r,\theta,\varphi)$ are the orthonormal vectors proportional to ∂_r, ∂_θ, ∂_φ, respectively, at the point (r,θ,φ). Therefore, (I.6.23) becomes

$$\langle \psi, \Delta\psi \rangle = -\int_0^\infty dr \int_0^\pi d\theta \int_0^{2\pi} d\varphi \left(\left| \partial_r \psi \right|^2 + \left| \frac{\partial_\theta \psi}{r} \right|^2 + \left| \frac{\partial_\varphi \psi}{r \sin\theta} \right|^2 \right) r^2 \sin\theta. \tag{I.6.25}$$

Integrating by parts,

$$\langle \psi, \Delta\psi \rangle = \int_0^\infty dr \int_0^\pi d\theta \int_0^{2\pi} d\varphi \, \bar\psi \left(\partial_r(r^2 \partial_r \psi) + \frac{\partial_\theta(\sin\theta \, \partial_\theta \psi)}{\sin\theta} + \frac{\partial_\varphi^2 \psi}{\sin^2\theta} \right) \sin\theta$$

$$= \left\langle \psi, \frac{\partial_r(r^2 \partial_r \psi)}{r^2} + \frac{\partial_\theta(\sin\theta \, \partial_\theta \psi)}{r^2 \sin\theta} + \frac{\partial_\varphi^2 \psi}{r^2 \sin^2\theta} \right\rangle. \tag{I.6.26}$$

Therefore, the Laplace operator in spherical coordinates reads as follows:

$$\Delta\psi = r^{-2}\partial_r(r^2 \partial_r \psi) + \frac{\partial_\theta(\sin\theta \, \partial_\theta \psi)}{r^2 \sin\theta} + \frac{\partial_\varphi^2 \psi}{r^2 \sin^2\theta} = r^{-2}\partial_r(r^2 \partial_r \psi) + r^{-2}\Lambda\psi; \tag{I.6.27}$$

here Λ is the following differential operator on the sphere $\mathbb{S}^2$ in the coordinates θ, φ:

$$\Lambda f(\theta,\varphi) = \frac{1}{\sin\theta}\partial_\theta(\sin\theta \, \partial_\theta f(\theta,\varphi)) + \frac{1}{\sin^2\theta}\partial_\varphi^2 f(\theta,\varphi). \tag{I.6.28}$$

Definition I.6.8. *The operator Λ is called the* spherical Laplace operator *(or Laplace–Beltrami operator).*

Exercise I.6.9. Check all the calculations (I.6.23)–(I.6.28).

Problem I.6.10. Verify the identity

$$\Lambda = -\mathbf{S}^2. \tag{I.6.29}$$

Hint: Both sides are second-order *spherically symmetric* elliptic operators.

I.6.3   Radial equation

Here we deduce Theorem I.6.1 from Theorem I.6.6 by substituting (I.6.21) into (I.6.3), taking into account (I.6.27), (I.6.29), and (I.6.20), and applying the Sommerfeld method of factorization. First, omitting $Y_l^m(\theta, \varphi)$ and multiplying by $-2m/\hbar^2$, we arrive at the *radial equation*

$$-\frac{2mE}{\hbar^2}R(r) = r^{-2}\partial_r r^2 \partial_r R(r) - \frac{l(l+1)}{r^2}R(r) + \frac{2me^2 Z}{\hbar^2 r}R(r), \quad r > 0. \quad (I.6.30)$$

If we let $r \to \infty$, this equation becomes

$$-\frac{2mE}{\hbar^2}R(r) \sim R_l''(r). \quad (I.6.31)$$

This suggests that $E < 0$ and that the asymptotics $R(r) \sim e^{-\gamma r}$ should hold as $r \to \infty$, where

$$\gamma = \sqrt{-2mE}/\hbar > 0. \quad (I.6.32)$$

Correspondingly, we set $R(r) = e^{-\gamma r} F(r)$ following the Sommerfeld *method of factorization*. Substitution into (I.6.30) gives

$$F'' + \left[\frac{2}{r} - 2\gamma\right]F' + \left[\frac{d}{r} - \frac{l(l+1)}{r^2}\right]F = 0, \qquad r > 0, \quad (I.6.33)$$

where

$$d = b - 2\gamma, \qquad b = 2me^2 Z/\hbar^2. \quad (I.6.34)$$

Finally, we introduce the new variable $\rho = 2\gamma r$. Then (I.6.33) becomes

$$f'' + \left[\frac{2}{\rho} - 1\right]f' + \left[\frac{\lambda - 1}{\rho} - \frac{l(l+1)}{\rho^2}\right]f = 0, \qquad \rho > 0, \quad (I.6.35)$$

where $f(\rho) = F(r)$ and $\lambda = b/(2\gamma)$. Now let us look for f in the form

$$f(\rho) = \rho^s(a_0 + a_1\rho + a_2\rho^2 + \dots) \equiv \rho^s L(\rho), \quad (I.6.36)$$

where $a_0 \neq 0$. Substituting (I.6.36) into (I.6.35), we obtain for $\rho > 0$:

$$\rho^2 L'' + \left[2s\rho + \left[\frac{2}{\rho} - 1\right]\rho^2\right]L' + \left[s(s-1) + \left[\frac{2}{\rho} - 1\right]s\rho\right.$$
$$\left. + \left[\frac{\lambda - 1}{r} - \frac{l(l+1)}{\rho^2}\right]\rho^2 L = 0. \right. \quad (I.6.37)$$

This gives

$$\rho^2 L'' + \rho[2(s+1) - \rho]L' + [\rho(\lambda - 1 - s) + s(s+1) - l(l+1)]L = 0, \quad \rho > 0. \quad (I.6.38)$$

Setting $\rho = 0$, we find that $s(s + 1) - l(l + 1) = 0$. Hence either $s = l$ or $s = -l - 1$. For $s = l$, equation (I.6.38) becomes

$$\rho L'' + [2(l + 1) - \rho]L' + [\lambda - 1 - l]L = 0, \qquad \rho > 0. \tag{I.6.39}$$

The case $s = -l - 1$ is forbidden, since the corresponding eigenfunction $\psi(\mathbf{x})$ is not a function of finite energy due to $\nabla\psi(\mathbf{x}) \notin L^2(\mathbb{R}^3)$.

Substituting $L(\rho) = a_0 + a_1\rho + a_2\rho^2 + \dots$ and equating the coefficients of terms with the same powers of ρ, we obtain the system

$$\begin{cases} \rho^0 : 2(l+1)a_1 + (\lambda - 1 - l)a_0 = 0 \\ \rho^1 : 2a_2 + 2(l + 1)2a_2 - a_1 + (\lambda - 1 - l)a_1 = 0 \\ \rho^2 : 3 \cdot 2a_3 + 2(l+1)3a_3 - 2a_2 + (\lambda - 1 - l)a_2 = 0 \\ \dots \\ \rho^k : (k+1)ka_{k+1} + 2(l+1)(k+1)a_{k+1} - ka_k + (\lambda - 1 - l)a_k = 0 \\ \dots \end{cases} \qquad , \tag{I.6.40}$$

which results in the recurrence equations

$$a_{k+1} = \frac{k - (\lambda - 1 - l)}{(k + 1)(k + 2l + 2)}a_k. \tag{I.6.41}$$

This shows that

$$a_{k+1}/a_k \sim 1/k, \qquad k \to \infty, \tag{I.6.42}$$

provided all $a_k \neq 0$. However, this asymptotics implies the bound

$$|L(\rho)| \geq Ce^\rho = Ce^{2\gamma r}, \qquad C > 0, \tag{I.6.43}$$

and then, by (I.6.36),

$$R(r) = F(r)e^{-\gamma r} = f(\rho)e^{-\gamma r} = \rho^s L(\rho)e^{-\gamma r} \to \infty, \qquad r \to \infty. \tag{I.6.44}$$

This, however, contradicts the fact that the energy of a quantum stationary state is finite. Therefore, $a_{\bar{k}+1} = 0$ and $a_{\bar{k}} \neq 0$ for some $\bar{k} = 0, 1, 2, \dots$. Hence $\bar{k} - (\lambda - 1 - l) = 0$, and so

$$\lambda = \frac{b}{2\gamma} = \bar{k} + l + 1 = n = 1, 2, \dots. \tag{I.6.45}$$

Substituting γ and b from (I.6.32) and (I.6.34), we finally arrive at

$$E = E_n = -\frac{me^4 Z^2}{2\hbar^2 n^2}, \qquad n = 1, 2, \dots, \tag{I.6.46}$$

which coincides with (I.6.5) and (I.6.6). This proves Theorem I.6.1.

Exercise I.6.11. Check all the calculations (I.6.33)–(I.6.46).

I.6.4 Eigenfunctions

From (I.6.21) and Theorem I.6.6, we obtain eigenfunctions corresponding to eigenvalues E_n. In spherical coordinates,

$$\varphi_{nlm} := Ce^{-r/r_n} P_{nl}(r) F_l^m(\theta) e^{im\varphi}, \quad n=1, 2, \ldots, \ 0 \le l \le n-1, \ m = -l, \ldots, l.$$
$$(I.6.47)$$

Here

$$r_n := 1/\gamma = \hbar/\sqrt{-2mE_n} = \hbar^2 n/(me^2) \tag{I.6.48}$$

and $P_{nl}(r) = \rho^l L(\rho)$ is a polynomial function of degree $\bar{k} + l = n - 1 \ge l$, according to (I.6.45). Now formulae (I.6.9) are proved, since the expressions for $\hat{\mathbf{L}}^2$ and $\hat{L}_3$ follow from (I.6.20).

Note that, by (I.6.41), the signs of the coefficients a_k alternate (if $a_{\bar{k}} \in \mathbb{R}$), since we have $k \le \bar{k} = \lambda - 1 - l$. Hence the (associated) *Laguerre polynomial* $L(\rho) = a_0 + \ldots + a_k \rho^k$ does not vanish for $\rho < 0$, and $n - 1 = k + l$ is the number of zeros of the radial function $P_{nl}(r)$ for $\rho \ge 0$.

The ground state φ_1 is defined as the eigenfunction corresponding to the lowest energy $E_1 = -2\pi\hbar cR = -me^4/(2\hbar^2)$:

$$\varphi_1(\mathbf{x}) = \varphi_{001} = C_1 e^{-|\mathbf{x}|/r_1}, \qquad r_1 = \frac{\hbar^2}{me^2} \sim 1\text{Å} = 10^{-8} \text{ cm}. \tag{I.6.49}$$

It is spherically symmetric, and is concentrated in a very small region of the *Bohr radius* r_1.

Exercise I.6.12. Calculate the multiplicity of the eigenvalues E_n.
Hint: calculate the sum $\sum_{0 \le l \le n-1} (2l + 1)$.

Problem I.6.13. Prove that all roots of the Laguerre polynomial $L(r)$ are real. **Hint:** see [105].

I.7 Spherical Eigenvalue Problem*

The proof of Theorem I.6.6 depends on the rotational of the Schrödinger operator and on the classification of irreducible representations of the Lie algebra $\mathfrak{so}(3)$.

I.7.1 The Hilbert–Schmidt argument

We start with the diagonalization of the operator $\mathbf{S}^2 = S_1^2 + S_2^2 + S_3^2$ defined in (I.6.16). Recall that the operators S_k from (I.6.13) act in the Hilbert space $\mathcal{E}_1 = L^2(\mathbb{S}^2)$.

Lemma I.7.1. *The operator $\mathbf{S}^2$ in the complex Hilbert space $\mathcal{E}_1$ is selfadjoint and admits the spectral resolution*

$$L^2(\mathbb{S}^2) = \oplus_{l=0}^{\infty} L(l), \qquad (\text{I.7.1})$$

where $L(l) \subset C^{\infty}(\mathbb{S}^2)$ are finite-dimensional orthogonal eigenspaces in $\mathcal{E}_1$, and $\mathbf{S}^2|_{L(l)} = \lambda_l$, where $\lambda_l \to \infty$ as $l \to \infty$.

Proof. Step i) Each operator S_k is selfadjoint in $\mathcal{E}_1$. Therefore, $\mathbf{S}^2$ is a selfadjoint nonnegative operator in $\mathcal{E}_1$.

Step ii) The operator $\mathbf{S}^2$ is a nonnegative elliptic second-order operator on $\mathbb{S}^2$. This follows from (I.6.29) and (I.6.28). Hence the operator

$$\mathbf{S}^2 + 1 : H^2(\mathbb{S}^2) \to H^0(\mathbb{S}^2) = L^2(\mathbb{S}^2)$$

is invertible by the theory of pseudodifferential operators [161], where $H^s(\mathbb{S}^2)$ denotes the corresponding Sobolev function space on the sphere $\mathbb{S}^2$. Accordingly, the operator $(\mathbf{S}^2 + 1)^{-1}$ is selfadjoint and compact on $\mathcal{E}_1$ by the Sobolev embedding theorem. Hence the resolution (I.7.1) holds by the Hilbert–Schmidt theorem [159, I, Theorem VI.16] applied to the operator $(\mathbf{S}^2 + 1)^{-1}$. $\qquad \square$

We can assume that $\lambda_l \neq \lambda_{l'}$ for $l \neq l'$. Then all the spaces $L(l)$ are invariant with respect to rotations of the sphere since $\mathbf{S}^2$ commutes with rotations. Similarly, we obtain

Corollary I.7.2. *All spaces $L(l)$ are invariant with respect to S_k with $k = 1, 2, 3$.*

Exercise I.7.3. Prove that the operator $\mathbf{S}^2$ is elliptic and selfadjoint in $L^2(\mathbb{S}^2)$, and $(\mathbf{S}^2 + 1)^{-1}$ is a compact operator in $L^2(\mathbb{S}^2)$.

---

*This section can be skipped at first reading.

I.7.2   The Lie algebra of quantum angular momenta

The linear span of S_1, S_2 and S_3 is a Lie algebra due to (I.6.15). Moreover, the commutation relations (I.6.15) imply the following two lemmas.

Lemma I.7.4. *All spaces $L(l)$ are invariant with respect to S_k for $k = 1$, 2, 3.*

Let us denote
$$S_\pm = S_1 \pm iS_2.$$

Lemma I.7.5. *All spaces $L(l)$ are invariant with respect to $S_\pm$. Moreover,*

$$[S_3, S_\pm] = \pm S_\pm, \tag{I.7.2}$$

$$\mathbf{S}^2 = S_+ S_- + S_3(S_3 - 1) = S_- S_+ + S_3(S_3 + 1), \tag{I.7.3}$$

$$\mathbf{S}^2 = \frac{1}{2}(S_+ S_- + S_- S_+) + S_3^2. \tag{I.7.4}$$

Exercise I.7.6. Prove Lemmas I.7.4 and I.7.5.

I.7.3   Irreducible representations

Here we give a complete classification of all possible triples of Hermitian operators satisfying the commutation relations (I.6.15).

Proposition I.7.7 (see [70]). *Let E be a nonzero finite-dimensional complex linear space with a Hermitian inner product. Suppose that*

i) *the linear Hermitian operators S_k, acting on E, $k = 1, 2, 3$, satisfy the commutation relations (I.6.15);*
ii) $\mathbf{S}^2 := S_1^2 + S_2^2 + S_3^2$ *is a scalar: $\mathbf{S}^2 = \alpha \geq 0$;*
iii) *the space E is an irreducible space, i.e., it does not contain any nontrivial subspace invariant under all the S_k.*

Then there exists a spin number $J \in \{0, \frac{1}{2}, 1, \frac{3}{2}, 2, \dots\}$ and an orthonormal basis for E, $\{e_m : m = -J, -J+1, \dots, J-1, J\}$, such that $\alpha = J(J+1)$ and

$$S_1 e_m = \frac{s^+_{Jm}}{2} e_{m+1} + \frac{s^-_{Jm}}{2} e_{m-1}, \quad S_2 e_m = \frac{s^+_{Jm}}{2i} e_{m+1} - \frac{s^-_{Jm}}{2i} e_{m-1}, \quad S_3 e_m = m e_m, \tag{I.7.5}$$

where

$$s^\pm_{Jm} = \sqrt{(J \mp m)(J \pm m + 1)}. \tag{I.7.6}$$

Proof. We set $S_\pm := S_1 \pm iS_2$ as before. Then all the relations (I.7.2)–(I.7.4) hold by (I.6.15), as in Lemma I.7.4. Since S_3 is a Hermitian operator, there exists at least one eigenvector e_M, and so we have $S_3 e_M = M e_M$ with a real eigenvalue $M \in \mathbb{R}$. The following lemma is obvious.

Lemma I.7.8. *Let e_m be an eigenvector of S_3 with eigenvalue m. Then either $e_\pm := S_\pm e_m = 0$ or $e_\pm$ is an eigenvector of S_3 with eigenvalue $m \pm 1$.*

By this lemma, each vector $e_{M-k} := (S_-)^k e_M$ with $k = 0, 1, \ldots$, is either an eigenvector of S_3 with eigenvalue $M - k$ or zero. Similarly, each vector $e_{M+k} := (S_+)^k e_M$ with $k = 0, 1, \ldots$, is either an eigenvector of S_3 with eigenvalue $M + k$ or zero.

Since E is finite-dimensional, both sequences of the eigenvectors must terminate by zero: $e_{M_-} \neq 0$, but $e_M = 0$ for $M < M_-$, and similarly, $e_{M_+} \neq 0$, but $e_M = 0$ for $M > M_+$.

Let us show that M_- coincides with the unique nonpositive solution of the equation $\alpha = M(M - 1)$, and similarly, M_+ coincides with the unique nonnegative solution of the equation $\alpha = M(M + 1)$. Indeed, identities (I.7.3) imply that $S_-^* S_- = \alpha - S_3(S_3 - 1)$ and $S_+^* S_+ = \alpha - S_3(S_3 + 1)$. Hence, for any $m = M_-, M_- + 1, \ldots, M$,

$$0 \leq \|S_- e_m\|^2 = \langle e_m, S_-^* S_- e_m \rangle = \langle e_m, (\alpha - S_3(S_3 - 1)) e_m \rangle$$
$$= [\alpha - m(m - 1)] \|e_m\|^2, \qquad (\text{I.7.7})$$

and similarly, for any $m = M, M + 1, \ldots, M_+$,

$$0 \leq \|S_+ e_m\|^2 = \langle e_m, S_+^* S_+ e_m \rangle = \langle e_m, (\alpha - S_3(S_3 + 1)) e_m \rangle$$
$$= [\alpha - m(m + 1)] \|e_m\|^2. \qquad (\text{I.7.8})$$

Hence, the expression $\alpha - m(m - 1)$ is nonnegative for $m = M_-, \ldots, M$, and vanishes at M_-, since $e_{M_- - 1} := S_- e_{M_-} = 0$. Similarly, the expression $\alpha - m(m + 1)$ is nonnegative for $m = M, \ldots, M_+$, and vanishes at M_+. Therefore, $M_+ = -M_- =: J$, where $2J = 1, 2, \ldots$, since m runs in integer steps from $-J$ to J. The vectors e_m, $m = -J, -J + 1, \ldots, J - 1, J$, constitute a basis of the space E, since E is irreducible.

Thus, $\alpha - J(J + 1) = 0$, and so (I.7.7) and (I.7.8) imply that

$$\|S_- e_m\|^2 = [J(J+1) - m(m-1)] \|e_m\|^2, \quad \|S_+ e_m\|^2 = [J(J+1) - m(m+1)] \|e_m\|^2.$$
$$(\text{I.7.9})$$

Therefore,

$$S_+ e_m = s_{Jm}^+ e_{m+1}, \qquad S_- e_m = s_{Jm}^- e_{m-1} \qquad (\text{I.7.10})$$

with suitable normalized vectors e_m. Now formulae (I.7.5) follow. $\qquad \square$

Denote by Z_{Jm}^+ the product $Z_{Jm}^+ = s_{J, -J}^+ \cdots s_{J, m-1}^+$.

Corollary I.7.9. *The vectors $Y_J^m := S_+^{J+m} e_{-J} / Z_{Jm}^+$, $m = -J, \ldots, J$, constitute an orthonormal basis of E.*

Definition I.7.10. *For $J = 0, \frac{1}{2}, 1, \frac{3}{2}, 2, \ldots$, denote by $D(J)$ the irreducible space E from Proposition I.7.7 with the operators S_k defined by (I.7.5).*

Example I.7.11. For $J = \frac{1}{2}$, the operators S_k are represented by the matrices $\hat{s}_k := \frac{1}{2}\sigma_k$ in the orthonormal basis $(Y_{1/2}^{-1/2}, Y_{1/2}^{1/2})$, where σ_k are the Pauli matrices:

$$\sigma_1 = \begin{pmatrix} 0 & 1 \\ 1 & 0 \end{pmatrix}, \qquad \sigma_2 = \begin{pmatrix} 0 & -i \\ i & 0 \end{pmatrix}, \qquad \sigma_3 = \begin{pmatrix} 1 & 0 \\ 0 & -1 \end{pmatrix}. \qquad (I.7.11)$$

Exercise I.7.12. Deduce (I.7.11) from (I.7.5).

Exercise I.7.13. Prove Lemma I.7.8.

I.7.4 Spherical harmonics. Proof of Theorem I.6.6

Now we can prove Theorem I.6.6. Recall that λ_l are distinct for distinct l. Proposition I.7.7 implies that each eigenspace $L(l)$ is isomorphic to the direct sum of $M(l, J)$ copies of the spaces $D(J)$ with the same values of

$$J = 0, \frac{1}{2}, 1, \frac{3}{2}, 2, \ldots$$

Indeed, distinct values of J are impossible, since the eigenvalue of $\mathbf{S}^2$ in $D(J)$ (which is equal to $J(J+1)$) is a strictly increasing function of $J \geq 0$. Hence Theorem I.6.6 i) will follow from the next lemma.

Lemma I.7.14. i) $M(n, J) = 0$ *for each n if J is half-integer.*

ii) *For every integer $J = 1, 2, 3, \ldots$, there exists a unique $l = l(J)$ such that $M(l, J) = 1$.*

Proof. i) Each eigenspace $L(l)$ is isomorphic to a direct sum of subspaces $D(J)$. Let us consider one of these subspaces. According to (I.6.13), in the spherical coordinates (I.6.22), $S_3 = -i\partial_\varphi$. Then, by (I.7.5),

$$S_3 e_m(\theta, \varphi) = -i\partial_\varphi e_m(\theta, \varphi) = m e_m(\theta, \varphi) \qquad (I.7.12)$$

for each eigenvector $e_m \in D(J)$ of S_3. Therefore,

$$e_m(\theta, \varphi) = F_J^m(\theta) e^{im\varphi}, \qquad (I.7.13)$$

an so m is an integer, because the function $e_m(\theta, \varphi)$ must be single-valued. Hence J is also an integer.

ii) Let us consider the least eigenvalue $m = -J$ in the subspace $D(J)$. By (I.7.13), the corresponding eigenfunction is $e_{-J}(\theta, \varphi) = F_J^{-J}(\theta) e^{-iJ\varphi}$. We have $S_- e_{-J}(\theta, \varphi) = 0$ by Lemma I.7.8. In spherical coordinates,

$$S_- = -e^{-i\varphi}[\partial_\theta - i \cot\theta \, \partial_\varphi], \qquad S_+ = e^{i\varphi}[\partial_\theta + i \cot\theta \, \partial_\varphi]. \qquad (I.7.14)$$

Hence, $(\partial_\theta - J \cot \theta)F_J^{-J}(\theta) = 0$ and $F_J^{-J} = C \sin^J \theta$. This means that $-J$ is a simple eigenvalue. Hence $M(l, J) \le 1$ for all l.

It remains to verify that $M(l, J) \ne 0$ for some l. This is equivalent to the existence of an eigenvector of $\mathbf{S}^2$ with eigenvalue $J(J+1)$. However, this eigenvector is given by above constructed function $e_{-J}(\theta, \varphi) = C \sin^J \theta e^{-iJ\varphi}$. Indeed, (I.7.3) implies that

$$\mathbf{S}^2 e_{-J} = S_+ S_- e_{-J} + S_3(S_3 - 1)e_{-J} = J(J+1)e_{-J},$$

since $S_- e_{-J} = 0$, and $S_3 e_{-J} = -Je_{-J}$, because $S_3 = -i\partial_\varphi$. $\qquad\square$

Now Theorem I.6.6 i) is proved, since the functions

$$Y_l^m(\theta, \varphi) := S_+^{l+m} e_{-l}/Z_{lm}^+, \qquad m = -l, \dots, l \qquad (\text{I.7.15})$$

form an orthonormal basis of the space $D(l)$ by Corollary I.7.9. Finally, Theorem I.6.6 ii) follows from (I.7.15), because

$$S_+^{l+m} e_{-l} = \left(e^{i\varphi}[\partial_\theta + i \cot \theta \; \partial_\varphi]\right)^{l+m}(\sin^l \theta e^{-il\varphi}) = G_l^m(\theta)e^{im\varphi}, \qquad (\text{I.7.16})$$

where $G_l^m(\theta)$ is a *real-valued function*.

Exercise I.7.15. Prove that $G_l^m(\theta)$ is a real-valued function. **Hint:** apply complex conjugation to (I.7.16) and then substitute $\varphi \mapsto -\varphi$.

I.7.5 Angular momentum in spherical coordinates

Let us prove (I.7.14). First, we rewrite (I.6.24) as follows:

$$\nabla \psi(r, \theta, \varphi) = \mathbf{e}_r \partial_r \psi + \mathbf{e}_\theta \frac{\partial_\theta \psi}{r} + \mathbf{e}_\varphi \frac{\partial_\varphi \psi}{r \sin \theta} = \mathbf{e}_1 \nabla_1 \psi + \mathbf{e}_3 \nabla_3 \psi + \mathbf{e}_3 \nabla_3 \psi; \quad (\text{I.7.17})$$

here $\mathbf{e}_1 := (1, 0, 0)$, etc. It is geometrically evident that

$$\begin{cases} \mathbf{e}_r = (\mathbf{e}_1 \cos \varphi + \mathbf{e}_2 \sin \varphi) \sin \theta + \mathbf{e}_3 \cos \theta \\ \mathbf{e}_\theta = (\mathbf{e}_1 \cos \varphi + \mathbf{e}_2 \sin \varphi) \cos \theta - \mathbf{e}_3 \sin \theta \\ \mathbf{e}_\varphi = \mathbf{e}_2 \cos \varphi - \mathbf{e}_1 \sin \varphi \end{cases} \qquad (\text{I.7.18})$$

Substituting this into (I.7.17), we obtain:

$$\begin{cases} \nabla_1 = \sin \theta \cos \varphi \partial_r + \cos \theta \cos \varphi \frac{\partial_\theta}{r} - \sin \varphi \frac{\partial_\varphi}{r \sin \theta} \\ \nabla_2 = \sin \theta \sin \varphi \partial_r + \cos \theta \sin \varphi \frac{\partial_\theta}{r} + \cos \varphi \frac{\partial_\varphi}{r \sin \theta} \\ \nabla_3 = \cos \theta \partial_r - \sin \varphi \frac{\partial_\theta}{r} \end{cases} \qquad (\text{I.7.19})$$

Finally, substituting (I.7.19) and (I.6.22) into $S_k = -i(\mathbf{x} \times \nabla)_k$, we can write

$$\left\{ \begin{array}{rcl} S_1 & = & i(\sin\varphi\partial_\theta + \cot\theta\,\cos\varphi\partial_\varphi) \\ S_2 & = & i(-\cos\varphi\partial_\theta + \cot\theta\,\sin\varphi\partial_\varphi) \\ S_3 & = & -i\partial_\varphi \end{array} \right| . \qquad (I.7.20)$$

Now (I.7.14) follows from the first two formulae of (I.7.20).

Exercise I.7.16. Check (I.7.18)–(I.7.20) and deduce (I.7.14).

Chapter II

Scattering of Light and Particles

The scattering of light and electron beams by a hydrogen atom can be described by the coupled Maxwell–Schrödinger equations. However, the coupled equations are nonlinear, and so the calculations can be done only by a perturbation procedure neglecting the selfaction, i.e., in the Born approximation. The corresponding scattering cross sections are similar to the classical ones given by the Thomson and Rutherford formulae, respectively.

The calculations rely on the *limiting amplitude principle* and the *limiting absorption principle*, which, in particular, are able to explain the Einstein rules for the photoelectric effect.

In this chapter, the Maxwell equations are presented in the *rationalized Gaussian units*, in order to facilitate the comparison of formulae with our references such as [48, p. 781] and [49].

II.1   Classical Scattering of Light

Scattering of light by matter is well known from everyday observations. It is clearly explained by the interaction of the electromagnetic wave with atomic electrons. The scattering by the nuclei is negligible, since they are relatively heavy. The incident electromagnetic wave is modeled by a plane wave satisfying the free Maxwell equations, while matter is composed of classical point-like oscillators subject to Lorentz forces. The inconsistency, inherent in the concept of point charged particles, is circumvented by neglecting the selfaction. Application of the Hertzian dipole radiation formula gives the Thomson formula for differential cross section.

II.1.1   Incident wave

In 1861, Maxwell identified light with electromagnetic waves. Incident light is described by a *plane wave*

$$\phi_{\text{in}}(\mathbf{x}, t) = 0, \qquad \mathbf{A}_{\text{in}}(\mathbf{x}, t) = A\mathbf{e}_3\Theta(ct - x^1)\sin k(x^1 - ct), \qquad \text{(II.1.1)}$$

where Θ is the Heaviside function, k is the wave number, and $\mathbf{e}_3 = (0, 0, 1)$ is the polarization. The incident wave is a solution of the free Maxwell equations with zero charge and current densities:

$$\Box\phi_{\text{in}}(\mathbf{x}, t) = 0, \qquad \Box\mathbf{A}_{\text{in}}(\mathbf{x}, t) = 0, \qquad (\mathbf{x}, t) \in \mathbb{R}^4, \qquad \text{(II.1.2)}$$

where $\Box := \frac{1}{c^2}\partial_t^2 - \Delta$ is the d'Alembert operator. According to (I.2.9), the corresponding Maxwell fields are expressed in terms of the potentials:

$$\mathbf{E}_{\text{in}}(\mathbf{x}, t) = -\nabla\phi_{\text{in}}(\mathbf{x}, t) - \frac{1}{c}\dot{\mathbf{A}}_{\text{in}}(\mathbf{x}, t), \quad \mathbf{B}_{\text{in}}(\mathbf{x}, t) = \text{curl}\,\mathbf{A}_{\text{in}}(\mathbf{x}, t). \quad \text{(II.1.3)}$$

For the potentials (II.1.1), we have:

$$\left\{ \begin{array}{lll} \mathbf{E}_{\text{in}}(\mathbf{x}, t) & & kA\mathbf{e}_3\Theta(ct - x^1)\cos k(x^1 - ct) \\ \\ \mathbf{B}_{\text{in}}(\mathbf{x}, t) & = & -kA\mathbf{e}_2\Theta(ct - x^1)\cos k(x^1 - ct) \end{array} \right. \Bigg| . \qquad \text{(II.1.4)}$$

Let us consider the scattering of the incident wave (II.1.4) by a classical electron. The energy flux in the Maxwell field, i.e., the Poynting vector, is defined by formula [48, (6.109)]. In the Gaussian units [48, p. 781], [49], this vector reads as (see [77, Section 45])

$$\mathbf{S}_{\text{in}}(\mathbf{x}, t) = \frac{c}{4\pi}\mathbf{E}_{\text{in}}(\mathbf{x}, t) \times \mathbf{B}_{\text{in}}(\mathbf{x}, t) = \frac{c\mathbf{E}_{\text{in}}^2}{4\pi}\mathbf{e}_1\Theta(ct - x^1)\cos^2 k(x^1 - ct), \text{ (II.1.5)}$$

where $\mathbf{E}_{\text{in}} = kA\mathbf{e}_3$. The energy flux is directed along $\mathbf{e}_1$, and its *time-averaged intensity* is given by

$$I_{\text{in}} := \lim_{T\to\infty}\frac{1}{T}\left|\int_0^T \mathbf{S}_{\text{in}}(\mathbf{x}, t)\,dt\right| = \frac{c\mathbf{E}_{\text{in}}^2}{8\pi} = \frac{ck^2A^2}{8\pi}, \qquad \mathbf{x} \in \mathbb{R}^3. \qquad \text{(II.1.6)}$$

Exercise II.1.1. Check (II.1.4)–(II.1.6).

II.1.2 The Thomson scattering

The Thomson scattering is described by the Maxwell equations for the fields coupled with the Lorentz equation for the electron's trajectory. In the case of point electron, the Maxwell equations (I.4.14) in the *rationalized* Gaussian units become

$$\begin{cases} \operatorname{div} \mathbf{E}(\mathbf{x}, t) = 4\pi e \delta(\mathbf{x} - \mathbf{x}(t)), & \operatorname{curl} \mathbf{E}(\mathbf{x}, t) = -\frac{1}{c}\dot{\mathbf{B}}(\mathbf{x}, t), \\ \operatorname{div} \mathbf{B}(\mathbf{x}, t) = 0, & \operatorname{curl} \mathbf{B}(\mathbf{x}, t) = \frac{1}{c}\dot{\mathbf{E}}(\mathbf{x}, t) + \frac{4\pi}{c}e\dot{\mathbf{x}}\,\delta(\mathbf{x} - \mathbf{x}(t)); \end{cases}$$
$$(\text{II.1.7})$$

see [48, p. 781] and [49]. The Lorentz equation (I.2.11) now reads

$$m\ddot{\mathbf{x}}(t) = e\left[\mathbf{E}(\mathbf{x}(t), t) + \frac{1}{c}\dot{\mathbf{x}}(t) \times \mathbf{B}(\mathbf{x}(t), t)\right], \qquad t \in \mathbb{R}. \qquad (\text{II.1.8})$$

The initial conditions for the electron do not matter. For example, take

$$\mathbf{x}(t) = 0, \quad \dot{\mathbf{x}}(t) = 0, \quad t < 0. \qquad (\text{II.1.9})$$

The initial condition for the fields depends on the incident wave (II.1.1), since

$$\mathbf{E}(\mathbf{x}, t) = \Theta(ct - x^1)\mathbf{E}_{\text{in}}(\mathbf{x}, t) - e\frac{\mathbf{x}}{|\mathbf{x}|^2}, \quad \mathbf{B}(\mathbf{x}, t) = \Theta(ct - x^1)\mathbf{B}_{\text{in}}(\mathbf{x}, t), \quad t < 0,$$
$$(\text{II.1.10})$$

where $-e\mathbf{x}/|\mathbf{x}|^2$ is the static Coulomb field generated by the electron with position (II.1.9). The functions (II.1.10), together with the electron trajectory (II.1.9), give a solution of the nonlinear system of equations (II.1.7), (II.1.8) for $t < 0$, since the right-hand side of (II.1.8) is identically zero for $t < 0$.

II.1.3 Neglecting the selfaction

Let us split the solution of (II.1.7) as

$$\begin{cases} \mathbf{E}(\mathbf{x}, t) &=& \Theta(ct - x^1)\mathbf{E}_{\text{in}}(\mathbf{x}, t) + \mathbf{E}_{\text{r}}(\mathbf{x}, t) \\ \mathbf{B}(\mathbf{x}, t) &=& \Theta(ct - x^1)\mathbf{B}_{\text{in}}(\mathbf{x}, t) + \mathbf{B}_{\text{r}}(\mathbf{x}, t) \end{cases}, \quad t \in \mathbb{R}, \qquad (\text{II.1.11})$$

where $\mathbf{E}_{\text{r}}(\mathbf{x}, t), \mathbf{B}_{\text{r}}(\mathbf{x}, t)$ are the outgoing *radiated waves*. Then the Maxwell equations (II.1.7) read as follows:

$$\begin{cases} \operatorname{div} \mathbf{E}_{\text{r}}(\mathbf{x}, t) = 4\pi e\delta(\mathbf{x} - \mathbf{x}(t)), & \operatorname{curl} \mathbf{E}_{\text{r}}(\mathbf{x}, t) = -\frac{1}{c}\dot{\mathbf{B}}_{\text{r}}(\mathbf{x}, t) \\ \operatorname{div} \mathbf{B}_{\text{r}}(\mathbf{x}, t) = 0, & \operatorname{curl} \mathbf{B}_{\text{r}}(\mathbf{x}, t) = \frac{1}{c}\dot{\mathbf{E}}_{\text{r}}(\mathbf{x}, t) + \frac{4\pi}{c}e\dot{\mathbf{x}}\,\delta(\mathbf{x} - \mathbf{x}(t)) \end{cases},$$
$$(\text{II.1.12})$$

since the incident wave (II.1.4) is a solution of the homogeneous Maxwell equations. From the initial conditions (II.1.10) it follows that

$$\mathbf{E}_{\mathrm{r}}(\mathbf{x}, t) = -e\frac{\mathbf{x}}{|\mathbf{x}|^2}, \qquad \mathbf{B}_{\mathrm{r}}(\mathbf{x}, t) = 0, \qquad t < 0. \tag{II.1.13}$$

The Lorentz equation (II.1.8) can now be written as

$$m\ddot{\mathbf{x}}(t) = e\left[\Theta(ct - x^1(t))\mathbf{E}_{\mathrm{in}}(\mathbf{x}(t), t) + \mathbf{E}_{\mathrm{r}}(\mathbf{x}(t), t)\right.$$

$$\left. + \frac{1}{c}\dot{\mathbf{x}}(t) \times (\Theta(ct - x^1(t))\mathbf{B}_{\mathrm{in}}(\mathbf{x}(t), t) + \mathbf{B}_{\mathrm{r}}(\mathbf{x}(t), t)\right], \quad t \in \mathbb{R}. \tag{II.1.14}$$

Unfortunately, problem (II.1.12), (II.1.14) is not well-posed. Indeed, it is clear from (II.1.12) and (II.1.13) that the solutions $\mathbf{E}_{\mathrm{r}}(\mathbf{x}, t), \mathbf{B}_{\mathrm{r}}(\mathbf{x}, t)$ of (II.1.12) are infinite at $(\mathbf{x}(t), t)$. Therefore, the right-hand side of equation (II.1.14) is not well-defined.

To make the problem well-posed, it is necessary to replace the point electron by an *extended electron*, as suggested by M. Abraham [1, 2]. For this model, the well-posedness is proved in [186]. Here we employ another traditional perturbation approach to make the problem well-posed. Namely, we neglect the interaction term, omitting the radiated waves on the right of (II.1.14), which gives

$$m\ddot{\mathbf{x}}(t) = e\left[\Theta(ct - x^1(t))\mathbf{E}_{\mathrm{in}}(\mathbf{x}(t), t) + \frac{1}{c}\dot{\mathbf{x}}(t) \times \Theta(ct - x^1(t))\mathbf{B}_{\mathrm{in}}(\mathbf{x}(t), t)\right], \quad t \in \mathbb{R}. \tag{II.1.15}$$

Now we substitute the obtained solution $\mathbf{x}(t)$ into the right-hand side of the Maxwell equations (II.1.12) to calculate the radiated waves $\mathbf{E}_{\mathrm{r}}(\mathbf{x}, t), \mathbf{B}_{\mathrm{r}}(\mathbf{x}, t)$.

Let us assume that the electron velocities are small as compared to the speed of light:

$$\beta := \max_{t \in \mathbb{R}} |\dot{\mathbf{x}}(t)|/c \ll 1. \tag{II.1.16}$$

Then we can neglect the contribution of the magnetic field to the right-hand side of (II.1.15). Thus, we obtain the equation

$$m\ddot{\mathbf{x}}(t) = e\mathbf{E}_{\mathrm{in}}(\mathbf{x}(t), t) = kA\mathbf{e}_3\Theta(ct - x^1(t))\cos k(x^1(t) - ct), \qquad t > 0. \tag{II.1.17}$$

Therefore, $x^1(t) \equiv 0 \equiv x^2(t)$ by the initial conditions (II.1.9), and so the equation becomes

$$m\ddot{\mathbf{x}}(t) = kA\mathbf{e}_3\cos kct, \qquad t > 0. \tag{II.1.18}$$

Now the initial conditions (II.1.9) define the trajectory $\mathbf{x}(t)$ uniquely:

$$\mathbf{x}(t) = \frac{eA}{mkc^2}\mathbf{e}_3(1 - \cos kct). \tag{II.1.19}$$

Note that condition (II.1.16) is equivalent to

$$\frac{|e|A}{mc^2} \ll 1. \qquad (II.1.20)$$

This relation means that the amplitude of oscillations $\frac{|e|A}{mkc^2}$ is small compared to the wavelength $\lambda = 2\pi/k$ of the incident wave.

Exercise II.1.2. Verify (II.1.19).

II.1.4 Dipole approximation

Our goal is to calculate the energy flux at infinity, i.e., to find the Poynting vector (cf. (II.1.5))

$$\mathbf{S}_r(\mathbf{x}, t) = \frac{c}{4\pi} \mathbf{E}_r(\mathbf{x}, t) \times \mathbf{B}_r(\mathbf{x}, t)$$

for large $|\mathbf{x}| \gg 1$. In order to determine the radiated waves $\mathbf{E}_r(\mathbf{x}, t)$ and $\mathbf{B}_r(\mathbf{x}, t)$, we must still solve the Maxwell equations (II.1.12). We will use the traditional *dipole approximation* to calculate the radiated waves; this leads to the *Thomson formula*.

To this end, let us expand the charge density in the Maxwell equations (II.1.12) in a formal Taylor series:

$$e\delta(\mathbf{x} - \mathbf{x}(t)) = e\delta(\mathbf{x}) + e\mathbf{x}(t) \cdot \nabla \delta(\mathbf{x}) + \frac{1}{2} e(\mathbf{x}(t) \cdot \nabla)^2 \delta(\mathbf{x}) + \dots, \quad t > 0. \qquad (II.1.21)$$

Here, the first term is static, and the corresponding radial Maxwell field does not contribute to the energy flux. The second term corresponds to the Hertzian dipole with dipole moment $\mathbf{p}(t) := e\mathbf{x}(t)$. The subsequent terms give small contributions to the energy flux at infinity, because $|\mathbf{x}(t)|$ is small by (II.1.20) ($k = 2\pi/\lambda \gg 1$ for visible light with $\lambda \sim 10^{-4}$ cm).

Therefore, we can use the Hertz formula for the dipole radiation [51, (12.124)]:

$$\mathbf{S}_r(\mathbf{x}, t) \sim \mathbf{n} \frac{\sin^2 \chi}{4\pi c^3 |\mathbf{x}|^2} \ddot{\mathbf{p}}^2(t - |\mathbf{x}|/c), \qquad |\mathbf{x}| \to \infty, \qquad (II.1.22)$$

where χ is the angle between $\ddot{\mathbf{p}}(t - |\mathbf{x}|/c) \sim \mathbf{e}_3$ and $\mathbf{n} := \mathbf{x}/|\mathbf{x}|$. By (II.1.18),

$$\ddot{\mathbf{p}}(t) = e\ddot{\mathbf{x}}(t) = \frac{e^2}{m} \mathbf{E}_{in} \cos kct, \qquad \mathbf{E}_{in} = kA\,\mathbf{e}_3. \qquad (II.1.23)$$

Hence,

$$\ddot{\mathbf{p}}^2(t) = \left(\frac{e^2}{m}\right)^2 \mathbf{E}_{in}^2 \cos^2 kct. \qquad (II.1.24)$$

Let θ be the angle between $\mathbf{n}$ and $\mathbf{e}_1$ and let φ be the *azimuthal angle* defined as the angle between $\mathbf{e}_3$ and the plane $(\mathbf{n}, \mathbf{e}_1)$. Then

$$\cos \chi = \cos \varphi \sin \theta, \qquad \sin^2 \chi = 1 - \cos^2 \varphi \sin^2 \theta. \qquad (II.1.25)$$

Therefore,

$$\mathbf{S}_r(\mathbf{x}, t) \sim \mathbf{n} \frac{1 - \cos^2 \varphi \sin^2 \theta}{4\pi c^3 |\mathbf{x}|^2} \left(\frac{e^2}{m}\right)^2 \mathbf{E}_{in}^2 \cos^2 kc(t - |\mathbf{x}|/c), \quad |\mathbf{x}| \to \infty. \qquad (II.1.26)$$

Hence the corresponding *time-averaged intensity* $I_r(\mathbf{x})$ is obtained if the expression $\cos^2 kc(t - |\mathbf{x}|/c)$ is replaced by $1/2$:

$$I_r(\mathbf{x}) := \lim_{T \to \infty} \frac{1}{T} \left| \int_0^T \mathbf{S}_r(\mathbf{x}, t) \, dt \right| \approx \frac{1 - \cos^2 \varphi \sin^2 \theta}{8\pi |\mathbf{x}|^2} \left(\frac{e^2}{mc^2}\right)^2 \mathbf{E}_{in}^2 \qquad (II.1.27)$$

$$= \left(\frac{e^2}{mc^2}\right)^2 \frac{\sin^2 \chi}{|\mathbf{x}|^2} I_{in}, \qquad (II.1.28)$$

where I_{in} is the intensity (II.1.6) of the incident wave. Therefore, the *mean intensity per unit angle* $i_r = \lim_{|\mathbf{x}| \to \infty} I_r(\mathbf{x})|\mathbf{x}|^2$ is given by

$$i_r(\varphi, \theta) \approx \left(\frac{e^2}{mc^2}\right)^2 (1 - \cos^2 \varphi \sin^2 \theta) I_{in}. \qquad (II.1.29)$$

Finally, the *differential cross section* $D(\varphi, \theta)$ is as follows:

$$D(\varphi, \theta) := \frac{i_r(\varphi, \theta)}{I_{in}} \approx \left(\frac{e^2}{mc^2}\right)^2 (1 - \cos^2 \varphi \sin^2 \theta); \qquad (II.1.30)$$

this is the *Thomson formula*.

Depending on φ, the differential cross section (II.1.30) is not invariant with respect to rotations around $\mathbf{e}_1$. This is due to the fact that incident wave is linearly polarized. Assume that the polarization of the incident light is random with uniform angular distribution. Then the differential cross section is given by (II.1.30) with $1/2$ instead of $\cos^2 \varphi$.

Finally, for the *total cross section* we have the following expression:

$$D_T := \frac{1}{I_{in}} \int i_r(\varphi, \theta) \, d\Omega = \int D(\varphi, \theta) \, d\Omega \approx \left(\frac{e^2}{mc^2}\right)^2 \int \sin^2 \chi \, d\Omega = \left(\frac{e^2}{mc^2}\right)^2 \frac{8\pi}{3}. \qquad (II.1.31)$$

Remark II.1.3. Now it is clear why we can neglect the scattering of light by nucleus. Indeed, the mass of the nucleus is about 1836 times the electron mass, so its classical scattering cross section (II.1.30), (II.1.31) should be about 1836^{-2} of the corresponding electron cross section.

Exercise II.1.4. Check (II.1.25) and (II.1.31).

II.2 Quantum Scattering of Light

Quantum scattering of light by a hydrogen atom is well described by the interaction of an electromagnetic wave with a Schrödinger wave field, because the scattering by the nucleus is negligible, as explained above.

Our aim here is to calculate the energy flux for the radiated Maxwell waves and find the corresponding differential cross section in the *first-order approximation* for small amplitudes of the incident wave.

II.2.1 Scattering problem

We will describe the scattering of the plane wave (II.1.4) by the hydrogen atom in its ground state. By (I.6.49), the hydrogen ground state energy is $E_1 = -2\pi\hbar cR = -\mathrm{me}^4/(2\hbar^2)$, and the corresponding eigenfunction is $\varphi_1(\mathbf{x}) = C_1 e^{-|\mathbf{x}|/r_1}$ (we assume that the atom is located at the origin). Then the corresponding solution of the Schrödinger equation (I.2.14) is as follows:

$$\psi_1(\mathbf{x}, t) = \varphi_1(\mathbf{x})e^{-i\omega_1 t}, \qquad \omega_1 = \frac{E_1}{\hbar} = -\frac{\mathrm{me}^4}{2\hbar^3}. \tag{II.2.1}$$

The scattering is described by the coupled Maxwell–Schrödinger equations (I.5.1) with the currents (I.5.2). We follow the perturbation approach neglecting the interaction terms. The external Maxwell potentials are replaced by the magnetic potential of the incident wave (II.1.1) and by the scalar Coulombic potential of the nucleus $\phi(\mathbf{x}) = -e/|\mathbf{x}|$. In the rationalized Gaussian units the equations read as

$$\begin{cases} \dfrac{1}{4\pi}\Box \mathbf{A}(\mathbf{x}, t) = \dfrac{\mathbf{j}(\mathbf{x}, t)}{c} = \dfrac{e}{\mathrm{mc}} \operatorname{Re}\left(\overline{\psi}(\mathbf{x}, t)\left[-i\hbar\nabla - \dfrac{e}{c}\mathbf{A}_{\mathrm{in}}(\mathbf{x}, t)\right]\psi(\mathbf{x}, t)\right) \\ (i\hbar\partial_t - e\phi(\mathbf{x}))\psi(\mathbf{x}, t) = \dfrac{1}{2\mathrm{m}}\left[-i\hbar\nabla - \dfrac{e}{c}\mathbf{A}_{\mathrm{in}}(\mathbf{x}, t)\right]^2 \psi(\mathbf{x}, t) \end{cases}. \tag{II.2.2}$$

see [48, p. 781] and [49]. In this model, the hydrogen nucleus is considered as fixed. This reflects the fact that the nucleus is much heavier than the electron.

II.2.2 Atomic form factor

We consider an incident Maxwell wave with small amplitudes $|A| \ll 1$, and suppose that the atom is in its ground state:

$$\psi(\mathbf{x}, t) = \varphi_1(\mathbf{x})e^{-i\omega_1 t}, \qquad t < 0. \tag{II.2.3}$$

In this section we neglect the incident wave in the Schrödinger equation from (II.2.2), i.e., we solve this equation in the *zero-order approximation* in the amplitude A. The first order approximations will be considered in the next three sections.

Thus, the corresponding approximation to the radiated waves is given by solutions of the Maxwell equation from (II.2.2) with $\psi(\mathbf{x}, t) = \psi_1(\mathbf{x}, t)$. It remains to solve this equation for $\mathbf{A}$ with the current

$$\frac{\mathbf{j}(\mathbf{x}, t)}{c} = \frac{e}{mc} \operatorname{Re} \left(\overline{\varphi}_1(\mathbf{x}) \left[-i\hbar \nabla \varphi_1(\mathbf{x}) - \frac{e}{c} \mathbf{A}_{\text{in}}(\mathbf{x}, t) \varphi_1(\mathbf{x}) \right] \right). \qquad (\text{II.2.4})$$

Here, the first term on the right is zero, since the corresponding eigenfunction $\varphi_1(\mathbf{x}) = C_1 e^{-|x|/r_1}$ is real. Therefore, the current reduces to

$$\frac{\mathbf{j}(\mathbf{x}, t)}{c} = -\frac{e^2}{mc^2} \mathbf{A}_{\text{in}}(\mathbf{x}, t) |\varphi_1(\mathbf{x})|^2. \qquad (\text{II.2.5})$$

Let us split the solution as follows:

$$\mathbf{A}(\mathbf{x}, t) = \mathbf{A}_{\text{in}}(\mathbf{x}, t) + \mathbf{A}_{\text{r}}(\mathbf{x}, t), \qquad (\text{II.2.6})$$

where $\mathbf{A}_{\text{r}}(\mathbf{x}, t)$ is the outgoing *radiated wave* with zero initial data

$$\mathbf{A}_{\text{r}}(\mathbf{x}, t_*) \equiv 0, \qquad x \in \mathbb{R}^3, \qquad (\text{II.2.7})$$

since the atom is located close to the origin, while the incident wave (II.1.1) vanishes there for $t < t_* < 0$. Now the Maxwell equation of (II.2.2) becomes

$$\begin{aligned} \Box \mathbf{A}_{\text{r}}(\mathbf{x}, t) &= - 4\pi \frac{e^2}{mc^2} A \mathbf{e}_3 \Theta(ct - x^1) \sin k(x^1 - ct) |\varphi_1(\mathbf{x})|^2 \\ &= - 4\pi \frac{e^2}{mc^2} \operatorname{Im} A \Theta(ct - x^1) e^{ik(x^1 - ct)} |\varphi_1(\mathbf{x})|^2 \mathbf{e}_3 =: f(\mathbf{x}, t), \quad (\text{II.2.8}) \end{aligned}$$

because $\mathbf{A}_{\text{in}}$ is a solution to the homogeneous equation. The initial condition (II.2.7) and the Kirchhoff formula for solutions of the wave equation imply that the radiated wave for large times approximately equals to the retarded potential (see Section 12.12 of [51])

$$\mathbf{A}_{\text{r}}(\mathbf{x}, t) \approx \int \frac{f(\mathbf{y}, t - |\mathbf{x} - \mathbf{y}|/c) \, d\mathbf{y}}{4\pi |\mathbf{x} - \mathbf{y}|}, \qquad t \to \infty. \qquad (\text{II.2.9})$$

Let us show that this integral representation implies the *limiting amplitude principle* which is expressed as

$$\mathbf{A}_{\text{r}}(\mathbf{x}, t) \sim \operatorname{Im} \left[a_{\text{r}}(\mathbf{x}) e^{-ikct} \right], \qquad t \to \infty. \qquad (\text{II.2.10})$$

Indeed,

$$\int \frac{f(\mathbf{y}, t - |\mathbf{x} - \mathbf{y}|/c)\, d\mathbf{y}}{4\pi|\mathbf{x} - \mathbf{y}|} = -\frac{e^2}{mc^2} A \operatorname{Im} \left[\int \frac{e^{ik(y^1 - c(t - |\mathbf{x} - \mathbf{y}|/c))} |\varphi_1(\mathbf{y})|^2\, d\mathbf{y}}{|\mathbf{x} - \mathbf{y}|} \right] \mathbf{e}_3$$

$$= -\frac{e^2}{mc^2} A \operatorname{Im} \left[e^{-ikct} \int \frac{e^{ik(y^1 + |\mathbf{x} - \mathbf{y}|)} |\varphi_1(\mathbf{y})|^2\, d\mathbf{y}}{|\mathbf{x} - \mathbf{y}|} \right] \mathbf{e}_3.$$
(II.2.11)

Let us find the asymptotics of the integral as $|\mathbf{x}| \to \infty$. For any fixed $\mathbf{y} \in \mathbb{R}^3$,

$$|\mathbf{x} - \mathbf{y}| = |\mathbf{x}| - \mathbf{y} \cdot \mathbf{n} + o(1), \qquad |\mathbf{x}| \to \infty \qquad (\text{II.2.12})$$

where $\mathbf{n} = \mathbf{n}(x) = \mathbf{x}/|\mathbf{x}|$. Since $y^1 = \mathbf{y} \cdot \mathbf{e}_1$, we have for large $|\mathbf{x}|$:

$$\mathbf{A}_r(\mathbf{x}, t) \sim -\frac{e^2}{mc^2} A \operatorname{Im} \left[e^{-ikct} \frac{e^{ik|\mathbf{x}|}}{|\mathbf{x}|} \int e^{ik\mathbf{y} \cdot (\mathbf{e}_1 - \mathbf{n})} |\varphi_1(\mathbf{y})|^2\, d\mathbf{y} \right] \mathbf{e}_3. \quad (\text{II.2.13})$$

Next, we evaluate the last integral. We set $K := k|\mathbf{e}_1 - \mathbf{n}|$ and denote by θ the angle between $\mathbf{n}$ and $\mathbf{e}_1$; then

$$K = K(k, \theta) = k\sqrt{(1 - n_1)^2 + (n^2)^2 + (n^3)^2}$$

$$= k\sqrt{2(1 - n^1)} = k\sqrt{2(1 - \cos\theta)} = 2k\sin\frac{\theta}{2}. \quad (\text{II.2.14})$$

Let α be the angle between $\mathbf{y}$ and $\mathbf{e}_1 - \mathbf{n}$, and let φ be the azimuthal angle around $\mathbf{e}_1 - \mathbf{n}$. Finally, let us take into account the fact that the ground state $\varphi_1(\mathbf{y}) = f(|\mathbf{y}|)$ is spherically symmetric. Now integral (II.2.13) becomes

$$\int_0^\infty |\mathbf{y}|^2 d|\mathbf{y}| \int_0^\pi \sin\alpha\, d\alpha \int_0^{2\pi} d\varphi\, e^{iK\cos\alpha|\mathbf{y}|} |\varphi_1(\mathbf{y})|^2$$

$$= 4\pi \int_0^\infty \frac{\sin K|\mathbf{y}|}{K|\mathbf{y}|} |f(|\mathbf{y}|)|^2 |\mathbf{y}|^2 d|\mathbf{y}| =: F_a(k, \theta), \quad (\text{II.2.15})$$

which is called the *atomic form factor* corresponding to the ground state $\varphi_1(\mathbf{x}) = C_1 e^{-|\mathbf{x}|/r_1}$. Since $F_a(k, \theta)$ is real, the asymptotics (II.2.13) becomes

$$\mathbf{A}_r(\mathbf{x}, t) \sim -\frac{e^2}{mc^2} A \frac{\sin k(|\mathbf{x}| - ct)}{|\mathbf{x}|} F_a(k, \theta) \mathbf{e}_3, \qquad |\mathbf{x}| \to \infty. \quad (\text{II.2.16})$$

Exercise II.2.1. Check (II.2.12) and (II.2.15).

II.2.3   Energy flux

We still have to calculate the Maxwell field and the Poynting vector corresponding to the last vector potential. To do this, it suffices to compare the expression (II.2.16) with the vector potential of the Hertzian dipole (formula (12.120) of [51]):

$$\mathbf{A}(\mathbf{x}, t) = \frac{1}{c} \frac{\dot{\mathbf{p}}(t - r/c)}{r}. \tag{II.2.17}$$

It is identical to (II.2.16) with $F_a(k, \theta) = 1$ if

$$\mathbf{p}(t) = -\frac{e^2}{mc^2 k} A\mathbf{e}_3 \cos kct, \tag{II.2.18}$$

where $\mathbf{e}_3 := (0, 0, 1)$. Therefore, the energy flux $\mathbf{S}(\mathbf{x}, t)$ corresponding to (II.2.16) is given, up to $\mathcal{O}(|\mathbf{x}|^{-3})$ as $|\mathbf{x}| \to \infty$, by the Hertz formula (II.1.22), with the additional factor $|F_a(k, \theta)|^2$. This follows from the fact that the angle θ is a homogeneous function of $\mathbf{x}$ of degree zero, hence any differentiation of the form factor $F_a(k, \theta)$ in $\mathbf{x}$ gives an additional factor with decay $\mathcal{O}(|\mathbf{x}|^{-1})$ as $|\mathbf{x}| \to \infty$. Finally, for the function (II.2.18), $\ddot{\mathbf{p}}(t)$ coincides with (II.1.23). In our case, (II.1.26) gives: for large $|\mathbf{x}|$,

$$\mathbf{S}(\mathbf{x}, t) \sim \mathbf{n} |F_a(k, \theta)|^2 \frac{1 - \cos^2 \varphi \sin^2 \theta}{4\pi c^3 |\mathbf{x}|^2} \left(\frac{e^2}{m}\right)^2 E_{\text{in}}^2 \cos^2 kc(t - |\mathbf{x}|/c). \tag{II.2.19}$$

Hence, the expressions for the intensity per unit angle and for the differential cross section also contain the additional factor $|F_a(k, \theta)|^2$. Finally, the value of differential cross section coincides with its value given by the Thomson formula (II.1.30) up to the atomic form factor, i.e., we have

$$D(k, \varphi, \theta) = |F_a(k, \theta)|^2 \left(\frac{e^2}{mc^2}\right)^2 (1 - \cos^2 \varphi \sin^2 \theta). \tag{II.2.20}$$

Remark II.2.2. The presence of the form factor leads to the key observation that quantum scattering is negligible for high frequencies, since

$$F_a(k, \theta) \to 0, \qquad k \to \infty. \tag{II.2.21}$$

Exercise II.2.3. Check (II.2.21).

II.3 Polarization and Dispersion

In the previous section, we studied the scattering of light of small amplitude A by the hydrogen atom and solved the Maxwell equations with unperturbed ground state. This means that we considered the Schrödinger equation (II.2.2) in the zero order approximation in the amplitude A. In this section and in the following two sections we will consider the first order approximations in the amplitude A.

More precisely, here we consider *nonresonant* light frequencies, which result in the corresponding modification of atomic charge and current densities; this means that *polarization* and *magnetization* of the atom take place. Thus we will obtain the corresponding *permittivity, electric susceptibility*, and the *refraction coefficient*.

The case of resonance with discrete and continuous spectrum will be considered in the next two sections.

II.3.1 First-order approximation

To calculate the first-order correction to the ground state, we expand the solution of the Schrödinger equation from the system (II.2.2) for small amplitudes:

$$\psi(\mathbf{x}, t) = \psi_1(\mathbf{x}, t) + A w(\mathbf{x}, t) + \mathcal{O}(A^2), \qquad |A| \ll 1. \qquad \text{(II.3.1)}$$

Here $\psi_1(\mathbf{x}, t) = \varphi_1(\mathbf{x}) e^{-i\omega_1 t}$, where $\varphi_1(\mathbf{x}) = C_1 e^{-|x|/r_1}$ is the ground state (I.6.49) which decays exponentially with the first derivatives:

$$|\nabla \varphi_1(\mathbf{x})| + |\varphi_1(\mathbf{x})| \le C e^{-\varepsilon|\mathbf{x}|}, \qquad \varepsilon = \frac{1}{r_1} > 0. \qquad \text{(II.3.2)}$$

Now the initial condition (II.2.3) implies that

$$w(\mathbf{x}, t) = 0, \qquad t < 0. \qquad \text{(II.3.3)}$$

Substituting (II.3.1) into the Schrödinger equation from (II.2.2), we obtain, in the first order in A,

$$A\big(i\hbar\partial_t - e\phi(\mathbf{x})\big) w(\mathbf{x}, t) = A\frac{1}{2\mathrm{m}}[-i\hbar\nabla]^2 w(\mathbf{x}, t) + \frac{i\hbar e}{\mathrm{m}c}\mathbf{A}_{\mathrm{in}}(\mathbf{x}, t) \cdot \nabla\psi_1(\mathbf{x}, t),$$
$$\text{(II.3.4)}$$

since $\psi_1(\mathbf{x}, t)$ is a solution of the Schrödinger equation (II.2.2) with $\mathbf{A}_{\mathrm{in}} = 0$. For $x^1 < ct$, we have by (II.1.1) and (II.2.1) that

$$\frac{i\hbar e}{\mathrm{m}c}\mathbf{A}_{\mathrm{in}}(\mathbf{x}, t) \cdot \nabla\psi_1(\mathbf{x}, t) = \frac{i\hbar e}{\mathrm{m}c} A \sin k(x^1 - ct)\mathbf{e}_3 \cdot \nabla\varphi_1(\mathbf{x}) e^{-i\omega_1 t}$$

$$= \frac{A\hbar e}{2\mathrm{m}c}[e^{ik(x^1 - ct)} - e^{-ik(x^1 - ct)}]e^{-i\omega_1 t}\nabla_3\varphi_1(\mathbf{x}), \quad \text{(II.3.5)}$$

where $\omega := kc$. Now we apply the *limiting amplitude principle* [113, 114, 152, 157]:

$$w(\mathbf{x}, t) = w_+(\mathbf{x})e^{-i(\omega_1+\omega)t} - w_-(\mathbf{x})e^{-i(\omega_1-\omega)t} + r(\mathbf{x}, t), \qquad \text{(II.3.6)}$$

where $w_\pm(\mathbf{x})$ are the *limiting amplitudes*, and the remainder $r(\cdot, t)$ tends to zero as $t \to \infty$ in an appropriate *weighted norms* (see Section 28 of [148] and Section 13.5 of [51]). This asymptotics holds in the *nonresonant case*, i.e., when

$$\omega_1 \pm \omega \neq \omega_l \qquad \forall l, \qquad \text{(II.3.7)}$$

where ω_l are the eigenvalues of the Schrödinger operator

$$H = \frac{1}{2\mathrm{m}}[-i\hbar\nabla]^2 + e\phi(\mathbf{x}). \qquad \text{(II.3.8)}$$

Moreover, in this section we will assume that

$$\omega_1 \pm \omega < 0, \qquad \text{(II.3.9)}$$

which means the absence of resonance with the continuous spectrum of the Schrödinger operator (II.3.8). For the hydrogen atom, we have

$$\omega_1 = -\frac{\mathrm{m}e^4}{2\hbar^3} \approx -20, 5 \cdot 10^{15} \text{ s}^{-1}$$

by (II.2.1), with values of the constants from (I.1.2). Hence the bound (II.3.9) holds for wave numbers $k < k_1 := |\omega_1|/c \approx 68 \cdot 10^7$ m^{-1}, or the light wavelengths

$$\lambda > 2\pi/k_1 = 0.91176 \cdot 10^{-5} \text{ cm} = 911.76 \,\text{Å}. \qquad \text{(II.3.10)}$$

The cases when conditions (II.3.7) and (II.3.9) are not satisfied will be considered in the next two sections. Thus, by (II.3.6),

$$w(\mathbf{x}, t) \sim w_+(\mathbf{x})e^{-i(\omega_1+\omega)t} - w_-(\mathbf{x})e^{-i(\omega_1-\omega)t}, \qquad t \to +\infty. \qquad \text{(II.3.11)}$$

The functions $w_\pm$ can be found from stationary equations

$$\left(\omega_1 \pm \omega + \frac{e^2}{\hbar|\mathbf{x}|}\right) w_\pm(\mathbf{x}) + \frac{\hbar}{2\mathrm{m}}\Delta w_\pm(\mathbf{x}) = \frac{e}{2\mathrm{mc}}e^{\pm ikx^1}\nabla_3\varphi_1(\mathbf{x}). \qquad \text{(II.3.12)}$$

The frequency bound (II.3.9) implies that the values $\hbar(\omega_1 \pm \omega)$ do not belong to the *continuous spectrum* of the Schrödinger operator (II.3.8). Therefore, equation (II.3.12) implies that

$$w_\pm \in L^2(\mathbb{R}^3). \qquad \text{(II.3.13)}$$

Actually, the amplitudes decay exponentially,

$$|w_\pm(\mathbf{x})| \leq Ce^{-\varepsilon_\pm |\mathbf{x}|}, \tag{II.3.14}$$

where $\varepsilon_\pm > 0$. Indeed, we can neglect the term with $\dfrac{e^2}{\hbar|\mathbf{x}|}$ in equation (II.3.12), since it is relatively small and decays at infinity. So, we obtain:

$$(\Delta + z_\pm)w_\pm(\mathbf{x}) = f_\pm(\mathbf{x}), \qquad \mathbf{x} \in \mathbb{R}^3, \tag{II.3.15}$$

where $z_\pm = 2\mathrm{m}(\omega_1 \pm \omega)/\hbar < 0$ and

$$|f_\pm(\mathbf{x})| \leq Ce^{-\varepsilon|\mathbf{x}|}, \tag{II.3.16}$$

with $\varepsilon > 0$ by (II.3.2). Solutions are given by the convolutions $w_\pm = E_\pm * f_\pm$, where $E_\pm(\mathbf{x})$ are the fundamental solutions $E_\pm(\mathbf{x}) = -e^{-\varkappa_\pm|\mathbf{x}|}/(4\pi|\mathbf{x}|)$ with $\varkappa_\pm := \sqrt{-z_\pm} > 0$:

$$w_\pm(\mathbf{x}) = -\int \frac{e^{-\varkappa_\pm|\mathbf{x}-\mathbf{y}|}}{4\pi|\mathbf{x}-\mathbf{y}|} f_\pm(\mathbf{y})\, d\mathbf{y}. \tag{II.3.17}$$

As a result, the decay indicated in (II.3.14) actually occurs.

Exercise II.3.1. Prove the decay (II.3.14) with $\varepsilon_\pm = \min(\varepsilon, \varkappa_\pm) > 0$.

II.3.2 Limiting amplitudes

Alternatively, we can calculate the limiting amplitudes using the spectral resolution of the Schrödinger operator (II.3.8). First, expanding the right-hand side, we have

$$\frac{e}{2mc} e^{\pm ikx^1} \nabla_3 \varphi_1(\mathbf{x}) = \sum_l a_l^\pm \varphi_l(\mathbf{x}), \tag{II.3.18}$$

where $\sum_l$ *by definition* includes the sum over the discrete spectrum and the integral over the continuous spectrum of the Schrödinger operator (II.3.8). Hence the solutions $w_\pm$ are of the form

$$w_\pm(\mathbf{x}) = \sum_l \frac{a_l^\pm \varphi_l(\mathbf{x})}{\omega_1 \pm \omega - \omega_l}. \tag{II.3.19}$$

Now (II.3.11) implies that for large times,

$$w(\mathbf{x}, t) \sim \sum_l \frac{a_l^+ \varphi_l(\mathbf{x})}{\omega_1 + \omega - \omega_l} e^{-i(\omega_1+\omega)t} - \sum_l \frac{a_l^- \varphi_l(\mathbf{x})}{\omega_1 - \omega - \omega_l} e^{-i(\omega_1-\omega)t}. \tag{II.3.20}$$

Let us calculate the coefficients $a_l^\pm$. Formally,

$$a_l^\pm = \frac{e}{2mc} \int \overline{\varphi}_l(\mathbf{x}) e^{\pm ikx^1} \nabla_3 \varphi_1(\mathbf{x})\, d\mathbf{x} \tag{II.3.21}$$

if the eigenfunctions of the discrete spectrum are orthogonal and normalized, and the eigenfunctions of the continuous spectrum obey the "delta-function normalization".

We assume that $kr_1 \ll 1$, where r_1 is the Bohr radius from (I.6.49). Then we can substitute $e^{\pm ikx^1} = 1$. Hence $a_l^{\pm}$ are approximately equal. Therefore, using formula [77, (44.20)], we obtain

$$a_l^{\pm} \approx a_l := \frac{e\omega_{1l}}{2ch} \int x^3 \overline{\varphi}_l(\mathbf{x}) \varphi_1(\mathbf{x}) \, d\mathbf{x} = \frac{e\omega_{1l}}{2ch} x_{1l}^3, \qquad (\text{II.3.22})$$

where $\omega_{1l} := \omega_1 - \omega_l$ and $x_{1l}^3 := \int x^3 \overline{\varphi}_l(\mathbf{x}) \varphi_1(\mathbf{x}) \, d\mathbf{x}$. Finally, (II.3.20) becomes

$$w(\mathbf{x}, t) \sim \sum_l a_l \varphi_l(\mathbf{x}) \left(\frac{e^{-i\omega t}}{\omega_{1l} + \omega} - \frac{e^{i\omega t}}{\omega_{1l} - \omega} \right) e^{-i\omega_1 t}, \qquad t \to \infty, \qquad (\text{II.3.23})$$

and so the expansion (II.3.1) for large times takes the form

$$\psi(\mathbf{x}, t) = \left(\varphi_1(\mathbf{x}) + A \sum_l a_l \varphi_l(\mathbf{x}) \left(\frac{e^{-i\omega t}}{\omega_{1l} + \omega} - \frac{e^{i\omega t}}{\omega_{1l} - \omega} \right) \right) e^{-i\omega_1 t} + \mathcal{O}(A^2)$$
$$= \left(\varphi_1(\mathbf{x}) + A\Sigma(\mathbf{x}, t) \right) e^{-i\omega_1 t} + \mathcal{O}(A^2), \qquad |A| \ll 1. \qquad (\text{II.3.24})$$

Problem II.3.2. Prove formula (II.3.22). **Hints:** The integral (II.3.21) with $k = 0$ can be written as

$$I := \int \overline{\varphi}_l(\mathbf{x}) \nabla_3 \varphi_1(\mathbf{x}) \, d\mathbf{x} = \frac{i}{\hbar} \langle \varphi_l, \hat{p}_3 \varphi_1 \rangle.$$

The key point is the commutation relation $\hat{p}_3 = -\frac{m}{i\hbar}[H, \hat{x}^3]$, which follows from the last formula of (I.3.19) since the operator (II.3.8) can be written as $H = \frac{\hat{\mathbf{p}}^2}{2m} + e\phi(\hat{\mathbf{x}})$. Now

$$I = \frac{i}{\hbar} \left\langle \varphi_l, -\frac{m}{i\hbar}[H, \hat{x}^3]\varphi_1 \right\rangle = -\frac{m}{\hbar^2} \left\langle \varphi_l, [H, \hat{x}^3]\varphi_1 \right\rangle = -\frac{m}{\hbar^2} \left\langle \varphi_l, (H\hat{x}^3 - \hat{x}^3 H)\varphi_1 \right\rangle.$$

It remains to note that $H\varphi_1 = \omega_1 \hbar \varphi_1$ and $H\varphi_l = \omega_l \hbar \varphi_l$.

II.3.3   The Kramers–Kronig formula

Let us calculate the electric dipole moment of the atom. First, the charge density is given by

$$\rho(\mathbf{x}, t) = e\overline{\psi}(\mathbf{x}, t)\psi(\mathbf{x}, t) = e(\overline{\varphi}_1(\mathbf{x}) + A\overline{\Sigma}(\mathbf{x}, t))(\varphi_1(\mathbf{x}) + A\Sigma(\mathbf{x}, t)) + \mathcal{O}(A^2)$$

$$= e|\varphi_1(\mathbf{x}, t)|^2 + eA\left[\Sigma^+ e^{i\omega t} + \Sigma^- e^{-i\omega t} \right] + \mathcal{O}(A^2), \quad |A| \ll 1, \quad (\text{II.3.25})$$

where

$$\Sigma^+ = \sum_l \left(\frac{\varphi_1 \overline{a}_l \overline{\varphi}_l}{\omega_{1l} + \omega} - \frac{a_l \varphi_l \overline{\varphi}_1}{\omega_{1l} - \omega} \right), \qquad \Sigma^- = \overline{\Sigma^+}. \tag{II.3.26}$$

Therefore, by definition, the electric dipole moment is

$$\mathbf{p}(t) := \int \mathbf{x}\rho(\mathbf{x}, t) \, d\mathbf{x} = e\mathbf{x}_{11} + \mathbf{P}(t) + \mathcal{O}(A^2), \qquad |A| \ll 1; \tag{II.3.27}$$

here $\mathbf{x}_{11} := \int \mathbf{x}|\varphi_1(\mathbf{x})|^2 \, d\mathbf{x} = 0$ by the spherical symmetry, and

$$\mathbf{P}(t) = eA \left[\sum_l \left(\frac{\overline{a}_l \mathbf{x}_{1l}}{\omega_{1l} - \omega} - \frac{a_l \mathbf{x}_{1l}}{\omega_{1l} + \omega} \right) e^{i\omega t} + \sum_l \left(\frac{a_l \mathbf{x}_{1l}}{\omega_{1l} - \omega} - \frac{\overline{a}_l \mathbf{x}_{1l}}{\omega_{1l} + \omega} \right) e^{-i\omega t} \right], \tag{II.3.28}$$

where $\mathbf{x}_{1l} := \int \mathbf{x}\overline{\varphi}_l(\mathbf{x})\varphi_1(\mathbf{x}) \, d\mathbf{x}$. By symmetry arguments, we can assume that the vector $\mathbf{P}(t)$ is directed along $\mathbf{e}_3$. Indeed, the invariance of $\mathbf{P}(t)$ with respect to the reflection $x^2 \mapsto -x^2$ is obvious. The invariance with respect to the reflection $x^1 \mapsto -x^1$ follows from (II.3.18)–(II.3.20), since we finally set $k = 0$. Therefore, substituting a_l from (II.3.22) and projecting $\mathbf{x}_{1l}$ onto $\mathbf{e}_3$, we obtain:

$$\mathbf{P}(t) \approx A\mathbf{e}_3 \frac{2\omega e^2}{c\hbar} \sum_l \frac{\omega_{1l}|x_{1l}^3|^2}{\omega_{1l}^2 - \omega^2} \cos \omega t, \tag{II.3.29}$$

with $x_{1l}^3 = \mathbf{x}_{1l}\mathbf{e}_3$. Finally, *averaging* (II.3.27) *with respect to all possible orientations* of the atom, we obtain, in the first-order approximation,

$$\overline{\mathbf{p}}(t) = A\mathbf{e}_3 \frac{2ke^2}{\hbar} \sum_l \frac{\omega_{1l}\overline{|x_{1l}^3|^2}}{\omega_{1l}^2 - \omega^2} \cos \omega t, \tag{II.3.30}$$

since $\mathbf{x}_{11} = 0$ and $\omega = kc$. Above, the overline indicates the average.

Now we can express the *permittivity* of hydrogen in its ground state φ_1. Let $\mathbf{E}(t)$ be the electric field (II.1.4) at the position $\mathbf{x} = 0$ of the atom. By (II.1.4), we have $\mathbf{E}(t) = kA\mathbf{e}_3 \cos \omega t$. Hence (II.3.30) can be written as

$$\overline{\mathbf{p}}(t) = \frac{2e^2}{3\hbar} \sum_l \frac{\omega_{1l}\overline{|x_{1l}^3|^2}}{\omega_{1l}^2 - \omega^2} \mathbf{E}(t). \tag{II.3.31}$$

Therefore, the permittivity of atomic hydrogen in its ground state is given by the *Kramers–Kronig formula*,

$$\chi_e(\omega) = N|\overline{\mathbf{p}}(t)|/|\mathbf{E}(t)| = N \frac{2e^2}{3\hbar} \sum_l \frac{\omega_{1l}\overline{|x_{1l}^3|^2}}{\omega_{1l}^2 - \omega^2}, \tag{II.3.32}$$

where N is the number of atoms per unit volume. It is worth noting that formula (II.3.32) has the same analytic structure as its analog [51, (14.37)] first established in the framework of the Old Quantum Theory by H.A. Kramers and W. Heisenberg [52]–[54].

More precisely, the formula (II.3.32) should be rewritten as the sum over the discrete spectrum and the integral over the continuous spectrum,

$$\chi_e(\omega) = N\frac{2e^2}{3\hbar} \left[\sum_{l=1}^{L} \frac{\omega_{1l}\overline{|x_{1l}^3|^2}}{\omega_{1l}^2 - (\omega + i0)^2} + \int_0^\infty \frac{\omega_{1\nu}\overline{|x_{1\nu}^3|^2}\,d\nu}{(\omega_1 - \nu)^2 - (\omega + i0)^2} \right], \quad \text{(II.3.33)}$$

where the addition of $i0$ is in the agreement with the *limiting absorption principle* (for more details, see [148] and [51, (13.17)]). This famous formula has many important consequences that agree with experimental observations (see details in [5]):

I. It allows one to express the electric susceptibility $\varepsilon = 1 + 4\pi\chi_e$, and hence the refraction coefficient $n = \sqrt{\varepsilon\mu} \sim \sqrt{\varepsilon}$ in the case when the magnetic susceptibility μ is close to 1;

II. It implies that $\varepsilon(\omega)$ has a singularity at the eigenvalues ω_k, which is in a good agreement with experimental observations;

III. It explains the anomalous dispersion near the eigenvalues ω_k, i.e., the fact that the polarizability coefficient $\varepsilon(\omega)$ is a decaying function of $|\omega|$ near $|\omega_k|$;

IV. The function (II.3.33) admits an analytic continuation from the real axis into the *upper complex half-plane* $\mathrm{Im}\,\omega > 0$. This fact implies the integral *dispersion relations* between real and imaginary parts of $\chi_e(\omega)$ and $\varepsilon(\omega)$ discovered by H.A. Kramers and R. Kronig [55, 56, 57]. This theory was intensively developed later in the framework of quantum field theory; see the survey in [159, Vol. III].

II.4 Photoelectric Effect

We now consider the case when the condition (II.3.9) is not satisfied, which corresponds to the *resonance* of the incident light with the *continuous spectrum* of the Schrödinger operator (II.3.8). In this case, the electron cloud of an atom is not modified, but rather completely destroyed, and the atom is *ionized*. The emitted electrons produce a *photocurrent* with specific angular distribution.

This ionization was first observed by H. Hertz in 1887 and studied experimentally in detail by P. Lenard during 1902–1905. The first theoretical explanation was given by A. Einstein in 1905, who suggested the corpuscular theory of light by introducing "photons", particles of light. In the framework of the Schrödinger theory, this effect was first described by G. Wentzel, who calculated the angular distribution of the photocurrent. The corresponding calculation relies on the perturbation procedure applied to the coupled Maxwell–Schrödinger equations.

H. Hertz discovered the discharge of a negatively charged *electroscope* under the electromagnetic radiation of very short wavelength (like visible light or ultraviolet radiation). This discharge was understood as the knocking out the electrons from metals due to the absorption of energy of the electromagnetic radiation.

P. Lenard systematically studied the behavior of "photoelectrons" (i.e., knocked-out electrons) in external electric and magnetic fields. Lenard's conclusions were the following:

L1. The saturation photocurrent is proportional to the intensity of incident light.

L2. The photocurrent is observed only for sufficiently small wavelength, that is, for high frequencies,

$$|\omega| > \omega_{\text{red}}, \tag{II.4.1}$$

where ω_{red} is called the *red bound* of the photoelectric effect. This red bound depends on the substance, but *it is independent of the intensity of light*.

L3. The photocurrent vanishes if the *stopping voltage* U_{stop} is applied; the minimal voltage U_{stop} also depends on the substance, but *it is independent of the intensity of light*. Moreover, the minimal stopping voltage U_{stop} increases with the decrease of the wavelength of incident light.

This independence of ω_{red} and of the minimal stopping voltage U_{stop} on the intensity of light was the main difficulty in the theoretical explanation of the Lenard observations. This independence seemed to constitute a new mysterious phenomenon, which never occurred in classical physics.

In 1905, A. Einstein proposed a revolutionary interpretation for the photoelectric effect stating that atoms absorb light in portions with energy $\hbar\omega$. In other words, light with frequency ω is similar to a beam of particles, called *photons*, with energy $\hbar\omega$. These ideas were inspired by Planck's treatment of the Kirchhoff black-body radiation law; see Section 1.3 of [51].

The *Einstein rules* for the photoelectric effect are as follows:

E1. *The flux of photons is proportional to the intensity of the incident light.*

E2. *The maximal kinetic energy of photoelectrons is given by (I.1.4):*

$$K_{max} = \frac{m v_{max}^2}{2} = \hbar\omega - W, \qquad (II.4.2)$$

where W is the work function *which depends on the metal. Hence the emission of electron is possible only if $\hbar\omega - W > 0$; therefore, the red bound $\omega_{red} = W/\hbar$ is independent of the intensity of light, in agreement with Lenard's observations!*

E3. *Accordingly, the stopping voltage must satisfy the inequality*

$$- e U_{stop} > \hbar\omega - W, \qquad (II.4.3)$$

where $e < 0$. Thus the minimal stopping voltage U_{stop} is also independent of the intensity of light.

The relation (II.4.2) formally represents energy conservation in the absorption of a photon by an electron. However, let us stress that (II.4.2) is a theoretical interpretation of formula (II.4.3), which is verified experimentally and gives the minimum stopping voltage $-(\hbar\omega - W)/e$. Moreover, formula (II.4.3) allows one to measure the Planck constant $\hbar$ with a very high precision.

Thus the Einstein rules **E1–E3** give a complete explanation of Lenard's observations. In 1922, Einstein was awarded the 1921 Nobel Prize in Physics for his theory of the photoelectric effect, relying on the revolutionary *corpuscular theory of light*.

In 1927, G. Wentzel calculated the angular distribution of the photocurrent applying the first-order perturbation approach (II.3.1) to the coupled Maxwell–Schrödinger equations (II.2.2). We will present Wentzel's calculations [82, Vol. II] justifying them by the application of the *limiting amplitude principle*. Such a justification allows us to explain Einstein's rules for the photoelectric effect. Namely, we will show that the photoelectric effect is caused by the slow spatial decay of the limiting amplitude at infinity in the case

$$|\omega| > |\omega_1|, \qquad (II.4.4)$$

where ω_1 is the minimal eigenvalue of the Schrödinger operator (II.3.8). This slow decay is caused by the *continuous spectrum* and results in a nonvanishing electric current to infinity. Thus, $\omega_{\mathrm{red}} = |\omega_1|$. Moreover, we will show that the maximal photoelectron energy is given by (II.4.2), and the stopping voltage satisfies (II.4.3).

However, such a perturbation approach is not self-consistent and should be considered, rather, as a hint to a nonlinear theory of the photoelectric effect. Let us note that recently a rigorous theory of *atomic ionization* was developed in [99]–[102] in the framework of the linear Schrödinger equation with time-periodic potentials. This theory implies that a complete atomic ionization occurs for any light frequency $\omega \neq 0$. On the other hand, the results [164] on AC-Stark effect suggest that the *rate of ionization* strongly depends on the integer $N = 1, 2, \ldots$ defined by the inequalities

$$(N - 1)|\omega| < |\omega_1| < N|\omega|. \tag{II.4.5}$$

Namely, the ionization rate decreases significantly with increasing N. The rate is maximal for $N = 1$, which corresponds to (II.4.4).

For second-quantized models, a *perturbative treatment* of atomic ionization and of relation (II.4.2) was given in [98, 103, 107].

To summarize, a genuine nonlinear dynamical nonperturbative explanation of Einstein's rules in the framework of the coupled Maxwell–Schrödinger equations (I.5.1) remains a challenging open problem.

II.4.1 Resonance with the continuous spectrum

Now we apply the perturbation approach (II.3.1) to the problem of the scattering of light with large frequencies; here

$$\omega_1 + |\omega| > 0, \tag{II.4.6}$$

in contrast to (II.3.9). Let us recall that this condition for the hydrogen atom means that the light wavelength λ is less than $911.76\,\text{Å}$; see (II.3.10).

For simplicity of notation, we assume that $\omega > 0$. Hence $\omega_1 - \omega < 0$, while

$$\omega_1 + \omega > 0. \tag{II.4.7}$$

Consequently, $\hbar(\omega_1 + \omega)$ belongs to the continuous spectrum of the Schrödinger operator (II.3.8). Hence, equation (II.3.12) implies that for the limiting amplitude one has $w_+(\mathbf{x}) \notin L^2$, unlike in (II.3.13). More precisely, as we will show in the next section, the limiting amplitude *slowly decays at infinity*:

$$|w_+(\mathbf{x})| \sim \frac{a(\mathbf{n}(\mathbf{x}))}{|\mathbf{x}|}, \qquad |\mathbf{x}| \to \infty, \tag{II.4.8}$$

where $\mathbf{n}(\mathbf{x}) := \mathbf{x}/|\mathbf{x}|$. Moreover, we will calculate the amplitude $a(\mathbf{n})$ and obtain the main term of the radiation in the form of the outgoing wave:

$$Aw_+(\mathbf{x})e^{-i(\omega_1+\omega)t} \sim A\frac{a(\varphi,\theta)}{|\mathbf{x}|}e^{i[k_r|\mathbf{x}|-(\omega+\omega_1)t]}, \qquad |\mathbf{x}| \to \infty. \qquad \text{(II.4.9)}$$

On the other hand, $\hbar(\omega_1 - \omega) < 0$ does not belong to the continuous spectrum of the Schrödinger equation. Hence, $w_-(x)$ decays exponentially, similarly to (II.3.14):

$$|w_-(\mathbf{x})| \leq Ce^{-\varepsilon_-|\mathbf{x}|}, \qquad \mathbf{x} \in \mathbb{R}, \qquad \text{(II.4.10)}$$

where $\varepsilon_- > 0$.

From (II.4.9) and (II.4.10), we will deduce the following asymptotics for the limiting stationary electric current at infinity:

$$\mathbf{j}(\mathbf{x},t) \sim A^2\frac{e\hbar k_r}{\mathrm{m}}\frac{a^2(\varphi,\theta)}{|\mathbf{x}|^2}\mathbf{n}(\mathbf{x}), \qquad |\mathbf{x}| \to \infty. \qquad \text{(II.4.11)}$$

This formula was obtained by Wentzel in 1927 (see [106]) with the angular distribution

$$a(\varphi,\theta) = C\sin\theta\,\cos\varphi, \qquad \text{(II.4.12)}$$

where $C \neq 0$. Hence, the formula (II.4.11) describes a nonzero photocurrent from the atom to infinity. Indeed, asymptotics (II.4.11) imply that the *total photocurrent to infinity does not vanish*, i.e.,

$$J_\infty := \lim_{R\to\infty}\int_{|\mathbf{x}|=R}\mathbf{j}(\mathbf{x},t)\,dS(\mathbf{x}) \neq 0. \qquad \text{(II.4.13)}$$

Exercise II.4.1. Deduce (II.4.13) from (II.4.11) and (II.4.12) with $C \neq 0$.

II.4.2   Limiting amplitude

Let us calculate the limiting amplitude $w_+(\mathbf{x})$. First, we rewrite equation (II.3.12) as follows:

$$[\nabla^2 + k_r^2(\omega)]w_+(\mathbf{x}) = \frac{e}{\hbar c}e^{ikx^1}\nabla_3\varphi_1(\mathbf{x}) - \frac{2e^2\mathrm{m}}{\hbar^2|\mathbf{x}|}w_+(\mathbf{x}), \qquad \text{(II.4.14)}$$

with the wave number $k_r(\omega)$ of the radiated wave $w_+(\mathbf{x})$ given by

$$k_r(\omega) := \sqrt{\frac{2\mathrm{m}(\omega_1 + \omega)}{\hbar}} > 0.$$

In the leading order approximation, we can neglect the last term in the right-hand side of (II.4.14), since it is small and decays at infinity. This gives us the Helmholtz equation of type (II.3.15),

$$[\Delta + k_r^2(\omega)]w_+(\mathbf{x}) = f_+(\mathbf{x}) := \frac{e}{\hbar c}e^{ikx^1}\nabla_3\varphi_1(\mathbf{x}). \qquad \text{(II.4.15)}$$

Hence the exponential decay (II.4.10) does not take place for $w_+(\mathbf{x})$. This is obvious in the Fourier space, where (II.4.15) becomes

$$\hat{w}_+(\mathbf{k}) = \frac{\hat{f}_+(\mathbf{k})}{-\mathbf{k}^2 + k_r^2(\omega)}. \tag{II.4.16}$$

Here the denominator vanishes on the sphere $|\mathbf{k}| = k_r(\omega)$, while

$$\hat{f}_+(\mathbf{k}) \sim k_3 \hat{\varphi}_1(k_1 + k, k_2, k_3)$$

is zero only for $k_3 = 0$.

Exercise II.4.2. Verify that $\hat{f}_+(\mathbf{k}) \neq 0$ for $k_3 \neq 0$. **Hint:** Calculate $\hat{\varphi}_1(\mathbf{k})$ in spherical coordinates.

Thus, the quotient (II.4.16) is singular on the sphere $|\mathbf{k}| = k_r(\omega)$. Hence, $w_+(\mathbf{x})$ cannot decay exponentially. Now the solution is given by convolution with the fundamental solution:

$$w_+(\mathbf{x}) = -\int \frac{e^{ik_r(\omega)|\mathbf{x}-\mathbf{y}|}}{4\pi|\mathbf{x}-\mathbf{y}|} f_+(\mathbf{y}) \, d\mathbf{y}. \tag{II.4.17}$$

Let us note that *the fundamental solution* $-\dfrac{e^{-ik_r(\omega)|\mathbf{x}-\mathbf{y}|}}{4\pi|\mathbf{x}-\mathbf{y}|}$ *is not suitable*. This *selection rule* for the fundamental solutions results from the *limiting absorption principle* (see [51, Section 13.5.2] and [148, Chapter 6]), because only the fundamental solution $\dfrac{e^{ik_r(\omega+i\varepsilon)|\mathbf{x}|}}{4\pi|\mathbf{x}|}$ is a tempered distribution for small $\varepsilon > 0$. This is obvious since $\operatorname{Im} k_r(\omega + i\varepsilon) > 0$ for the analytic continuation of the fixed branch for which $k_r(\omega) > 0$ at $\omega > 0$.

Now we can calculate the asymptotics (II.4.9). To do so, we substitute the expression (II.4.15) for f_+ into (II.4.17). Integrating by parts, we obtain:

$$w_+(\mathbf{x}) = -\frac{e}{\hbar c}\int \nabla_{y^3} \frac{e^{ik_r|\mathbf{x}-\mathbf{y}|}}{4\pi|\mathbf{x}-\mathbf{y}|} e^{iky^1} \varphi_1(\mathbf{y}) \, d\mathbf{y}$$

$$= \frac{ik_r e}{\hbar c}\int \frac{e^{ik_r|\mathbf{x}-\mathbf{y}|}(x^3 - y^3)}{4\pi|\mathbf{x}-\mathbf{y}|^2} e^{iky^1} \varphi_1(\mathbf{y}) \, d\mathbf{y} + \mathcal{O}(|\mathbf{x}-\mathbf{y}|^{-2}) \tag{II.4.18}$$

as $|\mathbf{x} - \mathbf{y}| \to \infty$. Recall that θ denotes the angle between $\mathbf{n} := \mathbf{x}/|\mathbf{x}|$ and $\mathbf{e}_1$, and φ stands for the azimuthal angle between $\mathbf{e}_3$ and the plane $(\mathbf{n}, \mathbf{e}_1)$. Hence, $x^3 = |\mathbf{x}| \sin\theta \cos\varphi$, and now (II.4.18) implies the asymptotics (II.4.9) with the angular distribution (II.4.12), because the ground state $\varphi_1(\mathbf{y})$ decays rapidly at infinity. The constant C in (II.4.12) is given by

$$C = C(k) = \frac{ik_r e}{4\pi\hbar c}\int e^{iky^1} \varphi_1(\mathbf{y}) \, d\mathbf{y} \neq 0. \tag{II.4.19}$$

Remark II.4.3. Let us stress that the limiting amplitude principle holds in both cases, for $|\omega| > |\omega_1|$, as in (II.4.4), and for $|\omega| < |\omega_1|$, as in (II.3.9). The crucial difference lies in different rate of spatial decay of the corresponding limiting amplitudes as $|\mathbf{x}| \to \infty$. Namely, in the case $|\omega| < |\omega_1|$, the amplitude decays exponentially according to (II.3.14), which results in the *zero total electric current to infinity* (II.4.13). On the other hand, in the case $|\omega| > |\omega_1|$, the amplitude decays slowly according to (II.4.9), which results in the *nonzero total electric current to infinity* (II.4.13).

Problem II.4.4. Check (II.4.19) and deduce (II.4.9), (II.4.12) from (II.4.18).

II.4.3   Angular distribution: the Wentzel formula

Our aim here is to deduce (II.4.11) from (II.4.9). The first equation of (II.2.2) implies that for small amplitudes A of the incident wave (II.1.1), the leading term of the photocurrent is given by

$$\mathbf{j}(\mathbf{x}, t) := \frac{e}{m} \operatorname{Re} \left\{ \overline{\psi}(\mathbf{x}, t)(-i\hbar \nabla \psi(\mathbf{x}, t)) \right\}. \qquad (\text{II.4.20})$$

Further, by (II.4.10), $w_-(\mathbf{x})$ decays exponentially at infinity, as well as the eigenfunction $\varphi_1(\mathbf{x})$. Therefore, (II.3.1) and (II.3.11) imply the asymptotics

$$\psi(\mathbf{x}, t) \sim A w_+(\mathbf{x}) e^{-i(\omega_1 + \omega)t}, \qquad |\mathbf{x}| \to \infty. \qquad (\text{II.4.21})$$

Substituting this into (II.4.20), and using the asymptotics (II.4.9), we obtain the Wentzel formula (II.4.11) with amplitude (II.4.12).

Exercise II.4.5. Deduce (II.4.11) from (II.4.20), (II.4.21), and (II.4.9).

II.4.4   Derivation of Einstein's rules

Now we can explain Lenard's observations and Einstein's rules for the photoelectric effect:

E1 By (II.4.11), the saturation photocurrent is proportional to A^2, which in turn is proportional to the intensity of incident light by (II.1.6).

E2 The asymptotics (II.4.9) imply that the energy per one photoelectron is given by the *Einstein formula* (II.4.2). Indeed, for large $|\mathbf{x}|$, the outgoing radiated wave (II.4.9) is locally close to the plane wave (I.3.43) with $\omega - |\omega_1|$ instead of ω. Hence the formula (I.3.48) implies that the energy per one photoelectron does not exceed $K = \hbar(\omega - |\omega_1|)$, which is equivalent to (II.4.2) with

$$W = \hbar |\omega_1|. \qquad (\text{II.4.22})$$

E3 The application of an external voltage $\phi_{\text{stop}}(\mathbf{x})$ is equivalent to the corresponding modification of the scalar potential in the Schrödinger equation of (II.2.2): $\phi(\mathbf{x}) \mapsto \tilde{\phi}(\mathbf{x}) = \phi(\mathbf{x}) + \phi_{\text{stop}}(\mathbf{x})$, where $\phi_{\text{stop}}(\mathbf{x})$ is a slowly varying potential, and $\phi_{\text{stop}}(\mathbf{x}) = U_{\text{stop}} > 0$ in a macroscopic region containing the atom.

Let us show that the ground state energy $\hbar\omega_1$ changes to $\hbar\tilde{\omega}_1$, where

$$\hbar\tilde{\omega}_1 \approx \hbar\omega_1 + eU_{\text{stop}} \tag{II.4.23}$$

with a high precision. Indeed, by the Courant minimax principle, we have

$$\hbar\omega_1 = \min_{\|\varphi\|=1} \langle \varphi, H\varphi \rangle, \tag{II.4.24}$$

where $\langle \cdot, \cdot \rangle$ is the inner product in $L^2(\mathbb{R}^3) \otimes \mathbb{C}$ (see (I.3.3)). We can assume that $0 \leq \phi_{\text{stop}}(\mathbf{x}) \leq U_{\text{stop}}$ for $\mathbf{x} \in \mathbb{R}^3$. Hence

$$\hbar\tilde{\omega}_1 = \min_{\|\varphi\|=1} \langle \varphi, (H + e\phi_{\text{stop}}(\mathbf{x}))\varphi \rangle \geq \hbar\omega_1 + eU_{\text{stop}}, \tag{II.4.25}$$

since $e < 0$. On the other hand, the unperturbed ground state $\varphi_1(\mathbf{x})$ is localized in a very small region of size approximately $1\,\text{Å} = 10^{-8}\,\text{cm}$, where $e\phi_{\text{stop}}(\mathbf{x}) = U_{\text{stop}}$. Hence

$$\langle \varphi_1, (H + e\phi_{\text{stop}}(\mathbf{x}))\varphi_1 \rangle \approx \hbar\omega_1 + eU_{\text{stop}}, \tag{II.4.26}$$

and (II.4.23) follows.

Finally, the applied voltage $\phi_{\text{stop}}(\mathbf{x})$ prevents the occurrence of photoelectric effect if the spectral condition (II.4.6) fails for the modified ground state, i.e., $0 < \omega < |\tilde{\omega}_1|$ or

$$\hbar\omega < |\hbar\omega_1 + eU_{\text{stop}}| = \hbar|\omega_1| - eU_{\text{stop}}, \tag{II.4.27}$$

since $\omega_1 < 0$, while $e < 0$ and we define $U_{\text{stop}} > 0$. In other words,

$$-eU_{\text{stop}} > \hbar\omega - \hbar|\omega_1|, \tag{II.4.28}$$

which is equivalent to (II.4.3) by (II.4.22). Thus, the condition (II.4.3) for the stopping voltage is proved.

Remark II.4.6. For eigenfunctions with higher numbers, localization becomes progressively worse, and the eigenfunctions of the continuous spectrum are not localized at all. Accordingly, the shift of higher eigenvalues gets smaller and smaller, while the continuous spectrum of the modified Schrödinger operator remains unchanged.

II.4.5   Further improvements

The calculations of G. Wentzel take into account the interaction of the Maxwell and Schrödinger fields in the first-order approximation. The second-order correction was obtained by A. Sommerfeld and G. Schur [104]. The corresponding corrected formula reads as follows (see [82, Vol. II]):

$$\mathbf{j}(\mathbf{x}, t) \sim \frac{\sin^2 \theta \, \cos^2 \varphi \, (1 + 4\beta \cos \theta)}{|\mathbf{x}|^2} \, \mathbf{n}(\mathbf{x}), \qquad |\mathbf{x}| \to \infty. \qquad (\text{II.4.29})$$

Here $\beta = \dfrac{v}{c}$, with v the velocity of photoelectrons. The formula implies that we have an *increment* of the scattering amplitude for angles $0 < \theta < \dfrac{\pi}{2}$ and a *decrement* of the scattering amplitude for angles $\dfrac{\pi}{2} < \theta < \pi$. This means that there is a "forward shift" of scattering amplitude due to pressure of the incident light on the outgoing photocurrent, as predicted by G. Wentzel [106].

J. Fisher and F. Sauter obtained the following formula, which is correct in each order of perturbation theory (see [82, Vol. II]):

$$\mathbf{j}(\mathbf{x}, t) \sim \frac{\sin^2 \theta \, \cos^2 \varphi}{(1 - \beta \cos \theta)^4 |\mathbf{x}|^2} \mathbf{n}(\mathbf{x}), \qquad |\mathbf{x}| \to \infty. \qquad (\text{II.4.30})$$

II.5 Classical Scattering of Charged Particles

In 1911, E. Rutherford experimentally studied the scattering of beams of *alpha-particles* using a very thin gold foil. The results proved to be very surprising, testifying to the concentration of a positive charge at a point nucleus. E. Rutherford calculated the corresponding *differential cross section*, which was found to be in agreement with the experimental data.

II.5.1 The Kepler problem

A homogeneous beam of classical particles, with a charge Q and mass M, falls onto a heavy nucleus with a positive charge $|e|Z > 0$. The mass of the nucleus being much larger than M, we may assume that the nucleus is fixed at the origin $\mathbf{x} = 0$. The trajectory of the particle $\mathbf{x}(t)$ satisfies the Newton equation. In the Gaussian units [48, p. 781], [49], the equation reads as

$$M\ddot{\mathbf{x}}(t) = \frac{Q|e|Z}{|\mathbf{x}(t)|^2} \frac{\mathbf{x}(t)}{|\mathbf{x}(t)|}, \qquad t \in R. \tag{II.5.1}$$

This equation holds both in the repulsive case, when $Q > 0$, and in the attractive case, when $Q < 0$. The repulsive case corresponds, for example, to the scattering of α-particles with $Q = 2|e|$, and the attractive case corresponds, for example, to the scattering of electrons with $Q = e < 0$.

For an incident particle moving along the trajectory

$$\mathbf{x} = \mathbf{x}(t), \qquad -\infty < t < \infty,$$

we assume that the particle comes from infinity with nonzero velocity, that is,

$$\lim_{t \to -\infty} |\mathbf{x}(t)| = \infty, \qquad \lim_{t \to -\infty} |\dot{\mathbf{x}}(t)| = v > 0. \tag{II.5.2}$$

This trajectory is known to be a hyperbola [3, 36] (see also Section 14.1 of [51]). Let us choose coordinates in space so that $x^3(t) \equiv 0$, and

$$\lim_{t \to -\infty} \dot{\mathbf{x}}(t) = (v, 0, 0), \qquad \lim_{t \to -\infty} x^1(t) = -\infty, \qquad \lim_{t \to -\infty} x^2(t) = b, \tag{II.5.3}$$

where b is a constant called the *impact parameter*. Let us further use polar coordinates in the plane $x^3 = 0$,

$$x^1 = r \cos\theta, \qquad x^2 = r \sin\theta, \tag{II.5.4}$$

and denote by $r(t)$, $\theta(t)$ the trajectory of a particle in these coordinates. From the initial scattering conditions (II.5.3), we have

$$\lim_{t \to -\infty} \theta(t) = \pi, \qquad \lim_{t \to -\infty} r(t) \sin\theta(t) = b. \tag{II.5.5}$$

II.5.2 Angle of scattering

The determination of the rules of motion of a particle in the Coulomb (or Newton) potential is the *Kepler problem* which was solved by I. Newton in 1687. The trajectories are defined from the conservation of energy and angular momentum. We will apply these conservation laws to calculate the final scattering angle

$$\bar{\theta} := \lim_{t \to \infty} \theta(t). \tag{II.5.6}$$

Lemma II.5.1. *The final scattering angle is given by the formula*

$$\cot \frac{\bar{\theta}}{2} = \frac{Mbv^2}{Q|e|Z}. \tag{II.5.7}$$

Proof. The conservation of the angular momentum and of energy imply that

$$r^2(t)\dot{\theta}(t) = bv, \qquad \frac{M}{2}(\dot{r}^2(t) + r^2(t)\dot{\theta}^2(t)) + \frac{Q|e|Z}{r(t)} = \frac{M}{2}v^2. \tag{II.5.8}$$

Let us write $r(t) = r_*(\theta(t))$. Then

$$\dot{r}(t) = \frac{dr}{dt} = \frac{dr_*}{d\theta}\frac{d\theta}{dt} = r'_*(\theta(t))\dot{\theta}, \tag{II.5.9}$$

and the energy conservation relation becomes

$$\frac{M}{2}\dot{\theta}^2(t)(|r'_*(\theta(t))|^2 + r^2_*(\theta)) + \frac{Q|e|Z}{r_*(\theta)} = \frac{M}{2}v^2. \tag{II.5.10}$$

Now let us introduce the *Clairaut substitution* $r_*(\theta) = 1/\rho(\theta)$, so $r'_* = -\rho'/\rho^2$. Note that the momentum conservation gives $\dot{\theta}(t) = vb\rho^2$. Therefore, (II.5.10) becomes

$$\frac{M}{2}b^2v^2(|\rho'|^2 + \rho^2) + Q|e|Z\rho = \frac{M}{2}v^2. \tag{II.5.11}$$

We differentiate this expression in θ and divide by ρ', arriving at the *Clairaut equation*

$$\rho'' + \rho = C := -\frac{Q|e|Z}{Mbv^2}. \tag{II.5.12}$$

The general solution of this equation is:

$$\rho(\theta) = A\cos\theta + B\sin\theta + C. \tag{II.5.13}$$

Now the initial scattering conditions (II.5.5) give

$$\lim_{\theta \to \pi} \rho(\theta) = 0, \qquad \lim_{\theta \to \pi} \frac{\rho(\theta)}{\sin\theta} = \frac{1}{b}. \tag{II.5.14}$$

If we substitute (II.5.13) into the previous formula, we obtain $-A + C = 0$ and $B = 1/b$, and hence

$$\rho(\theta) = C(1 + \cos\theta) + \frac{1}{b}\sin\theta. \tag{II.5.15}$$

Thus, for the final scattering angle, it follows from (II.5.6) that $\rho(\overline{\theta}) = 0$. Hence

$$C(1 + \cos\overline{\theta}) + \frac{1}{b}\sin\overline{\theta} = 0. \tag{II.5.16}$$

This implies (II.5.7).   □

Remark II.5.2. The solution $\overline{\theta} \in (-\pi, \pi)$ of equation (II.5.7) exists and is unique. For $b > 0$, the repulsive case corresponds to $Q > 0$ and $\overline{\theta} \in (0, \pi)$, while the attractive case corresponds to $Q < 0$ and $\overline{\theta} \in (-\pi, 0)$.

II.5.3   The Rutherford scattering

Now let us assume that the incident particles constitute a beam with flux density of n particles per cm^2·s in the direction $\mathbf{e}_1 := (1, 0, 0)$. Let $N = N(b, b+db)$ be the number of incident particles per second with impact parameter within the interval $[b, b+db]$. By axial symmetry, we have, for an infinitesimal interval db,

$$N(b, b + db) = n 2\pi b\, db. \tag{II.5.17}$$

The particles are scattered in the spatial angle $d\Omega = 2\pi \sin\overline{\theta}\, d\overline{\theta}$.

Definition II.5.3. *The differential cross section of scattering is defined by*

$$D(\overline{\theta}) := \frac{N/d\Omega}{n} = \frac{b\, db}{\sin\overline{\theta}\, d\overline{\theta}}. \tag{II.5.18}$$

Let us calculate the cross section. Rewriting (II.5.7) as

$$b^2 = \left(\frac{Q|e|Z}{Mv^2}\right)^2 \cot^2\frac{\overline{\theta}}{2} \tag{II.5.19}$$

and differentiating, we find

$$2b\, db = \left(\frac{Q|e|Z}{Mv^2}\right)^2 2\cot\frac{\overline{\theta}}{2}\frac{1}{\sin^2\frac{\overline{\theta}}{2}}\frac{d\overline{\theta}}{2}. \tag{II.5.20}$$

Substituting this into (II.5.18), we arrive at the *Rutherford formula*

$$D(\overline{\theta}) = \frac{\left(\dfrac{Q|e|Z}{Mv^2}\right)^2}{4\sin^4\frac{\overline{\theta}}{2}}. \tag{II.5.21}$$

II.6   Quantum Scattering of Electrons

The quantum analog of the Rutherford scattering is the scattering of an electron beam by the hydrogen atom in its ground state. The incident electron beam is described by a plane wave of type (I.3.43):

$$\psi_{\text{in}}(\mathbf{x}, t) = A e^{i(\mathbf{k} \cdot \mathbf{x} - \omega t)}, \qquad \mathbf{k} = (k, 0, 0) \neq 0, \tag{II.6.1}$$

moving in the direction $\mathbf{e}_1$ (if $k > 0$). This incident wave is a solution of the free Schrödinger equation (I.2.5). Hence, similarly to (I.3.44)),

$$\hbar\omega = \frac{\hbar^2}{2\text{m}}\mathbf{k}^2 > 0. \tag{II.6.2}$$

The corresponding stationary electric current density is given by (I.3.45):

$$\mathbf{j}_{\text{in}} := \frac{e}{\text{m}} \, \text{Re} \left(\overline{\psi}_{\text{in}}(\mathbf{x}, t)[-i\hbar\nabla\psi_{\text{in}}(\mathbf{x}, t)] \right) = \frac{e\hbar\mathbf{k}}{\text{m}}|A|^2. \tag{II.6.3}$$

II.6.1   Radiated outgoing wave

The unperturbed atom is in its ground state $\psi_1(\mathbf{x}, t) = \varphi_1(\mathbf{x})e^{-i\omega_1 t}$, as given by (II.2.1). Let us decompose the total wave field into three terms:

$$\psi(\mathbf{x}, t) = \psi_{\text{in}}(\mathbf{x}, t) + \psi_1(\mathbf{x}, t) + \psi_{\text{r}}(\mathbf{x}, t), \tag{II.6.4}$$

where $\psi_{\text{r}}(\mathbf{x}, t)$ is the outgoing *radiated wave*. We will assume that the incident plane wave is a small perturbation of the ground state, that is,

$$|A| \ll C_1 \tag{II.6.5}$$

(see (I.6.49)). Hence the radiated wave is also small, so the total wave field is approximately equal to the solution of the Schrödinger equation

$$[i\hbar\partial_t - e\phi(\mathbf{x})]\psi(\mathbf{x}, t) = \frac{1}{2\text{m}}[-i\hbar\nabla]^2\psi(\mathbf{x}, t). \tag{II.6.6}$$

Here $\phi(\mathbf{x}) = -e/|\mathbf{x}|$ is the Coulomb potential; its decay at infinity is rather slow, resulting in divergence in the intermediate calculations. Let us approximate $\phi(\mathbf{x})$ as follows:

$$\phi_\varepsilon(\mathbf{x}) = -\frac{e}{|\mathbf{x}|}e^{-\varepsilon|\mathbf{x}|}, \tag{II.6.7}$$

where $0 < \varepsilon \ll 1$. Afterwards we will let ε tend to zero. Substituting (II.6.4) into the Schrödinger equation (II.6.6) with ϕ_ε replacing ϕ, we obtain, approximately,

$$(i\hbar\partial_t - e\phi_\varepsilon(\mathbf{x}))(\psi_{\text{in}}(\mathbf{x}, t) + \psi_{\text{r}}(\mathbf{x}, t)) = \frac{1}{2\text{m}}[-i\hbar\nabla]^2(\psi_{\text{in}}(\mathbf{x}, t) + \psi_{\text{r}}(\mathbf{x}, t)), \tag{II.6.8}$$

since $\psi_1(\mathbf{x}, t)$ is the exact solution. We rewrite this equation as follows:

$$\left(i\hbar\partial_t - e\phi_\varepsilon(\mathbf{x}) - \frac{1}{2m}[-i\hbar\nabla]^2\right)\psi_r(\mathbf{x}, t) = e\phi_\varepsilon(\mathbf{x})\psi_{\text{in}}(\mathbf{x}, t) = e\phi_\varepsilon(\mathbf{x})Ae^{i(\mathbf{k}\cdot\mathbf{x}-\omega t)}.$$
(II.6.9)

Since $\omega > 0$, it follows by (II.6.2) that the frequency ω lies in the continuous spectrum of the Schrödinger operator (II.3.8). Now we apply the *limiting amplitude principle* (see Section 28 of [148]) to obtain the long-time asymptotics for $\psi_r(\mathbf{x}, t)$:

$$\psi_r(\mathbf{x}, t) \sim \varphi_r(\mathbf{x})e^{-i\omega t}, \qquad t \to \infty;$$
(II.6.10)

here the contribution of the discrete spectrum is neglected, since the eigenfunctions rapidly decay at infinity, unlike $\varphi_r(\mathbf{x})$.

Substituting these asymptotics into equation (II.6.9), we obtain the stationary equation for the limiting amplitude:

$$\left(\hbar\omega - e\phi_\varepsilon(\mathbf{x}) - \frac{1}{2m}[-i\hbar\nabla]^2\right)\varphi_r(\mathbf{x}) = eA\phi_\varepsilon(\mathbf{x})e^{i\mathbf{k}\cdot\mathbf{x}}.$$
(II.6.11)

Neglecting the term $e\phi_\varepsilon(\mathbf{x})$ on the left, we finally arrive at the stationary Helmholtz equation

$$[\Delta + \mathbf{k}^2]\varphi_r(\mathbf{x}) = \frac{2meA}{\hbar^2}\phi_\varepsilon(\mathbf{x})e^{i\mathbf{k}\cdot\mathbf{x}},$$
(II.6.12)

because $\mathbf{k}^2 = \dfrac{2m\omega}{\hbar}$ by (II.6.2). This is an equation of the same type as (II.4.15). Therefore, the solution is given by a convolution similar to (II.4.17):

$$\varphi_r(\mathbf{x}) = -\frac{2meA}{\hbar^2}\int\frac{e^{ik|\mathbf{x}-\mathbf{y}|}}{4\pi|\mathbf{x}-\mathbf{y}|}\phi_\varepsilon(\mathbf{y})e^{i\mathbf{k}\cdot\mathbf{y}}\,d\mathbf{y}.$$
(II.6.13)

This convolution is similar to the last integral in (II.2.11). Evaluating it by the method used in (II.2.12)–(II.2.15) and sending $\varepsilon \to 0$, we obtain:

$$\varphi_r(\mathbf{x}) \sim -A\frac{e^{ik|\mathbf{x}|}}{|\mathbf{x}|}f(k, \theta), \qquad |\mathbf{x}| \to \infty,$$
(II.6.14)

where θ is the polar angle between $\mathbf{x}$ and $\mathbf{e}_1$, and

$$f(k, \theta) = \frac{2me}{\hbar^2}\int_0^\infty\frac{\sin K|\mathbf{y}|}{K|\mathbf{y}|}\phi(\mathbf{y})|\mathbf{y}|^2\,d|\mathbf{y}|, \qquad K := 2k\sin\frac{\theta}{2}.$$
(II.6.15)

Now (II.6.10) becomes, for large $|\mathbf{x}|$,

$$\psi_r(\mathbf{x}, t) \sim -A\frac{e^{ik|\mathbf{x}|}}{|\mathbf{x}|}f(k, \theta)e^{-i\omega t}, \qquad t \to \infty.$$
(II.6.16)

The limiting amplitude ψ_r decays slowly as $|\mathbf{x}| \to \infty$, its L^2-norm being infinite. This corresponds to the fact that the frequency $\omega > 0$ lies in the continuous spectrum. Physically, this means that the electron flux goes off to infinity, as we shall see in the next section.

II.6.2   Differential cross section

The asymptotics (II.6.16) gives us the stationary electric current corresponding to the outgoing radiated wave $\psi_r(\mathbf{x}, t)$ for large t and $|\mathbf{x}|$:

$$\mathbf{j}_r(\mathbf{x}) := \frac{e}{m} \operatorname{Re}\left(\overline{\psi}_r(\mathbf{x}, t)\left[-i\hbar\nabla\psi_r(\mathbf{x}, t)\right]\right) \sim \frac{e\hbar k\mathbf{n}(\mathbf{x})}{m} \frac{|f(k, \theta)|^2}{|\mathbf{x}|^2}|A|^2, \quad \text{(II.6.17)}$$

where $\mathbf{n}(\mathbf{x}) := \mathbf{x}/|\mathbf{x}|$. The *angular distribution* of the outgoing electric current is defined by

$$j_r(\mathbf{n}) := \lim_{R\to\infty} R^2\mathbf{j}_r(R\mathbf{n})\cdot\mathbf{n}, \qquad \mathbf{n}\in\mathbb{R}^3, \qquad |\mathbf{n}| = 1. \qquad \text{(II.6.18)}$$

The current (II.6.17) is radial at infinity, hence the corresponding angular density is given by

$$j_r(\mathbf{n}) = \frac{e\hbar k}{m}|f(k, \theta)|^2|A|^2. \qquad \text{(II.6.19)}$$

In the quantum case, Definition II.5.3 should be modified as follows.

Definition II.6.1. *The* differential cross section *of the quantum scattering of electrons is defined by*

$$D(\mathbf{n}) := \frac{j_r(\mathbf{n})}{|\mathbf{j}_{in}|}, \qquad \mathbf{n}\in\mathbb{R}^3, \quad |\mathbf{n}| = 1. \qquad \text{(II.6.20)}$$

Formulae (II.6.19) and (II.6.3) imply that

$$D(\mathbf{n}) = |f(k, \theta)|^2. \qquad \text{(II.6.21)}$$

Further, substituting (II.6.7) into (II.6.15), we obtain:

$$f(k, \theta) = \frac{2me^2}{K\hbar^2}\int_0^\infty \sin K|\mathbf{y}|e^{-\varepsilon|\mathbf{y}|}d|\mathbf{y}| = \frac{2me^2}{\hbar^2(K^2 + \varepsilon^2)}; \qquad \text{(II.6.22)}$$

here $K = 2k\sin\dfrac{\theta}{2} \neq 0$ if $\theta \neq 0$ and $k \neq 0$. Hence, we can drop ε^2 as $\varepsilon\to 0$ and write

$$f(k, \theta) = \frac{2me^2}{\hbar^2 K^2} = \frac{me^2}{2\hbar^2 k^2\sin^2\dfrac{\theta}{2}}. \qquad \text{(II.6.23)}$$

We rewrite this expression using the wave-particle duality (I.3.46):

$$f(k, \theta) = \frac{e^2}{2m\mathbf{v}^2\sin^2\dfrac{\theta}{2}}. \qquad \text{(II.6.24)}$$

Now (II.6.21) reads

$$D(\mathbf{n}) = \frac{\left(\dfrac{e^2}{m\mathbf{v}^2}\right)^2}{4\sin^4\dfrac{\theta}{2}}, \tag{II.6.25}$$

which coincides with the classical Rutherford formula (II.5.21) with $Q=e$ and $Z=1$.

Remark II.6.2. The agreement of the cross section (II.6.25) with its classical analogue (II.5.21) was considered by M. Born as a crucial justification of Schrödinger's quantum mechanics and of its *probabilistic interpretation* [11] (see Section V.4).

II.7   Electron Diffraction

In this section, we calculate the limiting amplitude of diffraction of an electron beam and calculate the corresponding current.

II.7.1   Introduction

The diffraction of electron beams was first observed by C. Davisson and L. Germer in 1924–1927 [112] (see also [116] and [51, Section 5.2]). In these first experiments, the electron beam was scattered by a crystal of nickel, and the reflected beam was registered on a film. The resulting images are similar to X-ray diffraction patterns ("Lauegrams"), first obtained in 1912 by the method of Laue. These experiments were the main motivation for Born's introduction of the probabilistic interpretation of the wave function.

Later on, such experiments were also carried out with transmitted electron beams scattered by thin gold and platinum crystalline films (G. P. Thomson, the 1937 Nobel Prize). Only recently R. Bach with collaborators first carried out the double-slit diffraction of electrons [108, 109].

We shall consider the diffraction of a transmitted electron beam for a general aperture in a plane screen. The incident electron beam is described by a plane wave. We will calculate the limiting diffraction amplitude applying Fresnel–Kirchhoff theory [12, 149] in the framework of the Schrödinger equation. We check that this approach agrees satisfactorily with the recent experiments of R. Bach *et al.* [108, 109] in the particular case of a double-slit aperture.

II.7.2   Electron diffraction

The experiment on the diffraction of a transmitted electron beam is shown schematically on Fig. II.1. The incident beam of electrons with energy E falls from the left on the scattering plane S with an aperture Q. We choose coordinates setting $x^3 = 0$ on the plane S and $x^3 < 0$ on the left of S. The wave function satisfies the free Schrödinger equation

$$i\hbar \partial_t \psi(\mathbf{x}, t) = -\frac{\hbar^2}{2\mathrm{m}} \Delta \psi(\mathbf{x}, t), \qquad x^3 \neq 0. \qquad (\text{II.7.1})$$

The diffraction pattern is observed on the observation screen P; see Fig. II.1.

The incident wave. The incident wave in the region $x^3 < 0$ to the left of the scattering plane $x^3 = 0$ describes the beam of free electrons,

$$\psi_{\text{in}}(\mathbf{x}, t) = a_{\text{in}} e^{i(kx^3 - \omega t)}, \qquad x^3 < 0, \qquad (\text{II.7.2})$$

where

$$k > 0, \qquad \omega = E/\hbar, \qquad \hbar\omega = \frac{\hbar^2 k^2}{2\mathrm{m}} \qquad (\text{II.7.3})$$

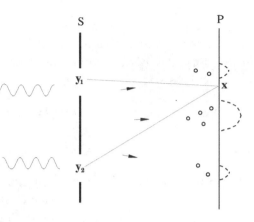

Figure II.1: Double-slit diffraction.

since ψ_{in} satisfies (II.7.1). In Section V.3.3, we will show, using the quasi-classical asymptotics (I.2.19), that such an incident wave is provided by an electron gun with voltage $V = -E/e$ (see (V.3.4) and (V.3.22), (V.3.23)).

The diffraction problem reduces to solving the free Schrödinger equation (II.7.1) in the region

$$\Omega = \{\mathbf{x} \in \mathbb{R}^3 : x^3 > 0\}, \tag{II.7.4}$$

with the boundary conditions defined by the incident wave (II.7.2). We assume the *limiting amplitude principle* to hold, i.e., we assume that

$$\psi(\mathbf{x}, t) \sim a_\infty(\mathbf{x})e^{-i\omega t}, \qquad t \to \infty. \tag{II.7.5}$$

This principle is well established for the Schrödinger equation with short-range potential and for boundary value problems; see [113, 114, 152, 157], and [148, Section 28], [51, Section 13.5]. The Schrödinger equation (II.7.1) implies that the limiting amplitude satisfies the stationary equation

$$H_0(k)a_\infty(\mathbf{x}) := [\Delta + k^2]a_\infty(\mathbf{x}) = 0, \quad \mathbf{x} \in \Omega := \{\mathbf{x} \in \mathbb{R}^3 : x^3 > 0\}. \tag{II.7.6}$$

The solution can be expressed in terms of its Cauchy data on $\partial\Omega = S$ by using the Green formula

$$a_\infty(\mathbf{x}) = \int_{y^3=0+} \left(G_0(\mathbf{x}, \mathbf{y})\partial_{y^3}a_\infty(\mathbf{y}) - \partial_{y^3}G_0(\mathbf{x}, \mathbf{y})a_\infty(\mathbf{y})\right) d\mathbf{y}, \quad \mathbf{x} \in \Omega, \tag{II.7.7}$$

where the Green function $G_0(\mathbf{x}, \mathbf{y})$ is the integral kernel of the inverse operator $H_0^{-1}(k + i0)$ on $\mathbb{R}^3$:

$$H_0(k)G_0(\cdot, \mathbf{y}) = \delta(\mathbf{x} - \mathbf{y}), \quad \mathbf{x}, \mathbf{y} \in \Omega; \quad G_0(\mathbf{x}, \mathbf{y}) = -\frac{e^{ik|\mathbf{x}-\mathbf{y}|}}{4\pi|\mathbf{x} - \mathbf{y}|}. \tag{II.7.8}$$

In (II.7.7) we have omitted the integral over the half-sphere $\{\mathbf{y} \in \Omega : |\mathbf{y}| = R\}$ with $R \to \infty$. This integral vanishes in the limit due to the Sommerfeld radiation condition [81, formulae (34.5)], which holds for this Green function and for the amplitude: as $|\mathbf{y}| \to \infty$,

$$\begin{cases} a_\infty(\mathbf{y}) = \mathcal{O}(|\mathbf{y}|^{-1}), & \partial_{|\mathbf{y}|} a_\infty(\mathbf{y}) = ika_\infty(\mathbf{y}) + o(|\mathbf{y}|^{-1}) \\ G_0(\mathbf{x}, \mathbf{y}) = \mathcal{O}(|\mathbf{y}|^{-1}), & \partial_{|\mathbf{y}|} G_0(\mathbf{x}, \mathbf{y}) = ikG_0(\mathbf{x}, \mathbf{y}) + o(|\mathbf{y}|^{-1}) \end{cases} . \quad \text{(II.7.9)}$$

Remarks II.7.1. This radiation condition obviously holds for the Green function (II.7.8), while it does not hold for the alternative Green function $\overline{G_0(\mathbf{x}, \mathbf{y})}$. We assume that the Sommerfeld condition also holds for the limiting amplitude, since it is well established for the Schrödinger equation with short-range potential and for boundary value problems outside a smooth bounded domains; see [113, 114, 152, 157].

The Kirchhoff approximation. The key point of the *Kirchhoff diffraction theory* [12] is the approximation of the Cauchy data for the limiting amplitude by the Cauchy data corresponding to the incident wave: (II.7.2)

$$a_\infty(\mathbf{y})|_{y^3=0+} \approx D_{\text{in}}(\mathbf{y}), \quad \partial_{y^3} a_\infty(\mathbf{y})_{y^3=0+} \approx N_{\text{in}}(\mathbf{y}), \quad \text{(II.7.10)}$$

where

$$D_{\text{in}}(\mathbf{y}) := \begin{cases} a_{\text{in}}, & \mathbf{y} \in Q, \\ 0, & \mathbf{y} \in S \setminus Q, \end{cases} \qquad N_{\text{in}}(\mathbf{y}) := \begin{cases} ika_{\text{in}}, & \mathbf{y} \in Q, \\ 0, & \mathbf{y} \in S \setminus Q. \end{cases} \quad \text{(II.7.11)}$$

For the Green function (II.7.8), the Cauchy data at $y^3 = 0+$ are given by

$$G_0(\mathbf{x}, \mathbf{y}) = -\frac{1}{4\pi} \frac{e^{iks}}{s}, \quad \partial_{y^3} G_0(\mathbf{x}, \mathbf{y}) = -\frac{ik}{4\pi} \frac{e^{iks}}{s} \cos\chi + \mathcal{O}(s^{-2}), \quad s \to \infty, \quad \text{(II.7.12)}$$

where $s := |\mathbf{x} - \mathbf{y}|$, $\mathbf{e}_3 := (0, 0, 1)$, and χ is the "Fresnel angle of diffraction" between the vectors $\mathbf{e}_3$ and $\mathbf{x} - \mathbf{y}$; see [12, Fig. 8.1]. Substituting this Cauchy data into the Green formula (II.7.7), we obtain a result similar to formula (18) of Section 8.3 of [12]:

$$a_\infty(\mathbf{x}) = -\frac{ika_{\text{in}}}{(4\pi)^2} \int_Q \frac{e^{ik|\mathbf{x}-\mathbf{y}|}}{|\mathbf{x} - \mathbf{y}|} (1 + \cos\chi) \, d\mathbf{y}, \qquad \mathbf{x} \in \Omega. \quad \text{(II.7.13)}$$

Exercise II.7.2. Deduce (II.7.7) from (II.7.6).

Exercise II.7.3. Verify (II.7.9) and (II.7.11)–(II.7.12).

Problem II.7.4. Deduce (II.7.13) from (II.7.7) and (II.7.10)–(II.7.12).

II.7.3 Limiting absorption principle

It is instructive to calculate the limiting amplitude $a_\infty(\mathbf{x})$ without reference to the Sommerfeld radiation condition. Let us show that this is possible with an application of the *limiting absorption principle*:

$$a_\infty^0(\mathbf{x}) = \lim_{\varepsilon \to 0+} a_\infty^0(\mathbf{x}, k + i\varepsilon), \tag{II.7.14}$$

where $a_\infty^0(\mathbf{x}, k + i\varepsilon)$ is the solution of the Cauchy problem for equation (II.7.6) with complex parameter $k + i\varepsilon$ replacing k and with the same Cauchy data (II.7.10), (II.7.11).

The key advantage of a complex parameter is that the corresponding Green function decays exponentially at infinity, so that a Sommerfeld-type condition is not necessary for obtaining the integral representation (II.7.7).

Indeed, to solve the stationary equation (II.7.6) with given Cauchy boundary conditions (II.7.10), (II.7.11), we extend the amplitudes $a_\infty(\mathbf{x}, z)$ by zero for $x^3 \le 0$ and set

$$a_\infty^0(\mathbf{x}, z) = \begin{cases} a_\infty(\mathbf{x}, z), & \mathbf{x} \in \Omega, \\ 0, & \mathbf{x} \in \mathbb{R}^3 \setminus \Omega, \end{cases} \tag{II.7.15}$$

with Ω from (II.7.4). Then the application of the operator $H_0(k) := \Delta + z^2$ *in the sense of distributions* gives

$$H_0(z)a_\infty^0(\mathbf{x}, z) = r(\mathbf{x}), \qquad \mathbf{x} \in \mathbb{R}^3, \tag{II.7.16}$$

where the right-hand side

$$r(\mathbf{x}) = a_\infty^0(x^1, x^2, 0+)\delta'(x^3) + \partial_{x^3} a_\infty^0(x^1, x^2, 0+)\delta(x^3) \tag{II.7.17}$$

is known from the Cauchy boundary conditions (II.7.10), (II.7.11). Applying the Fourier transform $a_\infty^0(\boldsymbol{\xi}, z) := F_{\mathbf{x} \to \boldsymbol{\xi}} a_\infty^0(\mathbf{x}, z)$, we obtain from (II.7.16) the equation

$$(-\boldsymbol{\xi}^2 + z^2)\hat{a}_\infty^0(\boldsymbol{\xi}, z) = \hat{r}(\boldsymbol{\xi}), \qquad \boldsymbol{\xi} \in \mathbb{R}^3 \tag{II.7.18}$$

in the sense of *tempered distributions*. The main problem in solving this equation with real parameter $z = k \in \mathbb{R}$ is that the quadratic form $-\boldsymbol{\xi}^2 + k^2$, which is the *symbol of the operator* $H_0(k)$, *vanishes on the sphere* $|\boldsymbol{\xi}| = |k|$. Hence, the *solution is not unique*, since every distribution $g(\boldsymbol{\xi})\delta(|\boldsymbol{\xi}| - |k|)$ is the solution of the homogeneous equation, and the space of these solutions is infinite-dimensional. On the other hand, for complex parameters $z \in \mathbb{C} \setminus \mathbb{R}$, the symbol does not vanish, and hence there is a unique solution

$$\hat{a}_\infty^0(\boldsymbol{\xi}; z) = \frac{\hat{r}(\boldsymbol{\xi})}{-\boldsymbol{\xi}^2 + z^2}, \qquad \boldsymbol{\xi} \in \mathbb{R}^3. \tag{II.7.19}$$

The key argument used to select the needed *unique solution* is the *limiting absorption principle* (II.7.14). Let us show that this principle leads to the same formula (II.7.7).

Indeed, applying the inverse Fourier transform to (II.7.19), we see that $a_\infty^0(\mathbf{x}, k + i\varepsilon)$ is given by the convolution

$$a_\infty^0(\,\cdot\,, k + i\varepsilon) = E_\varepsilon * r, \tag{II.7.20}$$

where r is the distribution (II.7.17), while E_ε is the fundamental solution

$$E_\varepsilon(\mathbf{x}) := F_{\xi \to \mathbf{x}}^{-1} \frac{1}{-\boldsymbol{\xi}^2 + (k + i\varepsilon)^2} = -\frac{e^{ik - \varepsilon|\mathbf{x}|}}{4\pi|\mathbf{x}|}, \qquad k, \varepsilon > 0. \tag{II.7.21}$$

Finally, this convolution reads as (II.7.7) with $E_\varepsilon(\mathbf{x} - \mathbf{y})$ replacing $G_0(\mathbf{x}, \mathbf{y})$, and taking the form (II.7.7) in the limit as $\varepsilon \to 0$ by (II.7.14), since

$$E_\varepsilon(\mathbf{x} - \mathbf{y}) \to G_0(\mathbf{x}, \mathbf{y}), \qquad \varepsilon \to 0+, \tag{II.7.22}$$

by (II.7.21) and (II.7.8).

Exercise II.7.5. Check (II.7.21). **Hint:** i) verify the equation

$$(\Delta + (k + i\varepsilon)^2)E_\varepsilon(\mathbf{x}) = \delta(\mathbf{x}), \qquad \mathbf{x} \in \mathbb{R}^3; \tag{II.7.23}$$

ii) use the spherical symmetry $E_\varepsilon(\mathbf{x}) = f(|\mathbf{x}|)$ and reduce (II.7.23) to an ODE for $f(r)$;

iii) take into account the fact that $E_\varepsilon(\mathbf{x})$ is a tempered distribution.

II.7.4   The Fraunhofer asymptotics

For a bounded aperture Q, the integral (II.7.13) admits the following asymptotics in each fixed direction $\mathbf{x}/|\mathbf{x}| = (\theta_1, \theta_2, \theta_3)$ as $|\mathbf{x}| \to \infty$ in the half-space $x^3 > 0$:

$$a_\infty(\mathbf{x}) \sim -\frac{ika_{\text{in}}(1 + \cos\chi_*)e^{ik|\mathbf{x}|}}{(4\pi)^2|\mathbf{x}|} \int_Q e^{-ik(\theta_1 y^1 + \theta_2 y^2)} \, d\mathbf{y}, \qquad |\mathbf{x}| \gg \text{diam}(Q),$$
$$\tag{II.7.24}$$

where $\cos\chi_* = \theta_3 > 0$ and $\text{diam}(Q)$ is the diameter of the aperture Q. This asymptotics is similar to *Fraunhofer diffraction* [12, Section 8.3.3, formula (28)]. Formula (II.7.24) implies that for any $C > 0$,

$$a_\infty(\mathbf{x}) \sim \frac{b(\theta_1, \theta_2)}{|\mathbf{x}|} e^{ikx^3} \quad \text{for } |(x^1, x^2)| \le C \quad \text{and} \quad x^3 \gg \text{diam}(Q), \tag{II.7.25}$$

where the amplitude $b(\theta_1, \theta_2)$ is a slowly varying function of the transversal variables. Hence, asymptotically,

$$\nabla a_\infty(\mathbf{x}) \sim ik\mathbf{e}_3 a_\infty(\mathbf{x}) \quad \text{for } |(x^1, x^2)| \le C \quad \text{and} \quad x^3 \gg \text{diam}(Q). \tag{II.7.26}$$

In particular, for one circular aperture of radius R centered at $y = 0$, the integral (II.7.13) was calculated in 1835 by Airy:

$$a_\infty(\mathbf{x}) \sim -\frac{ika_{\text{in}}(1 + \cos\chi_*)e^{ik|x|}}{(4\pi)^2|x|}\frac{2\pi r^2 J_1(kR\sin\chi_*)}{kR\sin\chi_*}, \qquad |\mathbf{x}| \to \infty, \quad (\text{II.7.27})$$

where $\sin\chi_* = |(\theta_1, \theta_2)|$.

Exercise II.7.6. Verify (II.7.24)–(II.7.26).

Problem II.7.7. Verify (II.7.27). Hint: see [12, Section 8.5.2, formula (13)].

II.7.5 Comparison with experiment

The double-slit diffraction of electrons was first observed experimentally first in 2013 (Bach *et al.* [108]), and our formula (II.7.13) is in satisfactory agreement with the results of these experiments. Indeed, in the particular case when the aperture Q consists of two small slits centered at the points $\mathbf{y}_1, \mathbf{y}_2 \in S$, formula (II.7.13) gives

$$a_\infty(\mathbf{x}) \sim \frac{e^{ik|\mathbf{x}-\mathbf{y}_1|}}{|\mathbf{x}-\mathbf{y}_1|}(1 + \cos\chi_1) + \frac{e^{ik|\mathbf{x}-\mathbf{y}_2|}}{|\mathbf{x}-\mathbf{y}_2|}(1 + \cos\chi_2), \qquad \mathbf{x} \in \Omega, \quad (\text{II.7.28})$$

with Ω from (II.7.4). In these experiments, the distance between the planes S and P was $D = 240\,\mu\text{m} = 240.000\,\text{nm}$, the distance between the centers of the slits was $2d = |\mathbf{y}_1 - \mathbf{y}_2| = 330\,\text{nm}$, and the wavelength $\lambda = 50\,\text{nm}$. So, for $x \in P$,

$$|\cos\chi_a - 1| \le |\mathbf{x} - \mathbf{y}_a|^2/D^2 \le 1/240^2, \qquad |\mathbf{x} - \mathbf{y}_a| < 1000\,\text{nm}, \qquad a = 1, 2.$$

Therefore, for such deviations, we can set $\cos\chi_a \approx 1$, and now formula (II.7.28) gives

$$|a_\infty(\mathbf{x})| \sim \frac{2}{D}\left|e^{ik[|\mathbf{x}-\mathbf{y}_1|-|\mathbf{x}-\mathbf{y}_2|]} + 1\right|. \qquad (\text{II.7.29})$$

Let the points $\mathbf{y}_1$, $\mathbf{y}_2$ be symmetric with respect to the origin. We choose the x^1-axis orthogonal to the plane of the drawing, so that $\mathbf{y}_1 = (0, d, 0)$ and $\mathbf{y}_2 = (0, -d, 0)$, where $d := |\mathbf{y}_1 - \mathbf{y}_2|/2$; see Fig. II.1. Then the amplitude (II.7.29) has maxima at the points $\mathbf{x} \in P$ with coordinates x^2 determined by the Bragg rule

$$\sqrt{(x^2 - d)^2 + D^2} - \sqrt{(x^2 + d)^2 + D^2} = n\lambda, \qquad n = 0, \pm 1, \ldots. \quad (\text{II.7.30})$$

We have $d \ll D$. Hence, expanding the square root in a Taylor series, we arrive at the equation

$$2x^2 d \approx n\lambda D. \qquad (\text{II.7.31})$$

For $n = 0$ we obtain the central maximum $x^2 = 0$, and for $n = 1$, the second maximum $x^2 = \lambda D/|\mathbf{y}_1 - \mathbf{y}_2| = \frac{50}{330}240 \approx 36,4\,\mu\text{m}$. The experimental value, as found in [108], is approximately $40\,\mu\text{m}$ [109, Fig. S2 a].

Remark II.7.8. Formula (II.7.29) satisfactorily reproduces the positions of the maxima, but not their magnitudes. On the other hand, the picture [109, Fig. S2 a] shows a perfect coincidence of the measured diffraction amplitude with the graph obtained by numerical integration over paths. This integration in the limit must coincide with the exact solution of the diffraction problem (II.7.1) with the incident wave (II.7.2).

Exercise II.7.9. Prove the Bragg rule (II.7.30) for the maxima of the amplitude (II.7.28).

Exercise II.7.10. Verify the approximation (II.7.31).

Chapter III

Atom in Magnetic Field

The Schrödinger equation in magnetic field depends on electron's angular momentum. The Zeeman experimental observations of 1897 demonstrated that the corresponding atomic spectra do not agree with this equation in many cases (*anomalous Zeeman effect*). This discrepancy suggests that the *orbital angular momentum* (I.3.14) must be modified. The suggestion was confirmed by the *Einstein–de Haas experiments* of 1915 and *Stern–Gerlach experiments* of 1922.

The modification of angular momentum was proposed by S. Goudsmit and G. Uhlenbeck in 1925 by introduction of *spin* which is an "intrinsic angular momentum" of electron. The corresponding modification of the Schrödinger equation, the *Pauli equation*, was introduced by W. Pauli in 1927.

This equation makes the electron dynamics covariant with respect to the *spinor representation* of the rotation group $SO(3)$. The total quantum angular momentum, corresponding to this representation in the sense of Definition III.3.3 (see below), includes the *spinor term*, which agrees with the Goudsmit–Uhlenbeck conjecture.

The Pauli equation, complemented with the *Russell-Saunders coupling term*, perfectly explains the Einstein–de Haas experiments and the anomalous Zeeman effect.

III.1   Normal Zeeman Effect

In 1895, P. Zeeman observed a splitting of the spectral lines (I.4.4) in a uniform magnetic field $\mathbf{B} = (0, 0, B)$.

III.1.1   Magnetic Schrödinger equation

The *normal Zeeman effect* is the splitting of a single spectral line into three separate lines (A.8.7), (A.8.7). This normal effect was successfully explained in the classical Lorentz model (Section A.2) as well as in the Old Quantum Theory (Section A.8).

The Schrödinger theory leads to the same result. Indeed, for an atom with one electron and nucleus charge $|e|Z$ in a magnetic field with potential $\mathbf{A}^{\mathrm{ext}}(\mathbf{x})$, the Schrödinger equation (I.2.14) reads as follows:

$$i\dot{\psi}(t) = H\psi(t), \qquad H = \frac{1}{2\mathrm{m}}\left[-i\hbar\nabla - \frac{e}{c}\mathbf{A}^{\mathrm{ext}}(\mathbf{x})\right]^2 - \frac{e^2 Z}{|\mathbf{x}|}. \tag{III.1.1}$$

In particular, for a uniform external magnetic field $\mathbf{B} = (0, 0, B)$, the vector potential has the form (I.3.26), and the Schrödinger equation (III.1.1) takes the form

$$i\dot{\psi}(t) = H_B\psi(t), \quad H_B = -\frac{\hbar^2}{2\mathrm{m}}\Delta - \frac{e^2 Z}{|\mathbf{x}|} - \frac{e}{2\mathrm{m}c}B\hat{L}_3 = H - \Omega_\Lambda\hat{L}_3 \tag{III.1.2}$$

if the quadratic terms with $\frac{e^2}{c^2}|\mathbf{B}|^2$ are neglected. Here

$$\Omega_\Lambda = \frac{eB}{2\mathrm{m}c} \tag{III.1.3}$$

is the *Larmor frequency*. Now (I.6.9) implies that

$$\begin{cases} H_B\varphi_{nlm} = E_{nm}\varphi_{nlm}, E_{nm} = E_n - \Omega_\Lambda\hbar m & \left| \begin{array}{l} n = 1, 2, \ldots \\ l = 0, 1, \ldots, n-1 \\ m = -l, \ldots, l \end{array} \right. \\ \hat{\mathbf{L}}^2\varphi_{nlm} = \hbar^2 l(l+1)\varphi_{nlm}, \hat{L}_3\varphi_{nlm} = \hbar m\varphi_{nlm} \end{cases}, \tag{III.1.4}$$

which coincides with the formula (A.8.6) from Appendix A except for the range of l. Thus, the spectral lines (I.4.4) are shifted by $-\Omega_\Lambda(m' - m)$ and are given by

$$\omega_{nn'} - \Omega_\Lambda(m' - m). \tag{III.1.5}$$

These spectral lines coincide with the result (A.8.7) in the Old Quantum Theory.

Note that the last term in (III.1.2) can be written as

$$\Omega_\Lambda\hat{L}_3 = \mathbf{B}\cdot\hat{\mathbf{M}}, \qquad \text{with} \quad \hat{\mathbf{M}} := \frac{e}{2\mathrm{m}c}\hat{\mathbf{L}}. \tag{III.1.6}$$

The classical formula (A.8.5) means that the magnetic moment **M** adds the value $-\mathbf{B} \cdot \mathbf{M}$ to the electron energy in the *uniform* magnetic field. Therefore, $\hat{\mathbf{M}}$ is the quantum observable of the magnetic moment, and the last relation in (III.1.6) agrees with (A.8.2). Now the operator (III.1.2) takes the form

$$H_B = H - B\hat{M}_3. \tag{III.1.7}$$

Exercise III.1.1. Check the expression (III.1.2) for H_B. **Hint:** use the definition of $\hat{\mathbf{L}}$ in (I.3.15).

III.1.2 Selection rules

Let us show that the selection rules (A.8.8) also hold in the Schrödinger theory in the sense that the intensity of the other lines vanishes. This follows from the formulae for the intensity of spectral lines. Indeed, the intensity of the line with frequency

$$\omega_{n'm'} - \omega_{nm} = (E_{n'm'} - E_{nm})/\hbar$$

is *proportional* to the sum

$$I_{nm\,n'm'} := |\omega_{n'm'} - \omega_{nm}|^4 \sum_{l=0}^{n-1} \sum_{l'=0}^{n'-1} |\mathbf{d}_{nlm\,n'l'm'}|^2, \tag{III.1.8}$$

where $\mathbf{d}_{nlmn'l'm'}$ are the *dipole momenta*

$$\mathbf{d}_{nlm\,n'l'm'} := \langle \varphi_{nlm}, \mathbf{x}\varphi_{n'l'm'} \rangle = \int \mathbf{x}\overline{\varphi}_{nlm}(\mathbf{x})\varphi_{n'l'm'}(\mathbf{x})\,d\mathbf{x}; \tag{III.1.9}$$

see [77, (45.13), (45.22)], [51, (7.48), (7.52)]. In spherical coordinates r, φ, θ (see (I.6.22)), all eigenfunctions of the operator $\hat{L}_3 = -i\hbar\partial_\varphi$ have the form $c(r, \theta)e^{im\varphi}$ with integer m and with the corresponding eigenvalues $\hbar m$. Hence, for a suitable choice of a multiplier, the eigenfunctions φ_{nlm} read

$$\varphi_{nlm} = c_{nlm}(r, \theta)e^{im\varphi} \tag{III.1.10}$$

and the corresponding eigenvalue equals $\hbar m$. The vector $\mathbf{x} = (x^1, x^2, x^3)$ contains the harmonics $e^{ik\varphi}$ only for

$$k = 0, \pm 1. \tag{III.1.11}$$

Hence, in spherical coordinates, the dipole moment (III.1.9) is proportional to the integral

$$\int_0^{2\pi} e^{i(m-m'+k)\varphi}\,d\varphi,$$

which vanishes for $|m' - m| > 1$. Thus, the contribution to the intensity of radiation (III.1.8) from the quantum numbers nlm and $n'l'm'$ with $|m'-m| > 1$ is absent. This fact is interpreted as the following *selection rule*:

$$m' - m = 0, \pm 1 \qquad \text{(III.1.12)}$$

for possible *quantum transitions $nlm \mapsto n'l'm'$*. A similar selection rule holds for the azimuthal quantum number l,

$$l' - l = \pm 1. \qquad \text{(III.1.13)}$$

The proof can be found in [11].

Thus, the normal Zeeman effect is fully explained in the Schrödinger theory. However, the anomalous Zeeman effect is not explained by this theory. That problem was solved only after introduction i) of *electron spin* and ii) of the corresponding *Russell–Saunders coupling* term in the Schrödinger equation (III.1.2); this modified equation describes the spin interaction with quantum angular momentum $\hat{\mathbf{L}}$.

Remark III.1.2. The Schrödinger operator (III.1.2) is generalized in a natural way to the case of a varying magnetic field $\mathbf{B}(\mathbf{x}, t)$:

$$H_B = -\frac{\hbar^2}{2\mathrm{m}}\Delta - \frac{e^2}{|x|} - \frac{e}{2\mathrm{m}c}\mathbf{B}(\mathbf{x}, t) \cdot \hat{\mathbf{L}}, \qquad \text{(III.1.14)}$$

where $\mathbf{x}(t)$ is the trajectory of atom in space.

Exercise III.1.3. Check that the operator (III.1.14) is obtained from the classical Hamiltonian (A.8.5) with $\mathbf{B}(\mathbf{x}, t)$ instead of $\mathbf{B}$ by the canonical quantization (I.2.6).

III.2 Intrinsic Magnetic Moment of Electrons

The introduction of electron spin was suggested by the experiments of Einstein–de Haas, Stern–Gerlach, by Landé's vector model and the Bohr–Pauli theory of the Mendeleev periodic table of chemical elements.

III.2.1 The Einstein–de Haas experiment

In 1915, the presence of the intrinsic magnetic moment of an electron was demonstrated in the experiments of Einstein–de Haas, who measured the ratio of the magnetic moment of the electrons of an atom to their angular ("mechanical") momentum. The result, very surprisingly, did not fit into the classical theory. The measurements were based on the observation of torsional vibrations of a ferromagnetic rod suspended on a thin filament inside a solenoid; see Fig. III.1.

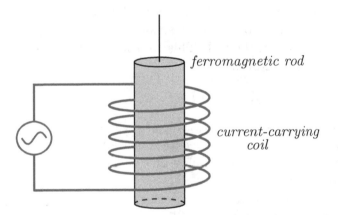

ferromagnetic rod

current-carrying coil

Figure III.1: Einstein–de Haas experiment.

The classical angular momentum and magnetic moment of the electrons of an atom are determined similarly to (A.8.1):

$$\mathbf{L} := m \sum_n \mathbf{x}_n \times \mathbf{v}_n, \qquad \mathbf{M} := \frac{e}{2c} \sum_n \mathbf{x}_n \times \mathbf{v}_n, \qquad \text{(III.2.1)}$$

where $\mathbf{x}_n$ are the positions of the atom's electrons and $\mathbf{v}_n$ are their velocities. Therefore,

$$\mathbf{M} = \frac{e}{2mc} \mathbf{L}, \qquad \text{(III.2.2)}$$

similarly to (A.8.2). Thus, the corresponding *gyromagnetic ratio* equals one:

$$g := \frac{|\mathbf{M}|}{\frac{e}{2mc}|\mathbf{L}|} = 1. \qquad \text{(III.2.3)}$$

In stationary states of the atom, one has $\overline{\mathbf{v}_n} = 0$, where the bar means the average over time. Therefore, formulae like (III.2.1) are suitable for the definition of the *time-averaged* values $\overline{\mathbf{M}}$ and $\overline{\mathbf{L}}$, regardless of the choice of the origin and position of the atom. The same is true for time-averaged values of the total angular momentum and the total magnetic moment $\overline{\mathbf{L}}_{\text{rod}}$ and $\overline{\mathbf{M}}_{\text{rod}}$ of all electrons of the rod. Thus,

$$\overline{\mathbf{M}}_{\text{rod}} = \frac{e}{2mc}\overline{\mathbf{L}}_{\text{rod}}. \tag{III.2.4}$$

When the current in the solenoid is turned on, the magnetic field orients the magnetic moments of all the atoms of the rod so that the total magnetic moment of these atoms almost instantly becomes maximal, $|\overline{\mathbf{M}}_{\text{rod}}|^{\max}$. According to (III.2.4), the total angular momentum of electrons also instantaneously increases to the maximal value $|\overline{\mathbf{L}}_{\text{rod}}|^{\max}$. But then the conservation of the total angular momentum of the rod (electrons together with the crystal lattice) leads to the opposite rotation with angular velocity $\omega^{\max} = |\overline{\mathbf{L}}_{\text{rod}}|^{\max}/I$ of the crystal lattice of the rod to which the suspension thread is attached; here I is the moment of inertia of the rod. This rotation is described by the following equation and initial conditions:

$$\ddot{\varphi}(t) = -I\varphi(t), \qquad \varphi(0) = 0, \qquad \dot{\varphi}(0) = \omega^{\max}. \tag{III.2.5}$$

The solution is $\varphi(t) = (\omega^{\max}/\omega_0)\sin\omega_0 t$, where $\omega_0 := \sqrt{I}$. Its maximal amplitude $\omega^{\max}/\omega_0$ was measured experimentally, giving the value of $\omega^{\max}$ and the corresponding maximal value of the angular momentum of the rod, $|\overline{\mathbf{L}}_{\text{rod}}|^{\max} = I\omega^{\max}$.

On the other hand, the remnant magnetic moment $|\overline{\mathbf{M}}_{\text{rod}}|^{\max}$ can be easily calculated from the frequency of oscillations of the rod in a given magnetic field (the magnetic field of Earth; see Fig. 8-15 of [63]). Thus relation (III.2.4) can be verified experimentally.

However, this plan is very difficult to carry out with satisfactory accuracy since the amplitude $\omega^{\max}/\omega_0$ of $\varphi(t)$ is very small. Einstein and de Haas followed a special strategy based on a detailed study of forced oscillations of the rod excited by an alternating magnetic field. The field was generated by an alternating current of varying frequency in the coil, and the entire impedance characteristics was analyzed; see [4, §18].

The substitution of the observed values $\overline{\mathbf{M}}_{\text{rod}}$ and $\overline{\mathbf{L}}_{\text{rod}}$ into (III.2.4) led to a contradiction! The observed result was

$$\overline{\mathbf{M}}_{\text{rod}} = g\frac{e}{2mc}\overline{\mathbf{L}}_{\text{rod}}, \qquad \text{with} \quad g > 1. \tag{III.2.6}$$

In later, more accurate experiments, it turned out that the gyromagnetic ratio satisfies

$$g \approx 2. \tag{III.2.7}$$

This led to the assumption that the electron itself has an *intrinsic* angular momentum and magnetic moment with this ratio; see Section III.2.4 below.

III.2.2   The Landé vector model

In 1921, A. Landé suggested a *phenomenological classical model* for the description of the *anomalous Zeeman effect* in the framework of the Old Quantum Theory. According to his model, the angular momentum and the magnetic moment are considered as suitable mixtures of two components which are due to the orbital motion and to the "spinning" of the electron with the corresponding gyromagnetic ratio $g = 1$ and $g = 2$; see [11, Section VI.2], [51, Section 14.7]. Landé's formula is in a good agreement with the experimental observations of the anomalous Zeeman effect.

III.2.3   The Stern–Gerlach experiment

In 1922, O. Stern and W. Gerlach conducted a crucial and clarifying experiment on the splitting of a beam of neutral silver atoms into *two component beams* in a *strongly inhomogeneous* magnetic field. Formula (A.8.5) with $\mathbf{B}(\mathbf{x}, t)$ instead of the uniform magnetic field $\mathbf{B}$ implies that the *inhomogeneous magnetic field* $\mathbf{B}(\mathbf{x}, t)$ acts on the atom with magnetic moment $\mathbf{M}$ and the trajectory $\mathbf{x}(t)$ with the force

$$\mathbf{F}(t) \approx (\mathbf{M} \cdot \nabla)\, \mathbf{B}(\mathbf{x}(t), t), \tag{III.2.8}$$

while the Lorentz force vanishes due to the neutrality of the atoms. Hence the double splitting of the atomic beam means that electrons in the atoms of these two beams are in states with different magnetic moments, since the magnetic moment of the nucleus is negligible ($\sim |\mathbf{M}|/1800$). Later, similar experiments were conducted with hydrogen atoms.

III.2.4   The Goudsmit–Uhlenbeck hypothesis

In 1925, S. Goudsmit and G. Uhlenbeck suggested an explanation for this fact by introducing the concept of "intrinsic magnetic moment" of the electron.

Namely, they suggested that the electron can be only in two states, with its magnetic moment parallel and anti-parallel to the magnetic field.

Particularly puzzling was the splitting into exactly two beams, since, according to the Old Quantum Theory, the splitting is possible only into an odd number of components. Indeed, according to (A.7.20), the projection of quantum angular momentum of an atomic electron with a fixed angular momentum $\hbar l$ onto the magnetic field takes $2l + 1$ values

$$L_{\mathbf{B}} = m\hbar, \qquad m = -l, \ldots, l, \qquad l = 0, 1, 2, \ldots. \tag{III.2.9}$$

On the other hand, according to (A.8.2), the projection of the magnetic moment at a fixed value of l takes $2l + 1$ values

$$M_{\mathbf{B}} = \frac{e\hbar}{2mc}m, \qquad m = -l, \ldots, l. \qquad (III.2.10)$$

Moreover, the same odd number and the same possible values were predicted by formulae (III.1.4) and (III.1.6) of the Schrödinger theory.

Accordingly, S. Goudsmit and G. Uhlenbeck interpreted the double splitting in the Stern–Gerlach experiment as the presence of electron states with $l = 1/2$ (so that $2l + 1 = 2$!), and with the corresponding quantum angular momentum $\pm\hbar/2$. At the same time, they kept the value $e\hbar/(2mc)$ for the magnetic moment (III.2.10) with the magnetic quantum numbers $m = \pm 1$:

> *The electron has an intrinsic "spin angular momentum"*
>
> *of magnitude $\pm\hbar/2$ and an intrinsic* $\qquad\qquad$ (III.2.11)
>
> *"spin magnetic moment" of magnitude $\pm e\hbar/(2mc)$.*

This conjecture yields $g = 2$ for the gyroscopic ratio (III.2.6) of the *intrinsic* angular momentum and magnetic moment of electrons; modern value

$$g = 2.0023193043617(15).$$

The experimentally measured value $g \approx 2$ is due to the mixture of the intrinsic and orbital angular momenta, which is the key point of the Landé model.

Exercise III.2.1. Check the formula (III.2.8). **Hint:** write the Hamiltonian equations corresponding to the Hamiltonian functional (A.8.5) with $\mathbf{B}(\mathbf{x}, t)$ instead of $\mathbf{B}$.

III.3 Spin and the Pauli Equation

After the emergence of the Schrödinger theory in 1926, a new contradiction arose with the Stern–Gerlach experiment, since silver atoms were prepared in a *spherically symmetric state*, as was known from spectroscopy. But the last term in the Schrödinger operator (III.1.14) cancels such states, since $\mathbf{B} \cdot \hat{\mathbf{L}}$ is a generator of rotations about the vector $\mathbf{B}$. Therefore, there should have been no first-order splitting!

III.3.1 Uniform magnetic field

In 1927, W. Pauli proposed to modify the Schrödinger equation by doubling the multiplicity of all stationary states in accordance with the additional *two-valued quantum number* introduced by him in the theory of the Mendeleev periodic table (see Section A.9). For this purpose, Pauli introduced the two-component wave functions

$$\psi(\mathbf{x}) = (\psi_1(\mathbf{x}), \psi_2(\mathbf{x})) \in \mathcal{E}^{\otimes} := L^2(\mathbb{R}^3) \otimes \mathbb{C}^2. \qquad (III.3.1)$$

Moreover, Pauli added a new *spin term* to the energy operator (III.1.7) with the coefficient corresponding to the Goudsmit–Uhlenbeck conjecture. Namely, he replaced the angular momentum $\hat{L}_3$ and the magnetic moment $\hat{M}_3 = \frac{e}{2mc}\hat{L}_3$ (see (III.1.6)), respectively, with

$$\hat{J}_3 := \hat{L}_3 + \hat{s}_3 \qquad \text{and} \qquad \hat{M}_3 = \frac{e}{2mc}(\hat{L}_3 + 2\hat{s}_3), \qquad (III.3.2)$$

where $\hat{s}_3 = \frac{\hbar}{2}\sigma_3$ and $\sigma_3 = \begin{pmatrix} 1 & 0 \\ 0 & -1 \end{pmatrix}$ is the Pauli matrix from (I.7.11).

Let us explain these modifications. First, the identification of $\hat{J}_3$ with the third component of the *total angular momentum* is justified by Lemma III.3.12 below. Second, for the Pauli two-component wave function (III.3.1), the contribution of the *spin angular momentum* $\hat{s}_3$ into the mean value $J_3 = \langle \psi, \hat{J}_3\psi \rangle$ is

$$\langle \psi, \hat{s}_3\psi \rangle = \frac{\hbar}{2}(\langle \psi_1, \psi_1 \rangle - \langle \psi_2, \psi_2 \rangle).$$

This contribution is $\hbar/2$ for $\psi_2 = 0$ and $-\hbar/2$ for $\psi_1 = 0$, which is in agreement with the conjecture (III.2.11). Finally, the contribution of $\hat{s}_3$ into the mean value $M_3 = \langle \psi, \hat{M}_3\psi \rangle$ is

$$\frac{e}{mc}\langle \psi, \hat{s}_3\psi \rangle = \frac{e\hbar}{2mc}(\langle \psi_1, \psi_1 \rangle - \langle \psi_2, \psi_2 \rangle),$$

which also agrees with the conjecture (III.2.11).

The substitution of the magnetic moment $\hat{M}_3$ from (III.3.2) into the Schrödinger operator (III.1.7) results in the *Pauli equation*

$$i\hbar\dot\psi(\mathbf{x}, t) = H_P\psi(\mathbf{x}, t), \quad H_P = H - B\hat{M}_3 = -\frac{\hbar^2}{2\mathrm{m}}\Delta - \frac{e^2 Z}{|\mathbf{x}|} - \Omega_\Lambda(\hat{L}_3 + 2\hat{s}_3). \quad \text{(III.3.3)}$$

Eigenfunctions of the Pauli operator H_P are obviously expressed in terms of
the eigenfunctions (III.1.4),

$$\varphi_{nlms} = \varphi_{nlm} \otimes e_s \text{for} \quad s = \pm\frac{1}{2}, \quad \text{where} \quad e_{+\frac{1}{2}} = (1, 0), \; e_{-\frac{1}{2}} = (0, 1).$$

The spin term $-2\Omega_\Lambda\hat{s}_3$ in (III.3.3) shifts the eigenvalues by $\pm\hbar\Omega_\Lambda$, and for
$B = 0$ the multiplicity of all stationary states of the operator H_P is doubled
as compared to the operator (III.1.2).

Thus (III.1.4) implies that the eigenvalues of the Pauli operator H_P are
numbered by the quantum numbers (A.9.1) together with the two-valued quantum number $s = \pm\frac{1}{2}$, which corresponds to the eigenvalues $\hbar s$ of the operator
$\hat{s}_3$:

$$\begin{cases} H_P\varphi_{nlms} = E_{nms}\varphi_{nlms} \\ \quad E_{nms} = E_n - \Omega_\Lambda\hbar(m + 2s) \\ \hat{\mathbf{L}}^2\varphi_{nlms} = \hbar^2 l(l+1)\varphi_{nlms} \\ \hat{L}_3\varphi_{nlms} = \hbar m\varphi_{nlms}, \; \hat{s}_3\varphi_{nlms} = \hbar s\varphi_{nlms} \end{cases} \left|\begin{array}{l} n = 1, 2, \ldots \\ l = 0, 1, \ldots, n-1 \\ m = -l, \ldots, l \\ s = \pm\frac{1}{2} \end{array}\right| ; \quad \text{(III.3.4)}$$

above, $\Omega_\Lambda = \dfrac{eB}{2\mathrm{mc}}$ is the Larmor frequency (III.1.3).

Remarks III.3.1. i) Formula (III.3.4) for E_{nms} demonstrates that for small
Ω_Λ (i.e., for a small magnetic field $\mathbf{B}$) each energy level E_n splits into two sub-
levels with $s = \pm 1/2$. This explains the double splitting in the Stern–Gerlach
experiments.

ii) For any fixed principal quantum number n, the sets of possible energies in
(III.3.4) and (A.8.6) coincide, because $m + 2s$ in (III.3.4) changes from $-n$
to n.

The Pauli equation (III.3.3) is still incapable of explaining the anomalous
Zeeman effect. The needed modification — the term describing the spin-orbital
interaction — was introduced by H.N. Russell and F.A. Saunders in 1925 in the
context of the Old Quantum Theory. In 1927, W. Pauli introduced this spin-
orbital term into the Pauli equation, which resulted in a perfect explanation
of the Einstein–de Haas experiments and of the anomalous Zeeman effect; see
Section III.4.

III.3.2   General Maxwell field

The Pauli equation (III.3.3) corresponds to a uniform magnetic field. For a general external Maxwell field, the Pauli equation reads

$$i\hbar\dot{\psi}(\mathbf{x}, t) = H_P(t)\psi(\mathbf{x}, t)$$
$$= \frac{1}{2\mathrm{m}}\left[\boldsymbol{\sigma}\cdot\left(-i\hbar\nabla - \frac{e}{c}\mathbf{A}^{\mathrm{ext}}(\mathbf{x}, t)\right)\right]^2\psi(\mathbf{x}, t) + eA_0^{\mathrm{ext}}(\mathbf{x}, t)\psi(\mathbf{x}, t), \tag{III.3.5}$$

where "·" stands for the inner product of three-dimensional vectors and

$$\boldsymbol{\sigma}\cdot\nabla = \sum_{k=1}^3 \sigma_k\partial_{x^k};$$

here $\boldsymbol{\sigma} := (\sigma_1, \sigma_2, \sigma_3)$ are the Pauli matrices (I.7.11) with cyclic product and commutation relations

$$\begin{cases} \sigma_1\sigma_2 = i\sigma_3, & \sigma_2\sigma_3 = i\sigma_1, & \sigma_3\sigma_1 = i\sigma_2, \\ [\sigma_1, \sigma_2] = 2i\sigma_3, & [\sigma_2, \sigma_3] = 2i\sigma_1, & [\sigma_3, \sigma_1] = 2i\sigma_2, \end{cases} \tag{III.3.6}$$

where the commutation relations are similar to (I.3.17) (and (III.3.14) below). Using these relations, we can rewrite the Pauli equation (III.3.5) in the following equivalent form:

$$i\hbar\dot{\psi}(\mathbf{x}, t) = H_P(t)\psi(\mathbf{x}, t)$$

$$= \frac{1}{2\mathrm{m}}\left[-i\hbar\nabla - \frac{e}{c}\mathbf{A}^{\mathrm{ext}}(\mathbf{x}, t)\right]^2\psi - \frac{e}{mc}\hat{\mathbf{s}}\cdot\mathbf{B}^{\mathrm{ext}}(\mathbf{x}, t)\psi + eA_0^{\mathrm{ext}}(\mathbf{x}, t)\psi, \tag{III.3.7}$$

where $\mathbf{B}^{\mathrm{ext}} = \mathrm{curl}\,\mathbf{A}^{\mathrm{ext}}$ and $\hat{\mathbf{s}} = \frac{\hbar}{2}\boldsymbol{\sigma}$ is the operator of the *spin angular momentum*. In particular, for the case of a uniform magnetic field $(0, 0, B)$, the last equation coincides with (III.3.3) with Ω_Λ given by (III.1.3) when the quadratic term with e^2/c^2 is neglected.

Exercise III.3.2. Check (III.3.6) and (III.3.7).

III.3.3   The Maxwell–Pauli equations

If we take into account electron's spin, then we must replace the Schrödinger equation in the Maxwell–Schrödinger system (I.5.1) with the corresponding Pauli equation which includes the Maxwell potentials $\mathbf{A}(\mathbf{x}, t)$ and $A_0(\mathbf{x}, t)$. Then we obtain the coupled nonlinear Maxwell–Pauli equations

$$\begin{cases} \frac{1}{c^2}\ddot{\mathbf{A}}(\mathbf{x}, t) = \Delta\mathbf{A}(\mathbf{x}, t) + \frac{1}{c}P\mathbf{j}(\mathbf{x}, t), \quad \Delta A_0(\mathbf{x}, t) = -\rho(\mathbf{x}, t) \\ i\hbar\dot{\psi}(\mathbf{x}, t) = \frac{1}{2\mathrm{m}}[\boldsymbol{\sigma}\cdot(-i\hbar\nabla - \frac{e}{c}(\mathbf{A}(\mathbf{x}, t) + \mathbf{A}^{\mathrm{ext}}(\mathbf{x}, t)))]^2\psi(\mathbf{x}, t) \\ \qquad\qquad + e(A_0(\mathbf{x}, t) + A_0^{\mathrm{ext}}(\mathbf{x}, t))\psi(\mathbf{x}, t), \end{cases} \tag{III.3.8}$$

where $A_0(\mathbf{x}, t) := (-\Delta)^{-1}\rho(\mathbf{x}, t)$ with $\rho(\mathbf{x}, t) := e|\psi(\mathbf{x}, t)|^2$.

III.3.4   Rotation group and angular momenta

Recall that we denote by $R_k(\theta_k)$ rotation of the space $\mathbb{R}^3$ by the angle θ_k about the x^k-axis for $k = 1, 2, 3$. The *generators* of these rotations are the matrices $r_k := \partial_{\theta_k} R_k(\theta_k)|_{\theta_k=0}$, which can be easily calculated:

$$r_1 = \partial_{\theta_1} \begin{pmatrix} 1 & 0 & 0 \\ 0 & \cos\theta_1 & -\sin\theta_1 \\ 0 & \sin\theta_1 & \cos\theta_1 \end{pmatrix}\Bigg|_{\theta_1=0} = \begin{pmatrix} 0 & 0 & 0 \\ 0 & 0 & -1 \\ 0 & 1 & 0 \end{pmatrix},$$

$$r_2 = \partial_{\theta_2} \begin{pmatrix} \cos\theta_2 & 0 & \sin\theta_2 \\ 0 & 1 & 0 \\ -\sin\theta_2 & 0 & \cos\theta_2 \end{pmatrix}\Bigg|_{\theta_2=0} = \begin{pmatrix} 0 & 0 & 1 \\ 0 & 0 & 0 \\ -1 & 0 & 0 \end{pmatrix}, \qquad \text{(III.3.9)}$$

$$r_3 = \partial_{\theta_3} \begin{pmatrix} \cos\theta_3 & -\sin\theta_3 & 0 \\ \sin\theta_3 & \cos\theta_3 & 0 \\ 0 & 0 & 1 \end{pmatrix}\Bigg|_{\theta_3=0} = \begin{pmatrix} 0 & -1 & 0 \\ 1 & 0 & 0 \\ 0 & 0 & 0 \end{pmatrix}.$$

A general rotation $R \in SO(3)$ can be written as

$$R = R(\boldsymbol{\theta}) := \exp(\boldsymbol{\theta} \cdot r), \qquad \boldsymbol{\theta} = (\theta_1, \theta_2, \theta_3) \in \mathbb{R}^3, \qquad \text{(III.3.10)}$$

where $r = (r_1, r_2, r_3)$ are the generators of rotations. The *regular representation* of the rotation group $SO(3)$ is defined by

$$T(\boldsymbol{\theta})f(\mathbf{x}) := f(R^{-1}(\boldsymbol{\theta})\mathbf{x}) = f(R(-\boldsymbol{\theta})\mathbf{x}), \qquad \boldsymbol{\theta} \in \mathbb{R}^3 \qquad \text{(III.3.11)}$$

for a suitable class of functions f on $\mathbb{R}^3$. It is easy to see that angular momenta (I.3.18) are proportional to the corresponding generators $T_k'(0) := \partial_{\theta_k} T(\boldsymbol{\theta})|_{\boldsymbol{\theta}=0}$ of the regular representation $T(\boldsymbol{\theta})$, i.e.,

$$\hat{L}_k = i\hbar T_k'(0). \qquad \text{(III.3.12)}$$

Hence, similarly to (III.3.10),

$$T(\boldsymbol{\theta}) = \exp\left(-\frac{i}{\hbar}\boldsymbol{\theta} \cdot \hat{\mathbf{L}}\right); \qquad \text{(III.3.13)}$$

see the details in [34, 89]. We know that the quadratic forms of all the $\hat{L}_k$ are conserved for solutions of the Schrödinger equation (I.2.14) when all the $\hat{L}_k$ commute with the Schrödinger operators; see (I.3.30). In this case, all the operators $T(\boldsymbol{\theta})$ also commute with the corresponding Schrödinger dynamics.

Conversely, for a broad class of *Hamiltonian systems* with dynamics commuting with a unitary representation of a Lie group, the Noether theory of invariants specifies the quadratic forms of the corresponding generators which are conserved in time (see Appendix B).

Definition III.3.3. *Let the quadratic forms of a generator $T'_k(0)$ be conserved in time for a representation $T(\boldsymbol{\theta})$ of the rotation group $SO(3)$ in the phase space of a certain dynamical system. Then this quadratic form is called the angular momentum of the dynamical system that* correspond to this representation.

The generators (III.3.9) satisfy commutation relations similar to (I.3.17):

$$[r_1, r_2] = r_3, \qquad [r_2, r_3] = r_1, \qquad [r_3, r_1] = r_2, \qquad \text{(III.3.14)}$$

which hold for generators of any representation of the rotation group $SO(3)$; see [34, 89].

Problem III.3.4. Prove that any rotation $R \in SO(3)$ admits the representation (III.3.10). **Hints:** i) prove that any rotation $R \in SO(3)$ leaves invariant a unitary vector $\mathbf{n} \in \mathbb{R}^3$; ii) check that $R(\boldsymbol{\theta})$ is the rotation about the vector $\boldsymbol{\theta}$ with the angle $|\theta|$: it suffices to consider $\boldsymbol{\theta} = (0, 0, \theta_3)$.

Exercise III.3.5. Check (III.3.9) and (III.3.14).

III.3.5 Rotational covariance

Let us show that the Pauli equation (III.3.5) is rotationally covariant. This means that the Pauli equation retains its form after rotations $\mathbf{x} \mapsto \mathbf{y} = R(\boldsymbol{\theta})\mathbf{x}$ with any $\boldsymbol{\theta} \in \mathbb{R}^3$, followed by a suitable transformation of the potentials, the wave function, and the basis in the space $\mathbb{C}^2$ of the values of the wave function. Let us define the transformations as follows:

$$\begin{cases} \tilde{\psi}(\mathbf{y}, t) := U(\boldsymbol{\theta})\psi(\mathbf{x}, t) = U(\boldsymbol{\theta})\psi(R(-\boldsymbol{\theta})\mathbf{y}, t) \\[2mm] \tilde{\mathbf{A}}^{\text{ext}}(\mathbf{y}, t) := R(\boldsymbol{\theta})\mathbf{A}^{\text{ext}}(R(-\boldsymbol{\theta})\mathbf{y}, t) \\[2mm] \tilde{A}_0^{\text{ext}}(\mathbf{y}, t) := A_0^{\text{ext}}(R(-\boldsymbol{\theta})\mathbf{y}, t) \\[2mm] \tilde{\mathbf{B}}^{\text{ext}}(\mathbf{y}, t) := R(\boldsymbol{\theta})\mathbf{B}^{\text{ext}}(R(-\boldsymbol{\theta})\mathbf{y}, t) \end{cases}, \qquad \text{(III.3.15)}$$

where

$$\begin{cases} R(\boldsymbol{\theta}) := \exp(\boldsymbol{\theta} \cdot \boldsymbol{r}), \qquad U(\boldsymbol{\theta}) := \exp\left(-i\dfrac{\boldsymbol{\theta} \cdot \boldsymbol{\sigma}}{2}\right) \\[3mm] \boldsymbol{\sigma}(\boldsymbol{\theta}) := U(\boldsymbol{\theta})\,\boldsymbol{\sigma}\,U(-\boldsymbol{\theta}), \quad \hat{s}(\boldsymbol{\theta}) := U(\boldsymbol{\theta})\,\hat{s}\,U(-\boldsymbol{\theta}) \end{cases}. \qquad \text{(III.3.16)}$$

Here $\boldsymbol{\sigma} = (\sigma_1, \sigma_2, \sigma_3)$ and $\boldsymbol{r} = (r_1, r_2, r_3)$, with r_k the generators (III.3.9). Note that the correspondence $R(\boldsymbol{\theta}) \mapsto U(\boldsymbol{\theta})$ is the *projective spinor representation* of the rotation group $SO(3)$. It is single-valued for small $|\theta|$, but its (real-)analytic continuation to all $\boldsymbol{\theta} \in \mathbb{R}^3$ is *two-valued*. This is evident from

$$R(\boldsymbol{\theta} + 2\pi\boldsymbol{\theta}/|\theta|) = R(\boldsymbol{\theta}), \quad U(\boldsymbol{\theta} + 2\pi\boldsymbol{\theta}/|\theta|) = -U(\boldsymbol{\theta}), \quad \boldsymbol{\theta} \in \mathbb{R}^3 \setminus 0. \quad \text{(III.3.17)}$$

Remark III.3.6. This doubling does not contradict the rotational covariance of the Pauli equation. Indeed, rotational covariance of a theory means that i) the *existence* for any rotation of a transformation which preserves the form of the theory, and ii) the *continuity* of such transformation under small perturbations of the rotation. Both these requirements hold for the transformations (III.3.15).

Let us show that for any solution $\psi(\mathbf{x}, t)$ to equation (III.3.5), the function $\tilde{\psi}(\mathbf{y}, t)$ is a solution to the same equation with

$$\mathbf{A}^{\text{ext}}(\mathbf{x}, t), \quad A_0^{\text{ext}}(\mathbf{x}, t), \quad \boldsymbol{\sigma}, \quad \hat{\mathbf{s}}, \quad \mathbf{B}^{\text{ext}}(\mathbf{x}, t)$$

replaced by

$$\tilde{\mathbf{A}}^{\text{ext}}(\mathbf{y}, t), \quad \tilde{A}_0^{\text{ext}}(\mathbf{y}, t), \quad \boldsymbol{\sigma}(\boldsymbol{\theta}), \quad \hat{\mathbf{s}}(\boldsymbol{\theta}), \quad \tilde{\mathbf{B}}^{\text{ext}}(\mathbf{y}, t),$$

respectively. Moreover, the basis in the space $\mathbb{C}^2$ of values of the wave functions $\tilde{\psi}(\mathbf{y}, t)$ must be changed. For the proof, we apply $U(\boldsymbol{\theta})$ to both sides of (III.3.5) and obtain, by (III.3.15),

$$i\hbar\dot{\tilde{\psi}}(\mathbf{y}, t) = \frac{1}{2\mathrm{m}}U(\boldsymbol{\theta})\left[\boldsymbol{\sigma} \cdot \left(-i\hbar\nabla - \frac{e}{c}\mathbf{A}^{\text{ext}}(\mathbf{x}, t)\right)\right]^2 \psi(\mathbf{x}, t) + e\tilde{A}_0^{\text{ext}}(\mathbf{y}, t)\tilde{\psi}(\mathbf{y}, t).$$

It remains to express the first term on the right-hand side via $\tilde{\psi}(\mathbf{y}, t)$. Since $\mathbf{x} = R(-\boldsymbol{\theta})\mathbf{y}$, we have

$$\partial_{y^k}\tilde{\psi}(\mathbf{y}) \;=\; \frac{\partial x^i}{\partial y^k}\partial_{x^i}U(\boldsymbol{\theta})\psi(\mathbf{x}) = U(\boldsymbol{\theta})R_{ik}(-\boldsymbol{\theta})\partial_{x^i}\psi(\mathbf{x})$$

$$\;=\; U(\boldsymbol{\theta})R_{ki}(\boldsymbol{\theta})\partial_{x^i}\psi(\mathbf{x}) = R_{ki}(\boldsymbol{\theta})U(\boldsymbol{\theta})\partial_{x^i}\psi(\mathbf{x}), \qquad \text{(III.3.18)}$$

since the matrix $R(-\boldsymbol{\theta})$ is orthogonal. Therefore,

$$\nabla_{\mathbf{x}} = U(-\boldsymbol{\theta})R(-\boldsymbol{\theta})\nabla_{\mathbf{y}}, \qquad \psi(\mathbf{x}) = U(-\boldsymbol{\theta})\tilde{\psi}(\mathbf{y}). \qquad \text{(III.3.19)}$$

Hence,

$$U(\boldsymbol{\theta})\left[\boldsymbol{\sigma} \cdot \left(-i\hbar\nabla_{\mathbf{x}} - \frac{e}{c}\mathbf{A}^{\text{ext}}(\mathbf{x}, t)\right)\right]^2 \psi(\mathbf{x})$$

$$= \left[\boldsymbol{\sigma}(\boldsymbol{\theta}) \cdot R(-\boldsymbol{\theta})\left(-i\hbar\nabla_{\mathbf{y}} - \frac{e}{c}R(\boldsymbol{\theta})\mathbf{A}^{\text{ext}}(R(-\boldsymbol{\theta})\mathbf{y}, t)\right)\right]^2 \tilde{\psi}(\mathbf{y})$$

$$= \left[\left(\boldsymbol{\sigma}(\boldsymbol{\theta}) \cdot R(-\boldsymbol{\theta})\left(-i\hbar\nabla_{\mathbf{y}} - \frac{e}{c}\tilde{\mathbf{A}}^{\text{ext}}(\mathbf{y}, t)\right)\right)\right]^2 \tilde{\psi}(\mathbf{y})$$

$$= \left[(R(\boldsymbol{\theta})\boldsymbol{\sigma}(\boldsymbol{\theta})) \cdot \left(-i\hbar\nabla_{\mathbf{y}} - \frac{e}{c}\tilde{\mathbf{A}}^{\text{ext}}(\mathbf{y}, t)\right)\right]^2 \tilde{\psi}(y), \qquad \text{(III.3.20)}$$

since the rotations $R(\boldsymbol{\theta})$ preserve the inner product "$\cdot$". It remains to note that the matrices

$$(\tilde{\sigma}_1, \tilde{\sigma}_2, \tilde{\sigma}_3) := R(\boldsymbol{\theta})\sigma(\boldsymbol{\theta}) \qquad (\text{III}.3.21)$$

satisfy the same commutation relations (III.3.6). This fact implies that the matrices have the same form (I.7.11) in an appropriate orthonormal basis of the space $\mathbb{C}^2$.

Exercise III.3.7. Check (III.3.18) and (III.3.20).

Exercise III.3.8. Check (III.3.17). **Hint:** It suffices to consider $\boldsymbol{\theta} = (0, 0, \theta_3)$.

Problem III.3.9. Check the commutation relations from (III.3.6) for the matrices (III.3.21). **Hints:** i) first, prove these commutation relations for the matrices

$$(\hat{\sigma}_1, \hat{\sigma}_2, \hat{\sigma}_3) := \sigma(\boldsymbol{\theta}); \qquad (\text{III}.3.22)$$

ii) it suffices to consider the matrices (III.3.22) and (III.3.21) with $\boldsymbol{\theta} = (0, 0, \theta_3)$.

Problem III.3.10. Check that the matrices satisfying the commutation relations from (III.3.6) coincide with the Pauli matrices (I.7.11) in appropriate orthonormal basis of $\mathbb{C}^2$. **Hint:** adapt the method from the proof of the Pauli theorem for the Dirac equation; see Section IV.2 below. See also [85, Lemma 2.25] and [134, Theorem VIII.7] (for the version of the Pauli theorem in any spatial dimension).

III.3.6 Conservation laws

Energy and momentum for the Pauli equation (III.3.7) are defined by the same formulae (I.3.2) and (I.3.14) as for the Schrödinger equation (with H replaced with H_P from (III.3.7)), and the proofs of their conservation remain unchanged.

However, the definition and proof of the conservation of the angular momentum require significant modifications.

Definition III.3.11. The angular momentum *for the Pauli equation is defined as*

$$\mathbf{J}(t) := \langle \psi(t), \hat{\mathbf{J}}\psi(t)\rangle, \qquad \hat{\mathbf{J}} := \hat{\mathbf{L}} + \hat{\mathbf{s}}. \qquad (\text{III}.3.23)$$

The commutation relations for the components of the angular momentum are similar to (I.3.17):

$$[\hat{J}_1, \hat{J}_2] = i\hbar\hat{J}_3, \qquad [\hat{J}_2, \hat{J}_3] = i\hbar\hat{J}_1, \qquad [\hat{J}_3, \hat{J}_1] = i\hbar\hat{J}_2. \qquad (\text{III}.3.24)$$

These relations follow from (I.3.17) and (III.3.6).

Lemma III.3.12. *Let the external potentials satisfy the rotational invariance condition* (I.3.23). *Then, for any solution of the Pauli equation* (III.3.7), *the component J_k of angular momentum* (III.3.23) *is conserved:*

$$J_k(t) = \text{const}, \qquad t \in \mathbb{R}. \tag{III.3.25}$$

Proof. It suffices to consider the case $k = 3$ and to prove the commutation

$$[\hat{J}_3, H_P(t)] = 0. \tag{III.3.26}$$

First, note that $H_P(t) = H(t) - \frac{e}{mc}\hat{\mathbf{s}} \cdot \mathbf{B}^{\text{ext}}(\mathbf{x}, t)$ by (III.3.7), where $H(t)$ is the Schrödinger operator (I.2.14). Due to the commutation (I.3.30), we have

$$[\hat{J}_3, H(t)] = [\hat{L}_3, H(t)] = 0.$$

Hence, it remains to verify that

$$[\hat{J}_3, \hat{\mathbf{s}} \cdot \mathbf{B}^{\text{ext}}(\mathbf{x}, t)] = 0. \tag{III.3.27}$$

We have

$$\left\{ \begin{aligned} [\hat{L}_3, \hat{\mathbf{s}} \cdot \mathbf{B}^{\text{ext}}(\mathbf{x}, t)] &= \sum_{k=1}^{3} \hat{s}_k [\hat{L}_3, B_k^{\text{ext}}(\mathbf{x}, t)] = \sum_{k=1}^{3} \hat{s}_k \hat{L}_3 B_k^{\text{ext}}(\mathbf{x}, t) \\ [\hat{s}_3, \hat{\mathbf{s}} \cdot \mathbf{B}^{\text{ext}}(\mathbf{x}, t)] &= i\hbar\hat{s}_2 B_1^{\text{ext}}(\mathbf{x}, t) - i\hbar\hat{s}_1 B_2^{\text{ext}}(\mathbf{x}, t) \end{aligned} \right. . \tag{III.3.28}$$

Note that the vector field $\mathbf{A}^{\text{ext}}(\mathbf{x}, t)$ is invariant under rotations $R_3(\theta_3)$ about the x^3-axis in the sense of (I.3.23), and so is $\mathbf{B}^{\text{ext}}(\mathbf{x}, t)$. Hence, $\mathbf{B}^{\text{ext}}(\mathbf{x}, t)$ admits the representation of type (I.3.25). In particular, the component $B_3^{\text{ext}}(\mathbf{x}, t)$ is invariant with respect to the rotations, so

$$\hat{L}_3 B_3^{\text{ext}}(\mathbf{x}, t) = 0.$$

Therefore, substituting $\hat{L}_3 = -i\hbar\partial_\varphi$ with $\varphi = \theta_3$ into (III.3.28) and summing up, we obtain:

$$[\hat{J}_3, \hat{\mathbf{s}} \cdot \mathbf{B}^{\text{ext}}(\mathbf{x}, t)]$$

$$= -i\hbar\hat{s}_1[\partial_\varphi B_1^{\text{ext}}(\mathbf{x}, t) + B_2^{\text{ext}}(\mathbf{x}, t)] - i\hbar\hat{s}_2[\partial_\varphi B_2^{\text{ext}}(\mathbf{x}, t) - B_1^{\text{ext}}(\mathbf{x}, t)]. \tag{III.3.29}$$

It remains to verify that the coefficients at $\hat{s}_1$ and $\hat{s}_2$ on the right-hand side vanish. The representation of type (I.3.25) for $\mathbf{B}^{\text{ext}}(\mathbf{x}, t)$ implies that the radial and angular components of the vector field $\tilde{\mathbf{B}}^{\text{ext}}(\mathbf{x}, t) := (B_1^{\text{ext}}(\mathbf{x}, t), B_2^{\text{ext}}(\mathbf{x}, t))$ do not depend on the angle φ:

$$\left\{ \begin{aligned} \tilde{B}_r^{\text{ext}}(\mathbf{x}, t) &= B_1^{\text{ext}}(\mathbf{x}, t)\cos\varphi + B_2^{\text{ext}}(\mathbf{x}, t)\sin\varphi = f(r, x^3, t) \\ \tilde{B}_\varphi^{\text{ext}}(\mathbf{x}, t) &= -B_1^{\text{ext}}(\mathbf{x}, t)\sin\varphi + B_2^{\text{ext}}(\mathbf{x}, t)\cos\varphi = g(r, x^3, t) \end{aligned} \right. . \tag{III.3.30}$$

Differentiating with respect to φ, we obtain:

$$
\begin{cases}
-B_1^{\text{ext}}(\mathbf{x},t)\sin\varphi + B_2^{\text{ext}}(\mathbf{x},t)\cos\varphi + \partial_\varphi B_1^{\text{ext}}(\mathbf{x},t)\cos\varphi + \partial_\varphi B_2^{\text{ext}}(\mathbf{x},t)\sin\varphi = 0 \\
-B_1^{\text{ext}}(\mathbf{x},t)\cos\varphi - B_2^{\text{ext}}(\mathbf{x},t)\sin\varphi - \partial_\varphi B_1^{\text{ext}}(\mathbf{x},t)\sin\varphi + \partial_\varphi B_2^{\text{ext}}(\mathbf{x},t)\cos\varphi = 0
\end{cases}
. \tag{III.3.31}
$$

This easily implies the annihilation of the coefficients in (III.3.29). $\qquad\square$

Remarks III.3.13. i) For equation (III.3.3) with the spherically symmetric Coulomb potential, the commutations (III.3.26) hold for all $k = 1, 2, 3$. Hence, the dynamics (III.3.3) commutes with the representation of the rotation group $SO(3)$ defined by

$$
T_P(\boldsymbol{\theta}) := e^{-\frac{i}{\hbar}\boldsymbol{\theta}\cdot\hat{\mathbf{J}}} = e^{-\frac{i}{\hbar}\boldsymbol{\theta}\cdot\hat{\mathbf{L}}}e^{-\frac{i}{\hbar}\boldsymbol{\theta}\cdot\hat{\mathbf{s}}} = T(\boldsymbol{\theta})U(\boldsymbol{\theta}), \qquad \boldsymbol{\theta} \in \mathbb{R}^3; \tag{III.3.32}
$$

see (III.3.13). Thus, the conserved angular momenta (III.3.23) correspond (up to a factor) to this representation in the sense of Definition III.3.3. In other words, the spin angular momentum is due to rotational covariance of electron dynamics.

ii) The representation $R(\boldsymbol{\theta}) \mapsto T_P(\boldsymbol{\theta})$ is single-valued for small $|\boldsymbol{\theta}|$, but its (real-)analytic continuation to all $\boldsymbol{\theta} \in \mathbb{R}^3$ is a two-valued (projective) representation, similarly to the spinor representation $U(\boldsymbol{\theta})$ (see (III.3.17)).

iii) The commutation relations (I.3.30) and (III.3.26) play a crucial role in the calculation of eigenvalues of quantum observables; for example, in the evaluation of the anomalous Zeeman effect (see Section III.4).

Exercise III.3.14. Check (III.3.24).

Exercise III.3.15. Verify (III.3.28). **Hint:** check that $[\partial_\varphi, f(\mathbf{x})] = \partial_\varphi f(\mathbf{x})$.

Exercise III.3.16. Check (III.3.30).

Exercise III.3.17. Deduce the annihilation of (III.3.29) from (III.3.31).

III.4   Anomalous Zeeman Effect

In this section, we justify the anomalous Zeeman splitting of spectral lines in a magnetic field using a suitable modification of the Pauli equation (III.3.3). Note that this equation itself is still insufficient for the justification. Indeed, formulae (A.9.1) give the splitting

$$E_{nms} - E_{nm's'} = \Omega_\Lambda \hbar[m' - m + 2(s' - s)], \qquad (III.4.1)$$

where $m' - m = 0, \pm 1$ according to the selection rule (III.1.12) and $s, s' = \pm 1/2$. This equidistant splitting is close to the normal splitting (III.1.5), and in many cases does not agree with experimental results.

An agreement in these more complicated situations was achieved due to H.N. Russell and F.A. Saunders's idea to take into account the interaction of the spin with the orbital angular momentum. Namely, the orbital motion of the (classical) electron results in a circular current, which generates a magnetic field acting on the spin magnetic moment. This interaction was suggested by the phenomenological *Landé vector model* in the Old Quantum Theory (see, e.g., [11] or Section 14.7 of [51]).

III.4.1   Spin-orbital coupling

The interaction between orbital and spin angular momenta was calculated in 1926 by L.H. Thomas [87] and Ya. Frenkel [33]. This interaction appears as a correction to the energy of a particle due to the interaction of the spin with the magnetic field $\mathbf{B}_*(\mathbf{x})$ arising in the comoving frame of the electron. The magnetic field is generated by the Coulombic electrostatic field via the Lorentz formulae (see [48, (11.150)] or [51, (12.74)]): in the first-order approximation in $\beta = \mathbf{v}/c$, we have

$$\mathbf{B}_*(\mathbf{x}) = \frac{1}{c}\mathbf{E}(\mathbf{x}) \times \mathbf{v} = -\frac{1}{c}\nabla\phi(|\mathbf{x}|) \times \mathbf{v} = -\frac{1}{mc}\phi'(|\mathbf{x}|)\frac{\mathbf{x}}{|\mathbf{x}|} \times \mathbf{p} = -\frac{1}{mc}\frac{\phi'(|\mathbf{x}|)}{|\mathbf{x}|}L.$$
$$(III.4.2)$$

The interaction of this magnetic field with the spin momentum $\mathbf{s}$ produces the energy correction $-\dfrac{e}{mc}\mathbf{s} \cdot \mathbf{B}_*(\mathbf{x})$. This corresponds to the correction

$$\frac{e}{mc}\hat{\mathbf{s}} \cdot \mathbf{B}_*(\mathbf{x})\psi(\mathbf{x}, t) = \frac{e}{m^2c^2}\frac{\phi'(|\mathbf{x}|)}{|\mathbf{x}|}\hat{\mathbf{s}} \cdot \hat{\mathbf{L}}\psi(\mathbf{x}, t)$$

of the right-hand side of the Pauli equation (III.3.3). However, this correction does not agree with experimental observations, which suggest that the factor 1/2 should be added. L.H. Thomas and Ya. Frenkel argued in favor of this additional factor by using a sophisticated analysis of the Larmor precession of the spin (see the details in [48]). A more direct argument for the *Thomas factor* $1/2$ comes from the relativistic Dirac theory (see Section IV.8). Finally, the modified Pauli equation (III.3.3) reads

$$i\hbar\partial_t\psi(\mathbf{x},t) = H_P\psi(\mathbf{x},t) := -\frac{1}{2m}\hbar^2\Delta\psi(\mathbf{x},t) + e\phi(|\mathbf{x}|)\psi(\mathbf{x},t)$$

$$+ \frac{e}{2m^2c^2}\frac{\phi'(|\mathbf{x}|)}{|\mathbf{x}|}\hat{\mathbf{s}}\cdot\hat{\mathbf{L}}\psi(\mathbf{x},t) - \Omega_\Lambda(\hat{L}_3 + 2\hat{s}_3)\psi(\mathbf{x},t). \quad \text{(III.4.3)}$$

The correction term with $\hat{\mathbf{s}}\cdot\hat{\mathbf{L}}$ was given the name *spin-orbital coupling* of Russell and Saunders after their papers (1925), which developed the Landé vector model for many-electron atoms.

As we will see in Section III.4.5 below, for equation (III.4.3) the spectral lines in the splitting are not equidistant, in contrast to (III.4.1). The result of the calculations perfectly describes the anomalous Zeeman effect, as well as the Einstein–de Haas experiment.

We can write the operator H_P in the form

$$H_P = H_P^0 - B\hat{M}_3, \quad \text{(III.4.4)}$$

where H_P^0 corresponds to $B = 0$, and the magnetic moment (III.3.2) is

$$\hat{M}_3 := \frac{e}{2mc}(\hat{L}_3 + 2\hat{s}_3) = \frac{e}{2mc}(\hat{J}_3 + \hat{s}_3) \quad \text{(III.4.5)}$$

according to (III.3.23).

III.4.2 Gyroscopic ratio

Due to the axial symmetry of the Einstein–de Haas experiment, the magnetic moment and angular momentum are given by

$$M_3 = \langle\varphi, \hat{M}_3\varphi\rangle, \qquad J_3 = \langle\varphi, \hat{J}_3\varphi\rangle, \quad \text{(III.4.6)}$$

where φ is the corresponding stationary state which is an eigenfunction of the operator H_P. Hence, the Landé factor g for such states is defined by

$$\frac{M_3}{J_3} = g\frac{e}{2mc}. \quad \text{(III.4.7)}$$

It is exactly this gyroscopic ratio that was measured in the Einstein–de Haas experiment.

We will calculate eigenvalues of the operators $\hat{M}_3$ and $\hat{J}_3$ which correspond to their common eigenfunctions φ. Then the ratio (III.4.7) equals to the quotient of the corresponding eigenvalues.

Exercise III.4.1. Check (III.4.2).

III.4.3   Quantum numbers

The operators H_P, $\hat{\mathbf{L}}^2$, $\hat{\mathbf{J}}^2$, $\hat{J}_3 := \hat{L}_3 + \hat{s}_3$ commute with each other. On the other hand, the momenta $\hat{L}_3$ and $\hat{s}_3$ do not commute with $\hat{\mathbf{s}} \cdot \hat{\mathbf{L}}$ (and hence with H_P). Thus, we cannot use a classification of type (III.3.4) for the quantum stationary states by "quantum numbers" n, l, m, s which correspond to eigenvalues of the operators H_P, $\hat{\mathbf{L}}^2$, $\hat{L}_3$, $\hat{s}_3$. We must choose other quantum numbers.

Exercise III.4.2. Verify that the operators H_P, $\hat{\mathbf{L}}^2$, $\hat{\mathbf{J}}^2$, $\hat{J}_3$ commute with each other. **Hints:**
i) Equation (III.4.3) implies that

$$H_P = -\frac{1}{2\mathrm{m}}\hbar^2\Delta + e\phi(|x|) + \frac{e}{2\mathrm{m}^2c^2}\frac{\phi'(|x|)}{|x|}\,\hat{\mathbf{s}}\cdot\hat{\mathbf{L}} - \Omega_\Lambda\left(\hat{L}_3 + 2\hat{s}_3\right). \qquad \text{(III.4.8)}$$

ii) $\hat{\mathbf{s}} \cdot \hat{\mathbf{L}}$ commutes with $\hat{\mathbf{L}}^2$, since each $\hat{L}_k$ and $\hat{s}_k$ commutes with $\hat{\mathbf{L}}^2$.

iii) $\hat{\mathbf{J}}^2$ commutes with $\hat{\mathbf{L}}^2$, since $\hat{\mathbf{J}}^2 = \hat{\mathbf{L}}^2 + 2\hat{\mathbf{s}} \cdot \hat{\mathbf{L}} + \hat{\mathbf{s}}^2$, with $\hat{\mathbf{s}}^2 = \frac{3}{4}$.

iv) $\hat{\mathbf{s}} \cdot \hat{\mathbf{L}}$ commutes with $\hat{\mathbf{J}}^2$ and $\hat{J}_3$, since $2\hat{\mathbf{s}} \cdot \hat{\mathbf{L}} = \hat{\mathbf{J}}^2 - \hat{\mathbf{L}}^2 - \hat{\mathbf{s}}^2$.

Thus, all the operators H_P, $\hat{\mathbf{L}}^2$, $\hat{\mathbf{J}}^2$, $\hat{J}_3$ commute with each other. Moreover, all these operators are selfadjoint in the Hilbert space $\mathcal{E}^{\otimes}$ (see (III.3.1)).

Let us consider an eigenvalue E of the operator H_P and let $X(E)$ denote the corresponding eigenspace. The operator H_P is the perturbation of the Laplace operator by the first order differential operator with coefficients which decay as $|\mathbf{x}| \to \infty$. While the Coulomb potential ϕ in (III.4.8) is singular at $\mathbf{x} = 0$, we expect that the actual potential is smooth since the nucleus is extended, not a point particle. Hence such a perturbation is relatively compact (see [19]), and then the eigenspace $X(E)$ is finite-dimensional.

The key observation is that the subspace $X(E)$ is invariant with respect to the operators $\hat{\mathbf{L}}^2$, $\hat{\mathbf{J}}^2$, and $\hat{J}_3$ since they all commute with H_P. Hence, there exists an orthonormal basis in $X(E)$ of common eigenvectors of all the operators H_P, $\hat{\mathbf{L}}^2$, $\hat{\mathbf{J}}^2$, $\hat{J}_3$.

By (III.1.4) and Proposition I.7.7, possible eigenvalues of $\hat{\mathbf{L}}^2$, $\hat{\mathbf{J}}^2$ and $\hat{J}_3$ are, respectively,

$$\left. \begin{array}{ll} \hbar^2 L(L+1), & L = 0,\, 1,\, 2,\, \ldots \\[2mm] \hbar^2 J(J+1), & J = 0,\, \frac{1}{2},\, 1,\, \frac{3}{2},\, \ldots \\[2mm] \hbar j_3, & j_3 = -J,\, \ldots,\, J \end{array} \right\}. \qquad \text{(III.4.9)}$$

Note that Proposition I.7.7 is applicable since the subspace $X(E)$ is *finite-dimensional*. The values (III.4.9) are allowed, however, only finitely many of these values correspond to eigenvectors from the subspace $X(E)$.

Exercise III.4.3. Prove that in $X(E)$ there is an orthonormal basis of common eigenvectors of all the operators H_P, $\hat{\mathbf{L}}^2$, $\hat{\mathbf{J}}^2$, $\hat{J}_3$. **Hint:** the operators are selfadjoint and commute with each other.

III.4.4 The Landé formula

Let $\varphi = |E, L, J, j_3\rangle$ be a common eigenvector corresponding to some eigenvalues E, $\hbar^2 L(L+1)$, $\hbar^2 J(J+1)$, $\hbar j_3$ of the operators H_P, $\hat{\mathbf{L}}^2$, $\hat{\mathbf{J}}^2$, $\hat{J}_3$, respectively.

Theorem III.4.4. *The Landé factor $g = g(\varphi)$ corresponding to the quantum stationary state*
$\varphi = |E, L, J, j_3\rangle$ *is given by*

$$g = \frac{3}{2} + \frac{3/4 - L(L+1)}{2J(J+1)}. \tag{III.4.10}$$

Proof. Below we write j instead of j_3. We have:

$$\hat{\mathbf{J}}^2 \varphi = \hbar^2 J(J+1)\varphi, \qquad \hat{J}_3 \varphi = \hbar j \varphi.$$

Therefore, $\varphi = e_j$ is an element of the canonical basis $e_{-J}, \ldots, e_J$, as constructed in Proposition I.7.7, and so all the basis vectors $e_{j'}$ are obtained from e_j by the application of the operators $\hat{J}_\pm := \hat{J}_1 \pm i\hat{J}_2$.

For a linear operator A in the space $\mathcal{E}^\otimes$, denote the corresponding matrix elements

$$A^{j,k} = \langle e_j, Ae_k \rangle. \tag{III.4.11}$$

In particular, $\hat{M}_3^{j,j}$ and $\hat{J}_3^{j,j} = \hbar j$ are eigenvalues of the operators $\hat{M}_3$ and $\hat{J}_3$, respectively, corresponding to their common eigenvector $\varphi = e_j$. Now the ratio (III.4.7) becomes

$$\frac{\hat{M}_3^{j,j}}{\hat{J}_3^{j,j}} = g \frac{e}{2mc}. \tag{III.4.12}$$

It remains to calculate these eigenvalues. First, (III.4.5) implies that

$$\hat{M}_3^{j,j} = \frac{e}{2mc}(\hat{J}_3^{j,j} + \hat{s}_3^{j,j}) = \frac{e}{2mc}(\hbar j + \hat{s}_3^{j,j}). \tag{III.4.13}$$

We will find the matrix element $\hat{s}_3^{j,j}$ and calculate g from definition (III.4.7). First let us collect the commutators

$$\left\{ \begin{array}{lll} [\hat{J}_1, \hat{s}_1] = 0, & [\hat{J}_1, \hat{s}_2] = i\hbar\hat{s}_3, & [\hat{J}_1, \hat{s}_3] = -i\hbar\hat{s}_2 \\ [\hat{J}_2, \hat{s}_2] = 0, & [\hat{J}_2, \hat{s}_3] = i\hbar\hat{s}_1, & [\hat{J}_2, \hat{s}_1] = -i\hbar\hat{s}_3 \\ [\hat{J}_3, \hat{s}_3] = 0, & [\hat{J}_3, \hat{s}_1] = i\hbar\hat{s}_2, & [\hat{J}_3, \hat{s}_2] = -i\hbar\hat{s}_1 \end{array} \right\}, \tag{III.4.14}$$

where the second and the third lines follow from the first one by cyclic permutations. We use (III.4.14) to calculate the commutators of $\hat{J}_\pm$ with $\hat{s}_\pm = \hat{s}_1 \pm i\hat{s}_2$:

$$[\hat{J}_-, \hat{s}_+] = -2\hbar\hat{s}_3, \qquad [\hat{J}_+, \hat{s}_+] = 0. \qquad (III.4.15)$$

The first formula implies the identity

$$\hat{J}_-^{j,j+1}\hat{s}_+^{j+1,j} - \hat{s}_+^{j,j-1}\hat{J}_-^{j-1,j} = -2\hbar\hat{s}_3^{j,j}, \qquad (III.4.16)$$

since all matrix elements $\hat{J}_-^{j',j''}$ with $j'' \neq j'+1$ vanish by (I.7.10). The nonzero matrix elements of $\hat{J}_\pm$ are known from (I.7.10) and (I.7.6):

$$\hat{J}_-^{j,j+1} = \hbar\sqrt{(J+j+1)(J-j)}, \quad \hat{J}_-^{j-1,j} = \hbar\sqrt{(J+j)(J-j+1)}. \qquad (III.4.17)$$

On the other hand, the matrix elements of $\hat{s}_+$ can be calculated using the second identity in (III.4.15): taking matrix element $(\cdot)^{j+1,j-1}$ of both sides, we obtain:

$$\hat{J}_+^{j+1,j}\hat{s}_+^{j,j-1} - \hat{s}_+^{j+1,j}\hat{J}_+^{j,j-1} = 0, \qquad (III.4.18)$$

where

$$\hat{J}_+^{j+1,j} = \hbar\sqrt{(J-j)(J+j+1)}, \quad \hat{J}_+^{j,j-1} = \hbar\sqrt{(J-j+1)(J+j)} \qquad (III.4.19)$$

by (I.7.10) and (I.7.6). Hence,

$$\frac{\hat{s}_+^{j+1,j}}{\sqrt{(J-j)(J+j+1)}} = \frac{\hat{s}_+^{j,j-1}}{\sqrt{(J-j+1)(J+j)}} =: a. \qquad (III.4.20)$$

Substituting this into (III.4.16) and using (III.4.17) for the matrix elements of $\hat{J}_-$, we have:

$$\hat{s}_3^{j,j} = aj. \qquad (III.4.21)$$

It remains to calculate a. We start with the identity

$$\hat{\mathbf{J}}^2 = (\hat{\mathbf{L}} + \hat{\mathbf{s}})^2 = \hat{\mathbf{L}}^2 + 2\hat{\mathbf{s}}\cdot\hat{\mathbf{L}} + \hat{\mathbf{s}}^2 = \hat{\mathbf{L}}^2 + 2\hat{\mathbf{s}}\cdot\hat{\mathbf{J}} - \hat{\mathbf{s}}^2. \qquad (III.4.22)$$

This implies that

$$(\hat{\mathbf{s}}\cdot\hat{\mathbf{J}})^{j,j} = \hbar^2\frac{J(J+1) - L(L+1) + 3/4}{2}, \qquad (III.4.23)$$

since $\hat{\mathbf{s}}^2 = 3/4$. On the other hand, the same matrix element can be expressed from a different identity, namely, from

$$2\hat{\mathbf{s}}\cdot\hat{\mathbf{J}} = \hat{s}_+\hat{J}_- + \hat{s}_-\hat{J}_+ + 2\hat{s}_3\hat{J}_3, \qquad (III.4.24)$$

which follows directly from the definitions of $\hat{s}_\pm$ and $\hat{J}_\pm$. Now

$$(\hat{\mathbf{s}} \cdot \hat{\mathbf{J}})^{j,j} = \frac{1}{2}\hat{s}_+^{j,j-1}\hat{J}_-^{j-1,j} + \frac{1}{2}\hat{s}_-^{j,j+1}\hat{J}_+^{j+1,j} + \hbar j \hat{s}_3^{j,j}. \qquad \text{(III.4.25)}$$

Substituting (III.4.23) into the left-hand side, we obtain:

$$\hbar^2[J(J+1) - L(L+1) + 3/4] = \hat{s}_+^{j,j-1}\hat{J}_-^{j-1,j} + \hat{s}_-^{j,j+1}\hat{J}_+^{j+1,j} + 2\hbar j \hat{s}_3^{j,j}. \quad \text{(III.4.26)}$$

We note that the matrix elements $\hat{s}_+^{j+1,j}$ and $\hat{s}_-^{j,j+1}$ are complex conjugates. Therefore, using (III.4.20), we see that

$$\hat{s}_+^{j+1,j} = a\sqrt{(J-j)(J+j+1)} = \hat{s}_-^{j,j+1}, \qquad \text{(III.4.27)}$$

since the constant a is real by (III.4.21). Finally, substituting (III.4.27), (III.4.17), (III.4.19), and (III.4.21) into (III.4.26), we obtain the equation for a whose solution is

$$a = \hbar \frac{J(J+1) - L(L+1) + 3/4}{2J(J+1)} = \hbar(g-1), \qquad \text{(III.4.28)}$$

where g is given by (III.4.10). Now, (III.4.21) and (III.4.13) imply that

$$\hat{M}_3^{j,j} = \frac{e}{2mc}(\hbar j + aj) = \frac{e}{2mc}g\hbar j = \frac{e}{2mc}g\hat{J}_3^{j,j}. \qquad \text{(III.4.29)}$$

Therefore, (III.4.12) is proved with g is given by (III.4.10). $\qquad\square$

Remarks III.4.5. i) Our proof follows the calculations from [61].
ii) A. Landé [59] was the first to obtain formula of type (III.4.10); he used the phenomenological vector model and the Bohr correspondence principle (see Section 14.7 of [51]).

Exercise III.4.6. Check (III.4.14)–(III.4.16).

Exercise III.4.7. Check (III.4.28).

III.4.5   Applications of the Landé formula

The Landé formula (III.4.10) has the following fundamental consequences.

The Einstein–de Haas experiment

Formula (III.4.10) for the Landé factor agrees with the experimental observations of A. Einstein and W.J. de Haas, and with later more accurate measurements (see [8, 11, 18] and [82, Vol. I]).

Anomalous Zeeman effect

The main virtue of the formula (III.4.10) is that it allows us to explain the multiplet structure in the anomalous Zeeman effect by the splitting of energy levels for operator (III.4.4) with small magnetic field B. Indeed, let

$$n = |E, L, J, j\rangle, \qquad n' = |E', L', J', j'\rangle$$

be stationary states of the unperturbed atom in the absence of magnetic field, i.e., when $B = 0$. According to (I.4.4),

$$\omega^0_{nn'} = \omega^0_{n'} - \omega^0_n \qquad \text{(III.4.30)}$$

is the spectral line radiated in transitions between these stationary states, where $\hbar\omega^0_n = E$ and $\hbar\omega^0_{n'} = E'$ are the eigenvalues of the unperturbed operator H^0_P from (III.4.4). For small $|B|$, the eigenvalues acquire the corresponding small corrections according to (III.4.29) and the perturbation formula (C.0.2) (see Appendix C) divided by $\hbar$:

$$\Delta\omega^0_n \approx -B\hat{M}_3^{j,j}/\hbar = -g(n)\Omega_\Lambda j, \qquad \text{(III.4.31)}$$

where $g(n)$ is the gyroscopic ratio (III.4.10) corresponding to the stationary state n. This approximation is accurate up to the order $\mathcal{O}(B^2)$ as $B \to 0$. Similarly, $\Delta\omega^0_{n'} \approx -g(n')\Omega_\Lambda j'$, and hence the split spectral lines are now given by

$$\begin{aligned} \omega_{nn'} &= \omega^0_{nn'} - [g(n')j' - g(n)j]\Omega_\Lambda \\ &= \omega^0_{nn'} - g(n')[j' - j]\Omega_\Lambda - [g(n') - g(n)]j\Omega_\Lambda. \end{aligned} \qquad \text{(III.4.32)}$$

Finally, we must take into account the selection rules

$$J \mapsto J' = J \pm 1, \qquad j \mapsto j' = j, j \pm 1, \qquad \text{(III.4.33)}$$

which follow similarly to (III.1.12) and (III.1.13), since the eigenfunctions have the structure similar to (I.6.21) with the same angular functions.

Now, in contrast to (III.4.1), the splitting (III.4.32) is not equidistant and depends on the quantum numbers L, J, j, L', J', j' by (III.4.10). The application of the formulae (III.4.32) and (III.4.33) perfectly agrees with experimental observations [8, 11, 18] and [82, Vol. I]. This agreement was one of the greatest successes of quantum theory.

Chapter IV

Relativistic Quantum Mechanics

In 1928, P. Dirac introduced a new wave equation, which was a relativistically covariant generalization of the Schrödinger equation. In the Dirac equation, the wave function has 4 complex components and the coefficients of the equation are 4×4 *Dirac matrices*.

P. Dirac calculated the first two approximations of this equation in the limit as $c \to \infty$. The approximation up to order $1/c$ coincides with the Pauli equation, while the second approximation displays, up to order $1/c^2$, the Russell–Saunders spin-orbital coupling, as well as some other effects.

The resulting equation admits Lagrangian and Hamiltonian formulations, which provide the corresponding conserved quantum observables and the continuity equation for the charge and current densities. The angular momentum automatically includes the spin component with factor 2, as predicted by S. Goudsmit and G. Uhlenbeck.

A few months later, W. Gordon and C. Darwin independently solved the Dirac spectral problem for hydrogen atom. Now the energies depend on the angular momentum, in contrast to the nonrelativistic case. This dependence perfectly explains the "fine structure" of the hydrogen spectrum.

The energy for the Dirac equation is not bounded from above and below, suggesting an instability of solutions. This issue was resolved in quantum electrodynamics by postulating anticommutation relations for the electron field.

IV.1   Free Dirac Equation

P. Dirac extended the Einstein Special Relativity Principle to quantum theory. Containing the first-order derivatives in time and the second-order derivatives in **x**, the Schrödinger equation is obviously noninvariant with respect to the Lorentz group. The latter consists of linear transforms of the *Minkowski space-time* of $(\mathbf{x}, t) \in \mathbb{R}^4$ that preserve the quadratic form $c^2 t^2 - \mathbf{x}^2$ (cf. (I.1.8)). One possible approach is to employ the Klein–Gordon equation (I.2.2),

$$\frac{\hbar^2}{c^2} \partial_t^2 \psi(\mathbf{x}, t) = \hbar^2 \Delta \psi(\mathbf{x}, t) - \mathrm{m}^2 c^2 \psi(\mathbf{x}, t), \qquad (\mathbf{x}, t) \in \mathbb{R}^4, \qquad \text{(IV.1.1)}$$

which is Lorentz-invariant, just like the wave equation [51, Section 12.3]. However, this approach results in negative energies. Indeed, after the Fourier transform,

$$\hat{\psi}(\mathbf{p}, t) := \int_{\mathbb{R}^3} e^{i \frac{\mathbf{p} \cdot \mathbf{x}}{\hbar}} \psi(\mathbf{x}, t)\, dx, \quad \mathbf{p} \in \mathbb{R}^3; \qquad \mathbf{p} \cdot \mathbf{x} := \sum_{k=1}^{3} p_k x^k, \qquad \text{(IV.1.2)}$$

the Klein–Gordon equation turns into the following ordinary differential equation with the parameter $\mathbf{p} \in \mathbb{R}^3$,

$$\frac{\hbar^2}{c^2} \partial_t^2 \hat{\psi}(\mathbf{p}, t) = -\mathbf{p}^2 \hat{\psi}(\mathbf{p}, t) - \mathrm{m}^2 c^2 \hat{\psi}(\mathbf{p}, t), \qquad t \in \mathbb{R}. \qquad \text{(IV.1.3)}$$

All solutions are linear combinations of $e^{-i\omega_\pm t}$, where $\omega_\pm = \pm \sqrt{\mathbf{p}^2 + \mathrm{m}^2 c^2}$. The solutions with $\omega_- = -\sqrt{\mathbf{p}^2 + \mathrm{m}^2 c^2} < 0$ seem to correspond to negative energies, since

$$\hat{E} e^{-i\omega_- t} = \hbar \omega_- e^{-i\omega_- t},$$

where $\hat{E} = i\hbar \partial_t$ is the energy operator (I.2.6). These negative energies are unbounded from below, and hence the physical interpretation of the Klein–Gordon equation requires some additional arguments.

This is why P. Dirac introduced in [22] a relativistically invariant wave equation of the first-order in time, just like the Schrödinger equation, trying to avoid negative roots. The relativistic invariance also required the first order in space; i.e.,

$$\gamma^\mu \hat{P}_\mu \psi(x) = \mathrm{m} c \psi(x), \qquad x \in \mathbb{R}^4, \qquad \text{(IV.1.4)}$$

where the summation over repeated indices $\mu = 0, \ldots, 3$ is assumed, and

$$\hat{P}_\mu := i\hbar \partial_\mu, \quad \mu = 0, \ldots, 3, \qquad \text{with } \partial_\mu = \frac{\partial}{\partial x^\mu}, \quad x^0 := ct. \qquad \text{(IV.1.5)}$$

The basic requirement is the following *correspondence principle*: equation (IV.1.4) should imply the Klein–Gordon equation (IV.1.1). Applying the operator $\gamma^\mu \hat{P}_\mu$ to both sides of (IV.1.4), we obtain:

$$[\gamma^\mu \hat{P}_\mu]^2 \psi(x) = \mathrm{m}^2 c^2 \psi(x), \qquad x \in \mathbb{R}^4. \qquad \text{(IV.1.6)}$$

On the other hand, the Klein–Gordon equation (IV.1.1) can be written as

$$[\hat{P}_0^2 - \hat{\mathbf{P}}^2]\psi = m^2 c^2 \psi, \qquad \hat{\mathbf{P}} := (\hat{P}_1, \hat{P}_2, \hat{P}_3).$$

Hence, the correspondence principle is equivalent to the algebraic identity for the linear form $\gamma(p) := \gamma^\mu p_\mu$,

$$[\gamma(p)]^2 = p_0^2 - \mathbf{p}^2, \qquad p = (p_0, \mathbf{p}) \in \mathbb{R}^4. \tag{IV.1.7}$$

Dirac's extra idea was the choice of the coefficients γ^μ in a matrix algebra, since scalar coefficients γ^μ satisfying (IV.1.7) do not exist: the existence of scalar coefficients would mean that the polynomial $p_0^2 - p_1^2 - p_2^2 - p_3^2$ is reducible, which is not true.

Exercise IV.1.1. *Verify that (IV.1.7) is impossible with scalar coefficients γ^μ.*

Lemma IV.1.2. *The matrices*

$$\gamma(p) = \begin{pmatrix} p_0 I_2 & \boldsymbol{\sigma} \cdot \mathbf{p} \\ -\boldsymbol{\sigma} \cdot \mathbf{p} & -p_0 I_2 \end{pmatrix} \tag{IV.1.8}$$

satisfy identity (IV.1.7), where $\boldsymbol{\sigma} := (\sigma_1, \sigma_2, \sigma_3)$ are the Pauli spin matrices (I.7.11) and I_2 is the 2×2 identity matrix.

Proof. Applying the multiplication rules for the Pauli matrices (III.3.6) to the multiplication of the 2×2 block matrices, we obtain:

$$\gamma^2(p) = \begin{pmatrix} (p_0^2 - (\boldsymbol{\sigma} \cdot \mathbf{p})^2) I_2 & 0 \\ 0 & (p_0^2 - (\boldsymbol{\sigma} \cdot \mathbf{p})^2) I_2 \end{pmatrix}. \tag{IV.1.9}$$

It remains to invoke the identity $(\boldsymbol{\sigma} \cdot \mathbf{p})^2 = \mathbf{p}^2$. $\square$

Now we can calculate the matrices $\gamma^\mu = \gamma(e_\mu)$, where $e_0 = (1, 0, 0, 0)$, etc. From (IV.1.8), we obtain:

$$\gamma^0 = \begin{pmatrix} 1 & 0 \\ 0 & -1 \end{pmatrix}, \qquad \gamma^k = \begin{pmatrix} 0 & \sigma_k \\ -\sigma_k & 0 \end{pmatrix} \quad \text{with } k = 1, 2, 3. \tag{IV.1.10}$$

Lemma IV.1.3. *The matrices γ^μ, $\mu = 0, \ldots, 3$, satisfy the relations*

$$(\gamma^0)^2 = 1; \quad (\gamma^k)^2 = -1, \quad k = 1, 2, 3; \quad \gamma^\mu \gamma^\nu + \gamma^\nu \gamma^\mu = 0, \quad \mu \neq \nu. \tag{IV.1.11}$$

Proof. Rewriting (IV.1.7) in the form

$$\gamma^2(p) = g(p) := p_0^2 - \mathbf{p}^2, \qquad p \in \mathbb{R}^4, \tag{IV.1.12}$$

we can write

$$\gamma(p)\gamma(q) + \gamma(q)\gamma(p) = 2g(p, q), \tag{IV.1.13}$$

where $g(p, q) = p_0 q_0 - \mathbf{p} \cdot \mathbf{q}$ is the corresponding symmetric bilinear form. In particular, for $p = e_\alpha$ and $q = e_\beta$, α, $\beta = 0$, ..., 3, we obtain:

$$\gamma^\mu \gamma^\nu + \gamma^\nu \gamma^\mu = 2g(e_\alpha, e_\beta),$$

which implies (IV.1.11). $\square$

The matrices (IV.1.10) are known as the *ordinary* (or *standard*) *representation* of the relations (IV.1.11). It is easily checked that these matrices are not unique solutions: for example, we can replace γ^α by $-\gamma^\alpha$ for certain indexes α. Below we shall prove the Pauli Theorem: the matrices γ^α are unique up to a change of the orthonormal basis e_α (for details, see Theorem IV.2.1 below).

Let us rewrite the *free Dirac equation* (IV.1.4) in the "Schrödinger form". The equation (IV.1.4) can be written as

$$i\hbar\gamma^0 \partial_t \psi = c(\mathrm{m}c - i\hbar\gamma^k \partial_{x^k})\psi. \tag{IV.1.14}$$

Multiplying by γ^0, we obtain:

$$i\hbar\partial_t \psi = \gamma^0(\mathrm{m}c^2 - ic\hbar\gamma^k \partial_{x^k})\psi. \tag{IV.1.15}$$

We set

$$\alpha^k := \gamma^0\gamma^k = \begin{pmatrix} 0 & \sigma_k \\ \sigma_k & 0 \end{pmatrix}, \quad k = 1, 2, 3; \qquad \beta := \gamma^0. \tag{IV.1.16}$$

Now (IV.1.15) reads as follows:

$$i\hbar\partial_t \psi = H_D^0 \psi := [\mathrm{m}c^2\beta - ic\hbar\alpha^k \partial_{x^k}]\psi = [\mathrm{m}c^2\beta + c\,\boldsymbol{\alpha} \cdot \hat{\mathbf{p}}]\psi, \tag{IV.1.17}$$

where $\hat{p}_k := -i\hbar\partial_{x^k}$ and $\boldsymbol{\alpha} = (\alpha^1, \ldots, \alpha^3)$. The operator H_D^0 is called the *free Dirac Hamiltonian*. The Dirac matrices α^k and β are Hermitian, and hence the operator H_D^0 is symmetric in the complex Hilbert space $L^2(\mathbb{R}^3, \mathbb{C}^4)$.

Exercise IV.1.4. Check that the operator H_D^0 is symmetric. **Hint:** α^k commute with $\hat{p}_k$.

IV.2 The Pauli Theorem

The relativistic covariance of the Dirac equation follows from the complete description of appropriate solutions of the commutation relations (IV.1.11). The following *Pauli theorem* states that all low-dimensional solutions of these relations are equivalent to the standard Dirac matrices (IV.1.10).

Theorem IV.2.1. *Let γ^μ, $\mu = 0, \ldots, 3$ be linear operators in a finite-dimensional complex vector space V of dimension $\dim V \leq 4$. Suppose that they satisfy relations (IV.1.11). Then $\dim V = 4$, and the operators γ^μ, for a suitable choice of basis in V, can be written in the matrix form (IV.1.10).*

See also [85]. The higher-dimensional analogue of this theorem is in e.g. [134].

Proof. Step i) The key idea of the proof is the following simple characterization of basis vectors. For the standard Dirac matrices (IV.1.10), the matrix $\gamma^1\gamma^2$ is diagonal, and hence it commutes with γ^0, which is also diagonal:

$$\gamma^0 = \begin{pmatrix} I_2 & 0 \\ 0 & -I_2 \end{pmatrix}, \qquad \gamma^1\gamma^2 = \begin{pmatrix} -i\sigma_3 & 0 \\ 0 & -i\sigma_3 \end{pmatrix}. \qquad \text{(IV.2.1)}$$

Therefore, the basis vectors $e_0, \ldots, e_3$ are common eigenvectors of the matrices γ^0 and $\gamma^1\gamma^2$ with the eigenvalues 1 and $-i$, 1 and i, -1 and $-i$, -1 and i, respectively.

Step ii) Now we apply this observation to general matrices γ^μ of Theorem IV.2.1. It follows from the anticommutation relations from (IV.1.11) that the matrices γ^0 and $\gamma^1\gamma^2$ commute with each other:

$$\gamma^0\gamma^1\gamma^2 = \gamma^1\gamma^2\gamma^0. \qquad \text{(IV.2.2)}$$

Hence there exists at least one common eigenvector v_1 for both of them (since V is a *complex vector space!*):

$$\gamma^0 v_1 = a v_1 \quad \text{and} \quad \gamma^1\gamma^2 v_1 = b v_1, \qquad \text{(IV.2.3)}$$

where a and b are suitable complex numbers.

Step iii) We have $a^2 = 1$, since $(\gamma^0)^2 = 1$, and similarly, $b^2 = -1$, since $(\gamma^1\gamma^2)^2 = -1$. Hence $a = \pm 1$ and $b = \pm i$. Let us check that all four combinations of the signs are possible for suitable eigenvectors v_1. Indeed,

$$\gamma^0\gamma^3 v_1 = -\gamma^3\gamma^0 v_1 = -a\gamma^3 v_1 \qquad \text{(IV.2.4)}$$

and

$$\gamma^1\gamma^2\gamma^3 v_1 = \gamma^3\gamma^1\gamma^2 v_1 = b\gamma^3 v_1. \qquad \text{(IV.2.5)}$$

Hence the vector $v_3 := \gamma^3 v_1$ is also a common eigenvector of γ^0 and $\gamma^1\gamma^2$, with eigenvalues $-a$ and b, respectively. Similarly, $v_2 := -\gamma^3\gamma^1 v_1$ and $v_4 := -\gamma^1 v_1$ are common eigenvectors with eigenvalues a, $-b$ and $-a$, $-b$, respectively. Since all four possible signs have occurred, we may permute the four vectors to ensure that $a = 1$ and $b = -i$.

Step iv) The vectors v_1, v_2, v_3, v_4 are linearly independent, and hence they span V. In the basis v_1, v_2, v_3, v_4, the operators γ^0 and $\gamma^1\gamma^2$ are block-diagonal matrices of the form of (IV.2.1). Moreover, in this basis, the operators γ^1 and γ^3 have the form

$$\gamma^1 = \begin{pmatrix} 0 & \sigma_1 \\ -\sigma_1 & 0 \end{pmatrix}, \qquad \gamma^3 = \begin{pmatrix} 0 & \sigma_3 \\ -\sigma_3 & 0 \end{pmatrix}, \qquad \text{(IV.2.6)}$$

which coincides with (IV.1.10). Hence $\gamma^2 = -\gamma^1(\gamma^1\gamma^2)$ also has the desired form.   □

Exercise IV.2.2. Verify that v_1, v_2, v_3, v_4 are linearly independent.
Hint: $\gamma^0 = 1$ in the linear span of v_1 and v_2 and $\gamma^0 = -1$ in the linear span of v_3 and v_4, while $\gamma^1\gamma^2 = -i$ in the span of v_1 and v_3 and $\gamma^1\gamma^2 = i$ in the span of v_2 and v_4.

Exercise IV.2.3. Check the formulae (IV.2.1)–(IV.2.6).

Let us recall the definition of Lorentz transformations.

Definition IV.2.4. *A real 4×4-matrix is a Lorentz transformation if it leaves invariant the quadratic form (IV.1.12).*

Equivalently, a matrix Λ is a Lorentz transformation if and only if

$$\Lambda g \Lambda^t = g, \qquad g := \mathrm{diag}(1, -1, -1, -1). \qquad \text{(IV.2.7)}$$

Corollary IV.2.5. *For any Lorentz transformation Λ (see, e.g., Section 12.3 of [51]), there exists a nondegenerate matrix $\Gamma(\Lambda) \in GL(4, \mathbb{C})$ such that*

$$\gamma(\Lambda p) = \Gamma(\Lambda)\gamma(p)\Gamma^{-1}(\Lambda), \qquad p \in \mathbb{R}^4. \qquad \text{(IV.2.8)}$$

Proof. From (IV.1.12), we have

$$\gamma^2(\Lambda p) = g(\Lambda p) = g(p), \qquad p \in \mathbb{R}^4, \qquad \text{(IV.2.9)}$$

since Λ is a Lorentz transformation. Hence the matrices $\gamma(\Lambda e_\mu)$ satisfy the relations (IV.1.11), and also $\gamma^\mu := \gamma(e_\mu)$. Therefore, by the Pauli theorem,

$$\gamma(\Lambda e_\mu) = \Gamma(\Lambda)\gamma(e_\mu)\Gamma^{-1}(\Lambda), \qquad \mu = 0, \ldots, 3, \qquad \text{(IV.2.10)}$$

where $\Gamma(\Lambda)$ is an invertible operator in $\mathbb{R}^4$ (which maps the vector e_μ to v_μ, $\mu = 0, \ldots, 3$). Now (IV.2.8) follows by linearity.   □

Exercise IV.2.6. Prove that a matrix Λ is a Lorentz transformation if and only if it satisfies (IV.2.7).

IV.3 The Lorentz Covariance

The Einstein postulate of Special Relativity (see, e.g., [51, (12.24)]) extends to quantum theory as follows:

The laws of quantum theory take identical form in all inertial frames.

The next theorem ensures the existence of the corresponding transform for wave functions, which leave the Dirac equation invariant. Thus the Dirac equation provides relativistically invariant quantum theory.

We consider two frames of reference related by the Lorentz transformation: $x' = \Lambda x$. For a function $\psi(x)$, define

$$\psi'(x') := \Gamma(\Lambda^{\#})\psi(\Lambda^{-1}x'), \qquad x' \in \mathbb{R}^4, \tag{IV.3.1}$$

where $\Lambda^{\#} := (\Lambda^t)^{-1}$ and Λ^t is the transposed matrix Λ.

Theorem IV.3.1. *Let $\psi(x)$ be a solution of the Dirac equation (IV.1.4). Then the function $\psi'(x')$ is also a solution to the Dirac equation.*

Proof. Let us apply the Fourier transform,

$$\tilde{\psi}(x) = \int_{\mathbb{R}^4} e^{\frac{p \cdot x}{\hbar}} \psi(x) \, dx, \qquad p \in \mathbb{R}^4; \quad px = p_\mu x^\mu. \tag{IV.3.2}$$

Then the Dirac equation (IV.1.4) takes the form

$$\gamma(p)\tilde{\psi}(p) = mc\tilde{\psi}(p), \qquad p \in \mathbb{R}^4. \tag{IV.3.3}$$

On the other hand, the Fourier transform translates (IV.3.1) into

$$\tilde{\psi}'(p') = \Gamma(\Lambda^{\#})\tilde{\psi}(p), \qquad p = \Lambda^t p', \quad p' \in \mathbb{R}^4. \tag{IV.3.4}$$

In other words,

$$\tilde{\psi}(p) = \Gamma^{-1}(\Lambda^{\#})\tilde{\psi}'(\Lambda^{\#}p), \qquad p \in \mathbb{R}^4. \tag{IV.3.5}$$

Now we can express (IV.3.3) in terms of the wave function $\tilde{\psi}'(p')$:

$$\gamma(p)\Gamma^{-1}(\Lambda^{\#})\tilde{\psi}'(\Lambda^{\#}p) = mc\Gamma^{-1}(\Lambda^{\#})\tilde{\psi}'(\Lambda^{\#}p), \qquad p \in \mathbb{R}^4. \tag{IV.3.6}$$

This identity is equivalent to the Dirac equation (IV.3.3) if and only if

$$\Gamma(\Lambda^{\#})\gamma(p)\Gamma^{-1}(\Lambda^{\#}) = \gamma(\Lambda^{\#}p), \qquad p \in \mathbb{R}^4. \tag{IV.3.7}$$

This last relation is equivalent to (IV.2.8) with $\Lambda^{\#}$ instead of Λ. It remains to note that (IV.2.8) holds for $\Lambda^{\#}$, since $\Lambda^{\#}$ also belongs to the Lorentz group. $\square$

This theorem implies that the transform (IV.3.1) leaves the Dirac equation unchanged. In other words, the Dirac equation is *covariant* with respect to the Lorentz group.

Exercise IV.3.2. Verify (IV.3.4). **Hint:** We formally have

$$\tilde{\psi}'(p') := \int e^{\frac{ip'\cdot x'}{\hbar}} \psi'(x')\, dx' = \int e^{\frac{ip'\cdot(\Lambda x)}{\hbar}} \Gamma(\Lambda^{\#})\psi(x)|\det\Lambda|\, dx$$

$$= \Gamma(\Lambda^{\#}) \int e^{\frac{i(\Lambda^{t}p')\cdot x}{\hbar}} \psi(x)\, dx = \Gamma(\Lambda^{\#})\tilde{\psi}(p), \qquad\qquad \text{(IV.3.8)}$$

because $|\det\Lambda| = 1$ for the Lorentz transformation Λ. These formal calculations are justified by the continuity of the Fourier transform in the space of tempered distributions.

Exercise IV.3.3. Verify that $\Lambda^{\#}$ is a Lorentz transformation for any $\Lambda \in L$.

IV.4 Angular Momentum

The conserved *orbital* angular momentum for the Schrödinger equation is defined by (I.3.14):

$$\mathbf{L} = \mathbf{L}(\psi) := \langle \psi, \hat{\mathbf{L}}\psi \rangle, \qquad \hat{\mathbf{L}} = \mathbf{x} \times \hat{\mathbf{p}}.$$

For solutions of the Dirac equation, the orbital angular momentum is generally not conserved, because the operator $\hat{\mathbf{L}}$ does not commute with the Dirac operator H_D^0. Hence for the Dirac equation, the definition of the angular momentum requires a certain modification.

Definition IV.4.1. *For the Dirac equation, the* angular momentum *is defined by*

$$\mathbb{J} = \mathbb{J}(\psi) = \langle \psi, \hat{\mathbb{J}}\psi \rangle, \qquad with \quad \hat{\mathbb{J}} := \begin{pmatrix} \hat{\mathbf{J}} & 0 \\ 0 & \hat{\mathbf{J}} \end{pmatrix}, \qquad (IV.4.1)$$

where $\hat{\mathbf{J}}$ is the operator (III.3.23).

Theorem IV.4.2. *For solutions of the free Dirac equation* (IV.1.17), *the angular momentum $\mathbb{J}$ is conserved:*

$$\mathbb{J}(\psi(\cdot, t)) = \text{const}, \qquad t \in \mathbb{R}. \qquad (IV.4.2)$$

Proof. Differentiating (IV.4.2) and using (IV.1.17), we obtain:

$$\frac{d}{dt}\mathbb{J}(\psi(t)) = \langle \dot{\psi}(t), \hat{\mathbb{J}}\psi(t) \rangle + \langle \psi(t), \hat{\mathbb{J}}\dot{\psi}(t) \rangle$$

$$= \left\langle \frac{1}{i\hbar} H_D^0 \psi(t), \hat{\mathbb{J}}\psi(t) \right\rangle + \left\langle \psi(t), \hat{\mathbb{J}} \frac{1}{i\hbar} H_D^0 \psi(t) \right\rangle$$

$$= -\frac{1}{i\hbar} \langle H_D^0 \psi(t), \hat{\mathbb{J}}\psi(t) \rangle + \frac{1}{i\hbar} \langle \psi(t), \hat{\mathbb{J}} H_D^0 \psi(t) \rangle$$

$$= \frac{1}{i\hbar} \langle \psi(t), [\hat{\mathbb{J}}, H_D^0] \psi(t) \rangle. \qquad (IV.4.3)$$

It remains to verify the commutation

$$[H_D^0, \hat{\mathbb{J}}] = 0. \qquad (IV.4.4)$$

We have:

$$\hat{\mathbb{J}}_k := \hat{\mathbb{L}}_k + \hat{\mathbb{S}}_k, \qquad k = 1, 2, 3, \qquad (IV.4.5)$$

where

$$\hat{\mathbb{L}}_k := \begin{pmatrix} \hat{L}_k & 0 \\ 0 & \hat{L}_k \end{pmatrix}, \quad \hat{\mathbb{S}}_k := \begin{pmatrix} \hat{s}_k & 0 \\ 0 & \hat{s}_k \end{pmatrix} = \frac{1}{2}\hbar\Sigma_k, \quad \Sigma_k := \begin{pmatrix} \sigma_k & 0 \\ 0 & \sigma_k \end{pmatrix}. \qquad (IV.4.6)$$

Recall that $H_D^0 = mc^2\beta + c\,\alpha^k\hat{p}_k$ by (IV.1.17), where we assume the Einstein summation convention over repeated indices, and by (IV.1.16) we have

$$\alpha^k = \begin{pmatrix} 0 & \sigma_k \\ \sigma_k & 0 \end{pmatrix}, \qquad k = 1,\,2,\,3.$$

To prove (IV.4.4), we first note that

$$[\hat{p}_i, \hat{\mathbb{L}}_j] = i\hbar \sum_{k=1}^{3} \varepsilon_{ijk}\hat{p}_k, \tag{IV.4.7}$$

where ε_{ijk} is a totally antisymmetric tensor with $\varepsilon_{123} = 1$. Therefore,

$$[H_D^0, \hat{\mathbb{L}}_j] = [mc^2\beta + c\,\alpha^i\hat{p}_i, \hat{\mathbb{L}}_j] = c\,\alpha^i[\hat{p}_i, \hat{\mathbb{L}}_j] = i\sum_{k=1}^{3} c\hbar\,\varepsilon_{ijk}\alpha^i\hat{p}_k, \tag{IV.4.8}$$

since the Dirac matrices α^k act only on the "spin variables", and thus commute with the "diagonal matrices" $\hat{p}_k$ and $\hat{\mathbb{L}}_k$. This shows that the angular momentum operators $\hat{\mathbb{L}}_k$ do not commute with H_D^0, and hence the orbital angular momentum $\mathbb{L}$ is generally not conserved.

It remains to calculate the commutators $[H_D^0, \mathbb{S}]$. The operators $\hat{p}_k$ commute with the Pauli matrices, while Σ_k is the diagonal matrix, so

$$[H_D^0, \Sigma_k] = [mc^2\beta + c\,\alpha^i\hat{p}_i, \Sigma_k] = c\sum_{i=1}^{3} \begin{pmatrix} 0 & [\sigma_i, \sigma_k] \\ [\sigma_i, \sigma_k] & 0 \end{pmatrix}\hat{p}_i. \tag{IV.4.9}$$

Finally, the commutation relations for the Pauli spin matrices (III.3.6) can be written in the form $[\sigma_i, \sigma_j] = 2i\sum_{k=1}^{3}\varepsilon_{ijk}\sigma_k$. Hence, we obtain:

$$[H_D^0, \Sigma_j] = 2ic\varepsilon_{ijk}\begin{pmatrix} 0 & \sigma_k \\ \sigma_k & 0 \end{pmatrix}\hat{p}_i = 2ic\,\varepsilon_{ijk}\alpha^k\hat{p}_i = 2ic\,\varepsilon_{kji}\alpha^i\hat{p}_k.$$

Multiplying by $\hbar/2$ and adding (IV.4.8), we get (IV.4.4) by the antisymmetry of ε_{ijk}. □

This theorem justifies the Goudsmit–Uhlenbeck conjecture on electron spin as an intrinsic property of relativistic dynamical equations. The terms $\mathbb{S}_k = \dfrac{1}{2}\hbar\Sigma_k$ in (IV.4.5), (IV.4.6) represent *spin*, the intrinsic angular momentum of the relativistic electron. Moreover, we will see in Section IV.8 that the coupling of the Dirac equation to the magnetic field in the nonrelativistic limit automatically provides the correct Landé factor $g = 2$ for the spin angular momentum.

These facts were a great triumph of Dirac's theory, suggesting that the electron spin is intrinsically connected to the relativistic invariance of the theory.

Remark IV.4.3. In Appendix B, we present a general theory of conservation laws for Hamiltonian systems with a Lie symmetry group. The angular momentum conservation (IV.4.2) is a consequence of that theory in the case of the group of rotations.

Exercise IV.4.4. Check (IV.4.7).

Exercise IV.4.5. Check the commutation relations of type (III.3.24) for angular momentum (IV.4.1). **Hint:** use the commutation relations (I.3.17) and (III.3.6).

IV.5   Negative Energies

The Dirac equation (IV.1.17) is a Hamiltonian system with the Hamiltonian functional

$$\mathcal{H}^0(\psi) := \langle \psi, H_D^0 \psi \rangle = \int \psi^*(\mathbf{x}) H_D^0 \psi(\mathbf{x}) \, d\mathbf{x}, \qquad \text{(IV.5.1)}$$

which is similar to the Schrödinger Hamiltonian (I.3.2). It is conserved, i.e.,

$$\mathcal{H}^0(\psi(\cdot, t)) = \text{const}, \qquad t \in \mathbb{R}, \qquad \text{(IV.5.2)}$$

for solutions of (IV.1.17) (or (IV.1.15)).

Exercise IV.5.1. Verify (IV.5.2). **Hint:** Take the time derivative of the left-hand side of (IV.5.2) and use the Dirac equation (IV.1.17) and the symmetry of the Dirac operator H_D^0.

Let us check that energy for the Dirac equation is not bounded from above and from below. Indeed, the quadratic form (IV.5.1) is not positive definite. To see this, let us split the wave function as

$$\varphi(\mathbf{x}) = \begin{pmatrix} \varphi_+(\mathbf{x}) \\ \varphi_-(\mathbf{x}) \end{pmatrix}, \qquad \text{(IV.5.3)}$$

where $\varphi_+ := \begin{pmatrix} \varphi_1 \\ \varphi_2 \end{pmatrix} \in \mathbb{C}^2$ and $\varphi_- := \begin{pmatrix} \varphi_3 \\ \varphi_4 \end{pmatrix} \in \mathbb{C}^2$, and let us apply the Fourier transform (IV.1.2):

$$\hat{\varphi}(\mathbf{p}) = \int e^{i \frac{\mathbf{p} \cdot \mathbf{x}}{\hbar}} \varphi(\mathbf{x}) \, d\mathbf{x}, \qquad \mathbf{p} \in \mathbb{R}^3. \qquad \text{(IV.5.4)}$$

Due to the Parseval–Plancherel identity, the quadratic form (IV.5.1) can be written as

$$\begin{aligned}
\mathcal{H}^0(\varphi) &= \frac{c}{(2\pi)^3} \left\langle \hat{\varphi}(\mathbf{p}), \gamma^0 \begin{pmatrix} mcI_2 & -\boldsymbol{\sigma} \cdot \mathbf{p} \\ \boldsymbol{\sigma} \cdot \mathbf{p} & mcI_2 \end{pmatrix} \hat{\varphi}(\mathbf{p}) \right\rangle \\
&= \frac{c}{(2\pi)^3} \left\langle \hat{\varphi}(\mathbf{p}), \begin{pmatrix} mcI_2 & -\boldsymbol{\sigma} \cdot \mathbf{p} \\ -\boldsymbol{\sigma} \cdot \mathbf{p} & -mcI_2 \end{pmatrix} \hat{\varphi}(\mathbf{p}) \right\rangle \\
&= \frac{c}{(2\pi)^3} \left(mc \langle \hat{\varphi}_+(\mathbf{p}), \hat{\varphi}_+(\mathbf{p}) \rangle - 2 \langle \hat{\varphi}_+(\mathbf{p}), \boldsymbol{\sigma} \cdot \mathbf{p} \, \hat{\varphi}_-(\mathbf{p}) \rangle \right. \\
&\qquad \left. - mc \langle \hat{\varphi}_-(\mathbf{p}), \hat{\varphi}_-(\mathbf{p}) \rangle \right). \qquad \text{(IV.5.5)}
\end{aligned}$$

In particular,

$$
\begin{cases}
\mathcal{H}^0 \begin{pmatrix} \varphi_+(\mathbf{x}) \\ 0 \end{pmatrix} = \dfrac{mc^2}{(2\pi)^3} \langle \hat{\varphi}_+(\mathbf{p}), \hat{\varphi}_+(\mathbf{p}) \rangle, \\[4mm]
\mathcal{H}^0 \begin{pmatrix} 0 \\ \varphi_-(\mathbf{x}) \end{pmatrix} = -\dfrac{mc^2}{(2\pi)^3} \langle \hat{\varphi}_-(\mathbf{p}), \hat{\varphi}_-(\mathbf{p}) \rangle.
\end{cases}
$$

Historically, it was expected that the negative energy can lead to an instability of the Dirac dynamics due to possible transitions of solutions from states with positive energy into states with negative energy, as the result of interaction with the Maxwell field (which is defined in the next section). On the other hand, this instability has never been proved.

Remark IV.5.2. Let us mention that the solitary wave solutions to the nonlinear Dirac equation — the Dirac equation with self action — can exhibit the so-called *spectral stability*; for details, see [134].

Dirac suggested that the transition of all particles into states with negative energies is forbidden by the Pauli exclusion principle, because almost all such states became occupied long time ago. On the other hand, by the Dirac theory, transitions for certain particles are possible, and the "negative states" can be interpreted as states with positive energy for *antiparticles* which are *positrons* (i.e., "positively charged electrons") [9]. Positrons were discovered experimentally in *cosmic rays* by C. Anderson in 1932; for this discovery, C. Anderson received the 1936 Nobel Prize in physics.

Note that in the context of the Dirac equation the states with positive and negative energies are entirely equivalent. The problem of negative energies was solved by Dirac in the framework of *quantum electrodynamics* by introduction of *anticommutation relations* in *second quantized* Dirac equation; see [25, 80].

IV.6   Coupling to the Maxwell Field

Interaction of the Dirac field with an external Maxwell field is defined by analogy with the Schrödinger equation so as to keep the gauge invariance; see Theorem IV.6.1 below. We shall obtain the corresponding Hamiltonian and Lagrangian formulations and the coupled Maxwell–Dirac equations. Finally, we will obtain the charge conjugation, which transforms the Dirac equation for electrons with negative charge $e < 0$ into a similar equation for *positrons* with positive charge $-e > 0$.

The free Dirac equation (IV.1.4) can be written as

$$\gamma(\hat{P})\psi(x) = mc\psi(x), \qquad x = (\mathbf{x}, t) \in \mathbb{R}^4, \tag{IV.6.1}$$

where $\gamma(\hat{P}) := \gamma^\mu \hat{P}_\mu$ and $\hat{P}_\mu := i\hbar\partial_\mu$. Recall the Schrödinger equation (I.2.14) with external Maxwell field (in the Gaussian and Heaviside–Lorentz units [48, p. 781], [49]):

$$\left(i\hbar\partial_t - eA_0(\mathbf{x}, t)\right)\psi(\mathbf{x}, t) = \frac{1}{2m}\left[-i\hbar\nabla - \frac{e}{c}\mathbf{A}(\mathbf{x}, t)\right]^2 \psi(\mathbf{x}, t). \tag{IV.6.2}$$

In the notation $\mathcal{A}_\mu = (A_0, -\mathbf{A})$ and (IV.1.5), we have:

$$c[\hat{P}_0 - \frac{e}{c}\mathcal{A}_0(\mathbf{x}, t)]\psi(\mathbf{x}, t) = \frac{1}{2m}\sum_{k=1}^{3}\left[\hat{P}_k - \frac{e}{c}\mathcal{A}_k(\mathbf{x}, t)\right]^2 \psi(\mathbf{x}, t). \tag{IV.6.3}$$

This suggests the following generalization of the Dirac equation (IV.6.1) for a relativistic electron in an external Maxwell field:

$$\gamma\left(\hat{P} - \frac{e}{c}\mathcal{A}(x)\right)\psi(x) = mc\psi(x), \qquad x \in \mathbb{R}^4. \tag{IV.6.4}$$

IV.6.1   Gauge invariance

The gauge transform of the Maxwell field is defined as follows (see [48]):

$$A_0(\mathbf{x}, t) \mapsto A_0(\mathbf{x}, t) + \frac{1}{c}\dot{\chi}(\mathbf{x}, t), \qquad \mathbf{A}(\mathbf{x}, t) \mapsto \mathbf{A}(\mathbf{x}, t) - \nabla\chi(\mathbf{x}, t), \tag{IV.6.5}$$

where $\chi(\mathbf{x}, t)$ is an arbitrary real-valued function. This transform does not change the Maxwell fields corresponding to the Maxwell potentials $A_0(\mathbf{x}, t)$ and $\mathbf{A}(\mathbf{x}, t)$. In our notation,

$$\mathcal{A}_\mu(x) \mapsto \mathcal{A}'_\mu(x) := \mathcal{A}_\mu(x) + \frac{1}{i\hbar}\hat{P}_\mu\chi(x). \tag{IV.6.6}$$

Theorem IV.6.1. *Let* $\psi(x)$ *be a solution of the Dirac equation (IV.6.4) with Maxwell potentials* $\mathcal{A}(x)$*. Then* $\psi'(x) := \exp\left(\dfrac{e\chi(x)}{i\hbar c}\right)\psi(x)$ *satisfies the equation with the potentials* $\mathcal{A}'(x)$*.*

Proof. This result follows from the differentiation

$$\hat{P} \exp\left(\frac{e\chi(x)}{i\hbar c}\right) = \frac{e\hat{P}\chi(x)}{i\hbar c} \exp\left(\frac{e\chi(x)}{i\hbar c}\right).$$

$\square$

IV.6.2   Antiparticles

There exists a one-to-one correspondence between solutions of the Dirac equations with charges e and $-e$, respectively. Indeed, consider the Dirac equation (IV.6.4) with $-e$ instead of e:

$$\gamma\left(\hat{P} + \frac{e}{c}\mathcal{A}(x)\right)\psi(x) = mc\psi(x), \qquad x \in \mathbb{R}^4. \tag{IV.6.7}$$

This equation describes the positron wave field corresponding to particles with positive charge $-e = |e|$. We will find an isomorphism between the solutions of (IV.6.4) and (IV.6.7).

Definition IV.6.2. *The* charge conjugation *is defined as the transform of each wave function $\psi(x)$ into $\psi_C(x) := \gamma^2 \boldsymbol{K}\psi(x)$, with $\boldsymbol{K}$ Wigner's (antilinear) operator of componentwise complex conjugation,*

$$\boldsymbol{K} : \mathbb{C}^{m\times n} \to \mathbb{C}^{m\times n}, \qquad (\boldsymbol{K}a)_{ij} = \overline{a_{ij}}. \tag{IV.6.8}$$

Theorem IV.6.3. *Let ψ satisfy the Dirac equation (IV.6.4). Then ψ_C satisfies the equation (IV.6.7).*

Proof. Step i) Note that for any vector $p \in \mathbb{C}^4$, we have

$$\gamma^2 \gamma(\bar{p})\gamma^2 = \boldsymbol{K}\gamma(p). \tag{IV.6.9}$$

Indeed, all the matrices γ^μ are real, except for γ^2, which is purely imaginary. Hence,

$$\boldsymbol{K}\gamma^\mu = \begin{cases} \gamma^\mu, & \mu \neq 2, \\ -\gamma^2, & \mu = 2. \end{cases} \tag{IV.6.10}$$

On the other hand, the anticommutation relations for the Dirac matrices imply that

$$\gamma^2 \gamma^\mu \gamma^2 = \begin{cases} \gamma^\mu, & \mu \neq 2, \\ -\gamma^2, & \mu = 2. \end{cases} \tag{IV.6.11}$$

This gives (IV.6.9).

Step ii) Applying the complex conjugation to the Dirac equation (IV.6.4) and using (IV.6.9), we obtain:

$$\boldsymbol{K}\left(\gamma\left(\hat{P} - \frac{e}{c}\mathcal{A}(x)\right)\psi\right) = \gamma^2 \gamma\left(\overline{\hat{P} - \frac{e}{c}\mathcal{A}(x)}\right)\gamma^2 \boldsymbol{K}\psi = mc\boldsymbol{K}\psi. \tag{IV.6.12}$$

Multiplying by γ^2 and taking into account that the vector $\dfrac{e}{c}\mathcal{A}(x)$ is real, while the differential operator $\hat{P}$ includes the imaginary factor i, we obtain, as claimed,

$$-\gamma\left(-\hat{P}-\frac{e}{c}\mathcal{A}(x)\right)\psi_C = \gamma\left(\hat{P}+\frac{e}{c}\mathcal{A}(x)\right)\psi_C = \mathrm{mc}\psi_C.$$

$\square$

Exercise IV.6.4. Check (IV.6.11).

IV.7 Charge and Current. Continuity Equation

The charge and current densities for the Dirac equation (IV.6.4) are defined by

$$\begin{cases} \rho(x) &= e\psi^*(x)\psi(x) = e|\psi(x)|^2 \\ j^k(x) &= e\psi^*(x)\alpha^k\psi(x), \quad k = 1, 2, 3 \end{cases} \quad \text{(IV.7.1)}$$

Let us check the charge continuity equation (I.3.11). Write (IV.6.4) as

$$i\partial_t\psi = \alpha^k\left[-i\partial_{x^k} + \frac{e}{\hbar c}A_k\right]\psi + \beta\frac{m}{\hbar}\psi + \frac{e}{\hbar c}A_0\psi,$$

where the summation in the repeated indices $k = 1, 2, 3$ is assumed. Then

$$\begin{aligned} \partial_t(\psi^*\psi) &= i\left(\alpha^k\left[-i\partial_{x^k} + \frac{e}{\hbar c}A_k\right]\psi + \beta\frac{m}{\hbar}\psi + \frac{e}{\hbar c}A_0\psi\right)^*\psi \\ &\quad -i\psi^*\left(\alpha^k\left[-i\partial_{x^k} + \frac{e}{\hbar c}A_k\right]\psi + \beta\frac{m}{\hbar}\psi + \frac{e}{\hbar c}A_0\psi\right) \\ &= i\left(\alpha^k\left[-i\partial_{x^k}\right]\psi\right)^*\psi - i\psi^*\left(\alpha^k\left[-i\partial_{x^k}\right]\psi\right) \\ &= -\partial_{x^k}[\psi^*\alpha^k\psi], \end{aligned}$$

which is exactly the charge continuity equation.

Exercise IV.7.1. Verify that the charge and current densities corresponding to the Dirac equation can be written as

$$j^\mu(x) = (\rho(x), \mathbf{j}(x)) = e\bar{\psi}(x)\gamma^\mu\psi(x), \quad \mu = 0, \ldots, 3,$$

where

$$\bar{\psi}(x) = \psi^*(x)\gamma^0$$

is the *Dirac conjugate* of $\psi(x)$ (which is not to be confused with the Hermitian conjugation $\psi^*(x)$ and with the componentwise complex conjugation $\mathbf{K}\psi(x)$ from (IV.6.8)).

IV.8   Nonrelativistic Limits

In his seminal paper [22], Dirac introduced equation (IV.6.4) for the electron in the Maxwell field and obtained the first two approximations in the non-relativistic limit as $c \to \infty$: up to terms of the order $1/c$ and of the order $1/c^2$. He discovered that, in the first approximation, the stationary Dirac equation reduces to the corresponding stationary Pauli equation, justifying the correct (maximal) Landé factor $g = 2$ for the spin. Moreover, in the next approximation, up to terms of the order $1/c^2$, there is a manifestation of the Russell–Saunders spin-orbital coupling with the correct Thomas factor $1/2$, and also of some other effects. These discoveries are regarded as a great triumph of the Dirac relativistic theory.

We consider the quantum stationary states for the Dirac equation (IV.6.4),

$$\psi(\mathbf{x}, t) = \varphi(\mathbf{x})e^{-iEt/\hbar},$$

in the case of static external potentials $\mathcal{A}(\mathbf{x}, t) = (\phi(\mathbf{x}), -\mathbf{A}(\mathbf{x}))$. First, we rewrite the Dirac equation (IV.6.4) similarly to (IV.1.17):

$$(i\hbar\partial_t - e\phi)\psi = \left(\mathrm{mc}^2\beta + c\boldsymbol{\alpha} \cdot \left[\hat{\mathbf{p}} - \frac{e}{c}\mathbf{A}\right]\right)\psi, \qquad \text{(IV.8.1)}$$

where $\phi = \phi(\mathbf{x})$, $\boldsymbol{\alpha} := (\alpha^1, \alpha^2, \alpha^3)$, and $\hat{\mathbf{p}} := -i\hbar\nabla$. Substituting the expressions (IV.1.16) for α^k and β and applying the splitting of the wave function (IV.5.3), we obtain the corresponding stationary eigenvalue problem

$$\begin{cases} (E - e\phi - \mathrm{mc}^2)\varphi_+(\mathbf{x}) = c\boldsymbol{\sigma} \cdot \left[\hat{\mathbf{p}} - \frac{e}{c}\mathbf{A}\right]\varphi_-(\mathbf{x}) \\ (E - e\phi + \mathrm{mc}^2)\varphi_-(\mathbf{x}) = c\boldsymbol{\sigma} \cdot \left[\hat{\mathbf{p}} - \frac{e}{c}\mathbf{A}\right]\varphi_+(\mathbf{x}) \end{cases}. \qquad \text{(IV.8.2)}$$

Our goal is to find the asymptotic expansion of the energy levels E in the nonrelativistic limit as $c \to \infty$. We consider these levels close to the free electron energy at rest. More precisely, we assume that

$$E \approx \mathrm{mc}^2, \qquad |e\phi(\mathbf{x})| \ll \mathrm{mc}^2. \qquad \text{(IV.8.3)}$$

The factor $E - e\phi + \mathrm{mc}^2$ in the second equation in (IV.8.2) is large. Hence, the component φ_+ is "big" compared to φ_- (the opposite conclusion would hold under the assumption $E \approx -\mathrm{mc}^2$). We can eliminate the small component φ_- from the first equation of (IV.8.2) by using the second one, getting

$$(E - e\phi - \mathrm{mc}^2)\varphi_+(\mathbf{x}) = \boldsymbol{\sigma} \cdot \left[\hat{\mathbf{p}} - \frac{e}{c}\mathbf{A}\right] \frac{c^2}{E - e\phi + \mathrm{mc}^2} \boldsymbol{\sigma} \cdot \left[\hat{\mathbf{p}} - \frac{e}{c}\mathbf{A}\right]\varphi_+(\mathbf{x}).$$
$$\text{(IV.8.4)}$$

Setting $E_0 = E - mc^2$, we are going to study the asymptotics of E_0 for large c. As a preliminary step, we expand the factor in the right-hand side of (IV.8.4):

$$\frac{c^2}{E - e\phi + mc^2} = \frac{c^2}{E_0 - e\phi + 2mc^2} = \frac{1}{2m} \frac{2mc^2}{2mc^2 + E_0 - e\phi}$$

$$= \frac{1}{2m} \left[1 - \frac{E_0 - e\phi}{2mc^2} + \cdots \right]. \qquad \text{(IV.8.5)}$$

IV.8.1 Order $1/c$

Keeping only the leading term in (IV.8.5), we find from (IV.8.4) that

$$(E_0 - e\phi)\varphi_+ = \frac{1}{2m} \left(\boldsymbol{\sigma} \cdot \left[\hat{\mathbf{p}} - \frac{e}{c}\mathbf{A} \right] \right)^2 \varphi_+, \qquad \text{(IV.8.6)}$$

neglecting the terms of the order $1/c^2$. Let us evaluate the right-hand side.

Lemma IV.8.1. *There is the identity*

$$\left(\boldsymbol{\sigma} \cdot \left[\hat{\mathbf{p}} - \frac{e}{c}\mathbf{A} \right] \right)^2 = \left[\hat{\mathbf{p}} - \frac{e}{c}\mathbf{A} \right]^2 - \frac{e}{c}\hbar\boldsymbol{\sigma} \cdot \mathbf{B}, \qquad \text{(IV.8.7)}$$

where $\mathbf{B} := \operatorname{curl} \mathbf{A}$ *is the magnetic field.*

Proof. The identities for the Pauli matrices (III.3.6) imply

$$\left(\boldsymbol{\sigma} \cdot \left[\hat{\mathbf{p}} - \frac{e}{c}\mathbf{A} \right] \right)^2 = \left[\hat{\mathbf{p}} - \frac{e}{c}\mathbf{A} \right] \cdot \left[\hat{\mathbf{p}} - \frac{e}{c}\mathbf{A} \right] + i\boldsymbol{\sigma} \cdot \left(\left[\hat{\mathbf{p}} - \frac{e}{c}\mathbf{A} \right] \times \left[\hat{\mathbf{p}} - \frac{e}{c}\mathbf{A} \right] \right).$$
$$\text{(IV.8.8)}$$

Note that the cross product of $\hat{\mathbf{p}} - \frac{e}{c}\mathbf{A}$ with itself does not vanish since the components do not commute. For example, let us calculate the first component of this cross product:

$$\left[\hat{p}_2 - \frac{e}{c}A_2 \right]\left[\hat{p}_3 - \frac{e}{c}A_3 \right] - \left[\hat{p}_3 - \frac{e}{c}A_3 \right]\left[\hat{p}_2 - \frac{e}{c}A_2 \right] = \left[\hat{p}_2 - \frac{e}{c}A_2, \hat{p}_3 - \frac{e}{c}A_3 \right].$$
$$\text{(IV.8.9)}$$

The commutator obviously reduces to

$$-\frac{e}{c}([\hat{p}_2, A_3] + [A_2, \hat{p}_3]) = i\frac{e\hbar}{c}(\partial_2 A_3 - \partial_3 A_2) = i\frac{e\hbar}{c}(\operatorname{curl}\mathbf{A})_1 = i\frac{e\hbar}{c}B_1, \quad \text{(IV.8.10)}$$

which implies (IV.8.7). $\qquad\qquad\square$

Now (IV.8.6) becomes

$$E_0\varphi_+ = \left(\frac{1}{2m}\left[\hat{\mathbf{p}} - \frac{e}{c}\mathbf{A} \right]^2 - \frac{e\hbar}{2mc}\boldsymbol{\sigma} \cdot \mathbf{B} + e\phi \right) \varphi_+, \qquad \text{(IV.8.11)}$$

which is the stationary eigenvalue problem corresponding to the Pauli equation (III.3.7). This agreement confirms the gyroscopic ratio $g = 2$ for the spin.

Exercise IV.8.2. Check (IV.8.8)–(IV.8.10).

IV.8.2   Order $1/c^2$

Equation (IV.8.11) neglects the terms of the order $1/c^2$ in (IV.8.4). Let us now take these terms into account for the case of zero magnetic potential $\mathbf{A} = 0$.

In this case, (IV.8.11) becomes the electrostatic problem with external static potential $\phi(\mathbf{x})$ (for example, the Coulomb potential of the hydrogen nucleus). Note that all terms of order $1/c$ vanish for the zero magnetic potential. We will show that E_0, up to the order $1/c^2$, is an eigenvalue of the problem

$$
E_0\varphi(\mathbf{x}) = \left[\frac{\hat{\mathbf{p}}^2}{2\mathrm{m}} + e\phi(\mathbf{x}) \right] \varphi(\mathbf{x})
$$

$$
+ \left[-\frac{\hat{\mathbf{p}}^4}{8\mathrm{m}^3c^2} - \frac{e\hbar^2}{8\mathrm{m}^2c^2}\nabla \cdot \mathbf{E}(\mathbf{x}) - \frac{e\hbar\boldsymbol{\sigma} \cdot (\mathbf{E} \times \mathbf{p})}{4\mathrm{m}^2c^2} \right] \varphi(\mathbf{x}), \qquad \text{(IV.8.12)}
$$

where $\varphi \in L^2(\mathbb{R}^3) \otimes \mathbb{C}^2$, $\mathbf{E} = -\nabla\phi$, and $\hat{\mathbf{p}}^4 := [\hat{\mathbf{p}}^2]^2$. Comparing this with (III.4.3) shows that the spin-orbital coupling appears precisely in the Russell–Saunders form, as conjectured by Ya. Frenkel and L.H. Thomas: for the *radial potential* $\phi(|\mathbf{x}|)$, we can write

$$
-\frac{e\hbar\boldsymbol{\sigma} \cdot (\mathbf{E} \times \hat{\mathbf{p}})}{4\mathrm{m}^2c^2} = \frac{e}{2\mathrm{m}^2c^2}\frac{\phi'(|\mathbf{x}|)}{|\mathbf{x}|}\hat{\mathbf{L}} \cdot \hat{\mathbf{s}}.
$$

This justification of the Russell–Saunders coupling was a great triumph of Dirac theory. The remaining terms are discussed in [8, Section 12], [47, Section 2.2.4], [75, Section 3-3], and [76], [77]. Note that the term with $\nabla \cdot \mathbf{E}$ is known as the *Darwin term*, which is responsible for the *Zitterbewegung*, while the term with $\hat{\mathbf{p}}^4$ is known to give the relativistic correction to the electron mass.

Proof of (IV.8.12) Let us keep the second term in the right-hand side of (IV.8.5). Then (IV.8.4) (with $\mathbf{A} = 0$) gives

$$
(E_0 - e\phi)\varphi_+ = \frac{1}{2\mathrm{m}}\boldsymbol{\sigma} \cdot \hat{\mathbf{p}} \left(1 - \frac{E_0 - e\phi}{2\mathrm{m}c^2} \right) \boldsymbol{\sigma} \cdot \hat{\mathbf{p}}\,\varphi_+. \qquad \text{(IV.8.13)}
$$

Collecting the terms involving E_0, we obtain:

$$
E_0 \left(1 + \frac{\hat{\mathbf{p}}^2}{4\mathrm{m}^2c^2} \right) \varphi_+ = \frac{1}{2\mathrm{m}}\boldsymbol{\sigma} \cdot \hat{\mathbf{p}} \left(1 + \frac{e\phi}{2\mathrm{m}c^2} \right) \boldsymbol{\sigma} \cdot \hat{\mathbf{p}}\varphi_+ + e\phi\varphi_+
$$

$$
= \frac{1}{2\mathrm{m}} \left(1 + \frac{e\phi}{2\mathrm{m}c^2} \right) \hat{\mathbf{p}}^2\varphi_+ + \frac{e}{4\mathrm{m}^2c^2}(\boldsymbol{\sigma} \cdot \hat{\mathbf{p}}\,\phi)\boldsymbol{\sigma} \cdot \hat{\mathbf{p}}\,\varphi_+ + e\phi\varphi_+
$$

$$
= \frac{1}{2\mathrm{m}} \left(1 + \frac{e\phi}{2\mathrm{m}c^2} \right) \hat{\mathbf{p}}^2\varphi_+ + \frac{ie\hbar}{4\mathrm{m}^2c^2}\boldsymbol{\sigma} \cdot \mathbf{E}\ \boldsymbol{\sigma} \cdot \hat{\mathbf{p}}\,\varphi_+ + e\phi\varphi_+, \qquad \text{(IV.8.14)}
$$

where we took into account that $\boldsymbol{\sigma} \cdot \hat{\mathbf{p}}\, \phi = -i\hbar\boldsymbol{\sigma} \cdot \mathbf{E} = i\hbar\boldsymbol{\sigma} \cdot \mathbf{E}$ as a consequence of the identities $\hat{\mathbf{p}} = -ih\nabla$ and $\mathbf{E} = -\nabla\phi$. By (IV.8.3), we may neglect the term with $\frac{e\phi}{2mc^2}$. Therefore, applying an identity of type (IV.8.8), we can write:

$$E_0 \left(1 + \frac{\hat{\mathbf{p}}^2}{4m^2c^2}\right)\varphi_+$$

$$= \frac{1}{2m}\hat{\mathbf{p}}^2\varphi_+ + \frac{ie\hbar}{4m^2c^2}\left(\mathbf{E} \cdot \hat{\mathbf{p}}\,\varphi_+ + i\boldsymbol{\sigma} \cdot (\mathbf{E} \times \hat{\mathbf{p}}\,\varphi_+)\right) + e\phi\varphi_+. \qquad \text{(IV.8.15)}$$

This equation has the form $E_0 B\varphi_+ = H\varphi_+$, where $B := 1 + \hat{\mathbf{p}}^2/(4m^2c^2)$ is a positive selfadjoint operator in the Hilbert space $L^2(\mathbb{R}^3) \otimes \mathbb{C}^2$. Hence,

$$E_0\varphi = \tilde{H}\varphi, \qquad \text{with} \quad \varphi := B^{1/2}\varphi_+, \quad \tilde{H} := B^{-1/2}HB^{-1/2}.$$

Writing $B^{-1/2} = 1 - \hat{\mathbf{p}}^2/(8m^2c^2) + \dots$, we approximate (IV.8.15) by

$$E_0\varphi = \left(1 - \frac{\hat{\mathbf{p}}^2}{8m^2c^2}\right)\left\{\frac{\hat{\mathbf{p}}^2}{2m} + \frac{ie\hbar}{4m^2c^2}\left(\mathbf{E} \cdot \hat{\mathbf{p}} + i\boldsymbol{\sigma} \cdot (\mathbf{E} \times \hat{\mathbf{p}})\right) + e\phi\right\}\left(1 - \frac{\hat{\mathbf{p}}^2}{8m^2c^2}\right)\varphi.$$

Neglecting the terms of higher orders $(1/c^4, \dots)$, this gives

$$E_0\varphi = \tilde{H}_1\varphi$$

$$:= \left(\frac{\hat{\mathbf{p}}^2}{2m} + e\phi\right)\varphi + \left[-\frac{\hat{\mathbf{p}}^4}{8m^3c^2} + \frac{e\hbar}{4m^2c^2}\left(\hbar\mathbf{E} \cdot \nabla - \boldsymbol{\sigma} \cdot (\mathbf{E} \times \hat{\mathbf{p}})\right)\right]\varphi. \qquad \text{(IV.8.16)}$$

Now we have to symmetrize the operator in the right-hand side. Indeed, the operator in (IV.8.14) is selfadjoint, while the term $\mathbf{E} \cdot \hat{\mathbf{p}}$ in (IV.8.15) is not. All other terms in (IV.8.15) are selfadjoint: in particular, the term $\boldsymbol{\sigma} \cdot (\mathbf{E} \times \hat{\mathbf{p}})$ is selfadjoint since $\operatorname{curl} \mathbf{E}(\mathbf{x}) = 0$. The symmetry breaking occurred because we neglected the term with the factor $\frac{e\phi}{2mc^2}$. Restoring this term in (IV.8.15), we obtain the symmetrized version involving the operator $(\tilde{H}_1 + \tilde{H}_1^*)/2$ instead of $\tilde{H}_1$. Finally, this version coincides with (IV.8.12), since

$$(\mathbf{E} \cdot \nabla + (\mathbf{E} \cdot \nabla)^*)/2 = -\nabla \cdot \mathbf{E}/2, \qquad \text{(IV.8.17)}$$

while the restored terms can again be neglected by (IV.8.3).

Exercise IV.8.3. Check (IV.8.13)–(IV.8.17).

IV.9   The Hydrogen Spectrum

In [22], Dirac introduced equation (IV.8.1) for the electron in the Maxwell field and obtained two nonrelativistic approximations which we presented in the previous section. However, he did not consider the corresponding spectral problem for the Coulomb potential; a few months later this spectral problem was solved independently by W. Gordon [39] and C. Darwin [20]. The solution relied on the separation of variables and the factorization method in solving the corresponding radial equation; this method was applied by Schrödinger in the nonrelativistic case (see Section I.6.3). For the Dirac equation, the reduction to the radial problem is more involved, requiring an application of the Clebsch–Gordan theorem to develop the Lie algebras arguments of Section I.7.

For the hydrogen atom, the corresponding four-potential of the nucleus is the Coulomb potential as in (I.6.1), with $Z = 1$. The case $Z \neq 1$ can be considered similarly. In the *rationalized* Gaussian units (see [48, p. 781], [49]), the potential reads as

$$\mathcal{A} = (\phi, 0, 0, 0), \qquad \phi = -\frac{e}{|\mathbf{x}|}. \qquad (\text{IV}.9.1)$$

Now the corresponding Dirac equation (IV.8.1) reads as follows:

$$i\hbar \partial_t \psi = H_D \psi := [mc^2 \beta + c\boldsymbol{\alpha} \cdot \hat{\mathbf{p}} + e\phi]\psi. \qquad (\text{IV}.9.2)$$

We are going to determine all the corresponding stationary orbits for the hydrogen atom, which are solutions of the form

$$\psi = \varphi(\mathbf{x})e^{-iEt/\hbar}, \qquad E \in \mathbb{R}. \qquad (\text{IV}.9.3)$$

However, now the requirement of finite energy, as in Definition I.4.1, is inadequate, since for the Dirac equation the energy is not bounded from below. On the other hand, the charge of the state is negative definite, since, according to (IV.7.1),

$$Q := \int_{\mathbb{R}^3} \rho(\mathbf{x})\, d\mathbf{x} = e \int |\varphi(\mathbf{x})|^2\, d\mathbf{x}. \qquad (\text{IV}.9.4)$$

Definition IV.9.1. Stationary orbits *(or* quantum stationary states*) for the Dirac equation (IV.6.4) with static potentials* $\mathcal{A}(\mathbf{x})$ *are solutions of the form (IV.9.3) with* finite charge *(IV.9.4).*

Thus, we are looking for solutions $\varphi \in L^2(\mathbb{R}^3, \mathbb{C}^4)$. Substituting (IV.9.3) into the Dirac equation (IV.9.2), we arrive at the corresponding eigenvalue problem

$$E\varphi(\mathbf{x}) = [mc^2 \beta + c\boldsymbol{\alpha} \cdot \hat{\mathbf{p}} + e\phi]\varphi(\mathbf{x}). \qquad (\text{IV}.9.5)$$

The splitting (IV.5.3) reduces it to coupled equations of type (IV.8.2),

$$\begin{cases} (E - mc^2 - e\phi)\varphi_+(\mathbf{x}) = c\boldsymbol{\sigma} \cdot \hat{\mathbf{p}}\,\varphi_-(\mathbf{x}) \\ (E + mc^2 - e\phi)\varphi_-(\mathbf{x}) = c\boldsymbol{\sigma} \cdot \hat{\mathbf{p}}\,\varphi_+(\mathbf{x}) \end{cases}. \qquad (IV.9.6)$$

The key argument for the solution is the spherical symmetry of the problem. Indeed, the Coulomb potential $\phi(\mathbf{x})$ is spherically symmetric. Hence, the angular momentum (IV.4.1) is conserved just like in the free case, since

$$[\hat{\mathbf{J}}, H_D] = 0. \qquad (IV.9.7)$$

This follows from the commutation $[\hat{\mathbf{J}}, \phi] = 0$, since $H_D = H_D^0 + e\phi$ and the commutation $[\hat{\mathbf{J}}, H_D^0] = 0$ was proved in (IV.4.4).

Further calculations follow the strategy of [41]. First, we will need an expression of the operator $\boldsymbol{\sigma} \cdot \hat{\mathbf{p}}$ from (IV.9.6) in terms of the orbital angular momentum and related operators.

Lemma IV.9.2. *There are the following relations:*

$$\boldsymbol{\sigma} \cdot \hat{\mathbf{p}} = |\mathbf{x}|^{-2} \boldsymbol{\sigma} \cdot \mathbf{x}(\mathbf{x} \cdot \hat{\mathbf{p}} + i\boldsymbol{\sigma} \cdot \hat{\mathbf{L}}), \qquad (IV.9.8)$$

$$\boldsymbol{\sigma} \cdot \mathbf{x}(\boldsymbol{\sigma} \cdot \hat{\mathbf{L}} + \hbar) + (\boldsymbol{\sigma} \cdot \hat{\mathbf{L}} + \hbar)\boldsymbol{\sigma} \cdot \mathbf{x} = 0. \qquad (IV.9.9)$$

Proof. From the formulae for products of spin matrices (III.3.6), we obtain:

$$(\boldsymbol{\sigma} \cdot \mathbf{x})(\boldsymbol{\sigma} \cdot \hat{\mathbf{p}}) = \mathbf{x} \cdot \hat{\mathbf{p}} + i\boldsymbol{\sigma} \cdot (\mathbf{x} \times \hat{\mathbf{p}}) = \mathbf{x} \cdot \hat{\mathbf{p}} + i\boldsymbol{\sigma} \cdot \hat{\mathbf{L}}. \qquad (IV.9.10)$$

Multiplying this equation on the left by $\boldsymbol{\sigma} \cdot \mathbf{x}$, we arrive at (IV.9.8). Now equation (IV.9.9) follows by multiplying the commutation relations

$$[\hat{L}_i, x^j] = i\hbar\varepsilon_{ijk}x^k$$

by $\sigma_i\sigma_j = 2\delta_{ij}I_2 - \sigma_j\sigma_i$ and simplifying. $\qquad \square$

Substituting (IV.9.8) into (IV.9.6) and using (IV.9.9), we obtain:

$$\begin{cases} (E - mc^2 - e\phi)\varphi_+ = c|\mathbf{x}|^{-2}\boldsymbol{\sigma} \cdot \mathbf{x}(\mathbf{x} \cdot \hat{\mathbf{p}} + i\boldsymbol{\sigma} \cdot \hat{\mathbf{L}})\varphi_-, \\ (E + mc^2 - e\phi)\varphi_- = c|\mathbf{x}|^{-2}\boldsymbol{\sigma} \cdot \mathbf{x}(\mathbf{x} \cdot \hat{\mathbf{p}} + i\boldsymbol{\sigma} \cdot \hat{\mathbf{L}})\varphi_+. \end{cases} \qquad (IV.9.11)$$

This system can be rewritten as

$$\begin{cases} (E - mc^2 - e\phi)\varphi_+ = c|\mathbf{x}|^{-2}(\mathbf{x} \cdot \hat{\mathbf{p}} - i(\boldsymbol{\sigma} \cdot \hat{\mathbf{L}} + \hbar))\boldsymbol{\sigma} \cdot \mathbf{x}\varphi_-, \\ (E + mc^2 - e\phi)\boldsymbol{\sigma} \cdot \mathbf{x}\varphi_- = c(\mathbf{x} \cdot \hat{\mathbf{p}} + i\boldsymbol{\sigma} \cdot \hat{\mathbf{L}})\varphi_+, \end{cases} \qquad (IV.9.12)$$

suggesting the notations

$$\Phi_+(\mathbf{x}) := \varphi_+(\mathbf{x}), \qquad \Phi_-(\mathbf{x}) := \frac{\boldsymbol{\sigma}\cdot\mathbf{x}}{|\mathbf{x}|}\varphi_-(\mathbf{x}). \qquad \text{(IV.9.13)}$$

Now we apply the operator $\dfrac{\boldsymbol{\sigma}\cdot\mathbf{x}}{|\mathbf{x}|}$ to the second equation of (IV.9.11) and use the fact that the potential ϕ is radial. As a result, equations (IV.9.11) in spherical coordinates take the following form:

$$\begin{cases} (E-mc^2 - e\phi)\Phi_+ = c\left(-i\hbar\dfrac{d}{dr} - ir^{-1}(\boldsymbol{\sigma}\cdot\hat{\mathbf{L}}+2\hbar)\right)\Phi_- \\[2mm] (E+mc^2 - e\phi)\Phi_- = c\left(-i\hbar\dfrac{d}{dr} + ir^{-1}\boldsymbol{\sigma}\cdot\hat{\mathbf{L}}\right)\Phi_+. \end{cases} , \quad r>0. \quad \text{(IV.9.14)}$$

Exercise IV.9.3. Check (IV.9.7).

Exercise IV.9.4. Check (IV.9.10) and deduce (IV.9.8).

Exercise IV.9.5. Check (IV.9.12) and (IV.9.14).

IV.9.1 Spinor spherical functions*

It is worth pointing out that we have solved the spectral problem for the nonrelativistic Schrödinger equation by employing the general strategy of separation of variables (Sections I.6). Now we are going to develop this strategy for the relativistic problem (IV.9.5) in a similar way, solving the corresponding spherical eigenvalue problem by developing the Lie algebra methods from Section I.7. In this case, the role of the orbital angular momentum $\hat{\mathbf{L}}$ is played by the total angular momentum $\hat{\mathbb{J}}$ defined in (IV.4.1). Hence, the strategy should now be modified accordingly:

I. (IV.9.7) implies that the operator H_D commutes with each $\hat{\mathbb{J}}_k$ and with $\hat{\mathbb{J}}^2 := \hat{\mathbb{J}}_1^2 + \hat{\mathbb{J}}_2^2 + \hat{\mathbb{J}}_3^2$:

$$[H_D, \hat{\mathbb{J}}_k] = 0 = [H_D, \hat{\mathbb{J}}^2]. \qquad (IV.9.15)$$

Hence, each eigenspace of the Dirac operator H_D is invariant with respect to each operator $\hat{\mathbb{J}}_k$ and $\hat{\mathbb{J}}^2$. Moreover, the commutation relations (III.3.24) imply that the operator $\hat{\mathbb{J}}^2$ also commutes with each operator $\hat{\mathbb{J}}_k$, similarly to (I.6.10). In particular,

$$[\hat{\mathbb{J}}^2, \hat{\mathbb{J}}_3] = 0. \qquad (IV.9.16)$$

II. The above arguments suggest that there is a basis of common eigenfunctions for the mutually commuting operators H_D, $\hat{\mathbb{J}}_3$, and $\hat{\mathbb{J}}^2$. In this section, we shall simultaneously diagonalize $\hat{\mathbb{J}}^2$ and $\hat{\mathbb{J}}_3$.

The finiteness of the charge (IV.9.4) means that we consider the eigenvalue problem (IV.9.5) in the Hilbert space $\mathcal{E} := L^2(\mathbb{R}^3) \otimes \mathbb{C}^4$. On the other hand, both operators $\hat{\mathbb{J}}_3$ and $\hat{\mathbb{J}}^2$ act only on spinor variables and angular variables in the spherical coordinates. Hence, the operators also act in the Hilbert space $\mathcal{E}_1 := L^2(\mathbb{S}^2) \otimes \mathbb{C}^4$, where $\mathbb{S}^2$ denotes the two-dimensional sphere $|\mathbf{x}| = 1$. We split the space $\mathcal{E}_1$ into the sum $\mathcal{E}_1 = \mathcal{E}_1^+ \oplus \mathcal{E}_1^-$ corresponding to the two-component representation (IV.5.3), where $\mathcal{E}_1^{\pm} = L^2(\mathbb{S}^2) \otimes \mathbb{C}^2$. By definition (IV.4.1), both operators $\hat{\mathbb{J}}_3$ and $\hat{\mathbb{J}}^2$ have the block form

$$\hat{\mathbb{J}}_3 = \begin{pmatrix} \hat{J}_3 & 0 \\ 0 & \hat{J}_3 \end{pmatrix}, \qquad \hat{\mathbb{J}}^2 = \begin{pmatrix} \hat{\mathbf{J}}^2 & 0 \\ 0 & \hat{\mathbf{J}}^2 \end{pmatrix}. \qquad (IV.9.17)$$

Hence, each space $\mathcal{E}_1^{\pm}$ is invariant with respect to both $\hat{\mathbb{J}}_3$ and $\hat{\mathbb{J}}^2$. Moreover, the action of $\hat{\mathbb{J}}_3$ on $\mathcal{E}_1^{\pm}$ coincides with $\hat{J}_3$, while the action of $\hat{\mathbb{J}}^2$ on $\mathcal{E}_1^{\pm}$ coincides with $\hat{\mathbf{J}}^2$. The eigenfunction expansion for these operators is given by the following particular version of the Clebsch–Gordan theorem [34, 37, 89]:

---

*This section can be skipped on the first reading.

Proposition IV.9.6. i) *In the space $L^2(\mathbb{S}^2) \otimes \mathbb{C}^2$, there exists an orthonormal basis of spinor spherical functions $\mathcal{S}_{j,k}(\theta, \varphi)$, which are common eigenfunctions of $\hat{J}_3$ and $\hat{\mathbf{J}}^2$:*

$$\hat{J}_3 \mathcal{S}_{j,k}(\theta, \varphi) = \hbar k \mathcal{S}_{j,k}(\theta, \varphi), \quad \hat{\mathbf{J}}^2 \mathcal{S}_{j,k}(\theta, \varphi) = \hbar^2 j(j+1) \mathcal{S}_{j,k}(\theta, \varphi), \quad \text{(IV.9.18)}$$

where $j = 1/2, 3/2, \ldots$ and $k = -j, -j+1, \ldots, j$.

ii) *The space of solutions to (IV.9.18) is two-dimensional for any fixed j, k.*

For the proof, let us first note that the operator $\hat{\mathbf{J}} = \hat{\mathbf{L}} + \hat{\mathbf{s}}$ acts on the space $L^2(\mathbb{S}^2) \otimes \mathbb{C}^2$, with the operator $\hat{\mathbf{L}}$ acting on $L^2(\mathbb{S}^2)$ and with $\hat{\mathbf{s}}$ acting on the second factor, $\mathbb{C}^2$. As a result, $\hat{\mathbf{L}}$ commutes with $\hat{\mathbf{s}}$. Hence,

$$\hat{\mathbf{J}}^2 = \hat{\mathbf{L}}^2 + 2\hat{\mathbf{L}} \cdot \hat{\mathbf{s}} + (3/4)\hbar^2$$

is a selfadjoint differential second-order elliptic operator on the sphere $\mathbb{S}^2$. Therefore, $L^2(\mathbb{S}^2) \otimes \mathbb{C}^2$ is an orthogonal sum of *finite-dimensional eigenspaces* of $\hat{\mathbf{J}}^2$; see [161]. Hence, Proposition I.7.7 is applicable to these eigenstates, so the commutation relations (III.3.24) imply formulae (IV.9.18) for the corresponding eigenvalues.

Let us check that only half-integer spins J are possible, with multiplicity two for $j \geq 3/2$ and multiplicity one for $j = 1/2$.

First, we know the spectral decomposition (I.7.1) of the operator $\hat{\mathbf{L}}^2$ in the space $L^2(\mathbb{S}^2)$:

$$L^2(\mathbb{S}^2) = \oplus_{l=0}^{\infty} L(l). \quad \text{(IV.9.19)}$$

Here $L(l)$ are the finite-dimensional orthogonal eigenspaces of $\hat{\mathbf{L}}^2$ corresponding to the eigenvalues

$$\hbar^2 l(l+1), \quad l = 0, 1, \ldots.$$

Second, we will use the structure of the rotation-invariant irreducible spaces $L(l)$, as established in the proof of Proposition I.7.7 and Lemma I.7.14. Indeed, in $L(l)$, there is an orthonormal basis $e_{-l}, \ldots, e_l$, where $e_m = \hat{S}_+^{m+l} e_{-l}$ (here $\hat{S}_+ := \hat{S}_1 + i\hat{S}_2$ with $\hat{S}_k := \hbar^{-1}\hat{L}_k$) and (cf. (IV.9.18))

$$\hat{L}_3 e_{lm} = \hbar m e_{lm}, \quad \hat{\mathbf{L}}^2 e_{lm} = \hbar^2 l(l+1) e_{lm}, \quad m = -l, \ldots, l. \quad \text{(IV.9.20)}$$

We recall that $e_{lm} = Y_l^m$, where $Y_l^m(\theta, \varphi)$ are the spherical harmonics (I.6.20). Similarly, in $\mathbb{C}^2$, there is an orthonormal basis $f_{-1/2}, f_{1/2}$, where $f_{1/2} = \hat{s}_+ f_{-1/2}$ and

$$\hat{s}_3 f_s = \hbar s f_s, \quad \hat{s}^2 f_s = \hbar^2 s(s+1) f_s, \quad s = -1/2, 1/2. \quad \text{(IV.9.21)}$$

Here $f_{-1/2} = \begin{pmatrix} 0 \\ 1 \end{pmatrix}$ and $f_{1/2} = \begin{pmatrix} 1 \\ 0 \end{pmatrix}$. Therefore, we have

$$L^2(\mathbb{S}^2) \otimes \mathbb{C}^2 = \oplus_{l=0}^{\infty} \mathcal{F}(l), \qquad \mathcal{F}(l) := L(l) \otimes \mathbb{C}^2. \tag{IV.9.22}$$

The tensor products $e_{lm} \otimes f_s$ with $m = -l, \ldots, l$ and $s = -1/2, 1/2$ form an orthonormal basis in $\mathcal{F}(l)$, so the dimension of $\mathcal{F}(l)$ is $4l+2$. Each space $\mathcal{F}(l)$ is invariant with respect to all operators $\hat{J}_n = \hat{L}_n + \hat{s}_n$ since $L(l)$ is invariant with respect to all operators $\hat{L}_n$ by Lemma I.7.4. Relations (IV.9.20), (IV.9.21) imply that

$$\hat{J}_3[e_{lm} \otimes f_s] = \hbar(m+s)e_{lm} \otimes f_s. \tag{IV.9.23}$$

So, only half-integer spins j are possible, and the multiplicity of the eigenvalue of $\hat{J}_3$ in the subspace $\mathcal{F}(l)$ is two if $|m+s| \leq l-1/2$ and one if $|m+s| = l+1/2$.

However, in general, the vectors $e_{lm} \otimes f_s$ are not eigenvectors for $\hat{\mathbf{J}}^2$. It turns out that $\hat{\mathbf{J}}^2$ has two different eigenvalues in $\mathcal{F}(l)$ if $l \geq 1$.

Lemma IV.9.7. i) *For $l = 0, 1, \ldots$, the space $\mathcal{F}(l)$ is the orthogonal sum of two eigenspaces $\mathcal{F}_{\pm}(l)$ of the operator $\hat{\mathbf{J}}^2$:*

$$\mathcal{F}(l) = \mathcal{F}_+(l) \oplus \mathcal{F}_-(l), \qquad \hat{\mathbf{J}}^2|_{\mathcal{F}_{\pm}(l)} = \hbar^2 j_{\pm}(j_{\pm}+1), \tag{IV.9.24}$$

where $j_{\pm} = j_{\pm}(l) := l \pm 1/2$ and $\dim \mathcal{F}_{\pm}(l) = 2j_{\pm}+1$.
ii) *For $l \geq 1$, in the space $\mathcal{F}_{\pm}(l)$ there exists a basis of functions*

$$\mathcal{S}_{j_{\pm},k}^{\pm}, \qquad k = -j_{\pm}, -j_{\pm}+1, \ldots, j_{\pm}, \tag{IV.9.25}$$

which satisfy equations (IV.9.18) with $j = j_{\pm}$.
iii) *For $l = 0$, one has $\mathcal{F}_-(0) = 0$, while in the space $\mathcal{F}_+(0)$ there is a basis of functions*

$$\mathcal{S}_{\frac{1}{2},k}^{+}, \qquad k = -1/2, 1/2 \tag{IV.9.26}$$

satisfying equations (IV.9.18) with $j = 1/2$.

Proof. The space $\mathcal{F}(l)$ is invariant with respect to all operators $\hat{J}_n$. Hence, $\mathcal{F}(l)$ is the finite sum of the orthogonal eigenspaces $\mathcal{F}_j(l)$ of the selfadjoint operator $\hat{\mathbf{J}}^2$ with the corresponding eigenvalues $\hbar^2 j(j+1)$. Finally, by Proposition I.7.7, each eigenspace $\mathcal{F}_j(l)$ is the finite sum of the irreducible subspaces isomorphic to $D(j)$ from Definition I.7.10. Each of these irreducible subspaces is generated by the vectors

$$\mathcal{S}_{j,k} = \hat{J}_+^{j+k}\mathcal{S}_{j,-j}, \qquad k = -j, -j+1, \ldots, j, \tag{IV.9.27}$$

where $\hat{J}_+ = \hat{J}_1 + i\hat{J}_2$ and $\hat{J}_3\mathcal{S}_{j,k} = \hbar k\mathcal{S}_{j,k}$. In particular, the dimension of $D(j)$ is $2j+1$, and $\mathcal{S}_{j,-j}$ is the eigenvector of $\hat{J}_3$ with the minimal eigenvalue $-\hbar j$ in this irreducible subspace. Note that

$$\mathcal{S}_{j_+,-j_+} := e_{-l} \otimes f_{-1/2}$$

is one of such vectors since it corresponds to the minimal eigenvalue

$$-\hbar(l+1/2) = -\hbar j_+$$

of $\hat{J}_3$ in the space $\mathcal{F}(l)$ in view of (IV.9.23). Therefore, we choose the space $\mathcal{F}_+(l)$ with the basis $\mathcal{S}_{j,k}^+$ defined as the right-hand side of (IV.9.27) with $j = l + 1/2 = j_+$. The dimension of this space is $2j_+ + 1 = 2l + 2$.

In the case $l = 0$, we have $\mathcal{F}(l) = \mathcal{F}_+(l)$, since the dimensions of both spaces are equal to 2. In the case $l \geq 1$, the orthogonal complement

$$\mathcal{F}_-(l) := \mathcal{F}(l) \ominus \mathcal{F}_+(l)$$

is a $\hat{J}_3$-invariant subspace of dimension $4l + 2 - 2l - 2 = 2l$, and it contains a nonzero eigenvector $\mathcal{S}_{j_-,-j_-}$ of $\hat{J}_3$ with eigenvalue $-\hbar(l - 1/2) = -\hbar j_-$ (see Exercise IV.9.10 below). Hence, the minimal eigenvalue of $\hat{J}_3$ in $\mathcal{F}_-(l)$ is $-\hbar(l - 1/2)$. This implies that $\mathcal{F}_-(l)$ is isomorphic to $D(l - 1/2)$ by Proposition I.7.7, since the dimension of $\mathcal{F}_-(l)$ is $2l = 2j_- + 1$. Thus, the basis $\mathcal{S}_{j,k}^-$ is defined as the right-hand side of (IV.9.27) with $j = l - 1/2 = j_-$. □

Proof of Proposition IV.9.6. It suffices to note that for all $l = 0, 1 \ldots$, we have $j_+(l) = j_-(l + 1)$. Hence, for any $j = 1/2, 3/2, \ldots$ and $k = -j, \ldots, j$, the eigenfunctions $\mathcal{S}_{j,k}^+$ and $\mathcal{S}_{j+1,k}^-$, as defined above, are orthogonal and satisfy identical equations (IV.9.18) with the same parameters j and k. □

It is worth noting that $\mathcal{F}_\pm(l)$ are also eigenspaces for $\boldsymbol{\sigma} \cdot \hat{\mathbf{L}}$; indeed, we have the following result.

Lemma IV.9.8. $\boldsymbol{\sigma} \cdot \hat{\mathbf{L}}$ *takes the value* $\hbar l$ *(respectively,* $-\hbar(l + 1)$*) when restricted onto the space* $\mathcal{F}_+(l)$ *(respectively, onto* $\mathcal{F}_-(l)$*).*

Proof. This follows from the second equation of (IV.9.18) with $j = j_\pm$ due to the identity

$$\hbar \boldsymbol{\sigma} \cdot \hat{\mathbf{L}} = \left(\hat{\mathbf{L}} + \frac{1}{2}\hbar\boldsymbol{\sigma}\right)^2 - \hat{\mathbf{L}}^2 - \frac{1}{4}\hbar^2\boldsymbol{\sigma}^2 = \hat{\mathbf{J}}^2 - \hat{\mathbf{L}}^2 - \frac{3}{4}\hbar^2, \qquad \text{(IV.9.28)}$$

since $j_\pm(j_\pm + 1) - l(l + 1) - 3/4$ is either l or $-(l + 1)$. □

Exercise IV.9.9. Check (IV.9.23).

Exercise IV.9.10. Prove that in the case $l \geq 1$ the orthogonal complement $\mathcal{F}_-(l) := \mathcal{F}(l) \ominus \mathcal{F}_+(l)$ contains a nonzero eigenvector v of $\hat{J}_3$ with eigenvalue $-\hbar(l - 1/2) = -\hbar j_-$.
Hints: i) the corresponding eigenspace of $\hat{J}_3$ in the space $\mathcal{F}(l)$ is two-dimensional, so v must be chosen as a vector orthogonal to $\mathcal{S}_{j_+,-j_-}^+$ in this eigenspace; ii) all eigenspaces of the selfadjoint operator $\hat{J}_3$ corresponding to different eigenvalues are mutually orthogonal.

Exercise IV.9.11. Check (IV.9.28). **Hint:** note that $\sigma_k^2 = 1$.

IV.9.2 Separation of variables

Now we can solve the coupled equations (IV.9.14), eliminating one of the components. The corresponding radial equation gives the energy levels by a suitable development of the Sommerfeld method of factorization from Section I.6.3. We will see that the energy levels, in contrast to the nonrelativistic case, depend on the angular momentum. This dependence perfectly explains Sommerfeld's "fine structure" of the spectrum.

Proposition IV.9.6 suggests that we can construct eigenfunctions of the Dirac operator H_D by separation of variables. Indeed, we will seek solutions to (IV.9.14) in the form

$$\begin{pmatrix} \Phi_+ \\ \Phi_- \end{pmatrix} = \begin{pmatrix} R_+^+(r)\mathcal{S}_{j,k}^+(\theta, \varphi) + R_+^-(r)\mathcal{S}_{j,k}^-(\theta, \varphi) \\ R_-^+(r)\mathcal{S}_{j,k}^+(\theta, \varphi) + R_-^-(r)\mathcal{S}_{j,k}^-(\theta, \varphi) \end{pmatrix}, \quad r > 0, \qquad \text{(IV.9.29)}$$

where $\mathcal{S}_{j,k}^+$ and $\mathcal{S}_{j,k}^-$ are arbitrary solutions of (IV.9.18). Each solution of the spectral problem (IV.9.14) is a finite sum of solutions of the particular form (IV.9.29), since the functions $\mathcal{S}_{j,k}^\pm$ constitute a basis in $\mathcal{E}_1$.

Let us construct all nonzero solutions of (IV.9.14) in the form (IV.9.29). For definiteness, let us assume that

$$R(r) := \begin{pmatrix} R_+^+(r) \\ R_-^+(r) \end{pmatrix} \not\equiv 0. \qquad \text{(IV.9.30)}$$

Other cases can be considered similarly (see Exercise IV.9.15). Let us substitute the expressions (IV.9.29) into (IV.9.14). Collecting the terms with $\mathcal{S}_{j,k}^+$ and using the fact that ϕ is spherically symmetric, we obtain the system

$$\begin{cases} (E - mc^2 - e\phi)R_+^+(r) = c\left(-i\hbar\dfrac{d}{dr} - ir^{-1}(\hbar l + 2\hbar)\right) R_-^+(r) \\[4mm] (E + mc^2 - e\phi)R_-^+(r) = c\left(-i\hbar\dfrac{d}{dr} + ir^{-1}\hbar l\right) R_+^+(r) \end{cases}. \qquad \text{(IV.9.31)}$$

Above, we used Lemma IV.9.8 with

$$l = j - \frac{1}{2} \qquad \text{(IV.9.32)}$$

since $\mathcal{S}_{j,k}^+$ is defined by (IV.9.25) with $j_+ = j$, and, moreover, $j_+ = l + 1/2$. The system (IV.9.31) is equivalent to the following equation for the *vector function* (IV.9.30):

$$(E - e\phi - mc^2\sigma_3)R(r) = -ic\hbar\left[\left(\frac{d}{dr} + \frac{1}{r}\right) + \frac{(l+1)}{r}\sigma_3\right]\sigma_1 R(r), \quad \text{(IV.9.33)}$$

where σ_k are the Pauli matrices (I.7.11). Finally,

$$(E - e\phi - mc^2\sigma_3)R(r) = -ic\hbar \left[\left(\frac{d}{dr} + \frac{1}{r}\right)\sigma_1 + i\frac{(l+1)}{r}\sigma_2\right] R(r), \quad (IV.9.34)$$

because $\sigma_3\sigma_1 = i\sigma_2$.

Exercise IV.9.12. Check (IV.9.31) and (IV.9.33).

IV.9.3   Factorization method

As in the nonrelativistic case, we substitute $R(r) = e^{-\varkappa r}P(r)$ with an unknown parameter $\varkappa > 0$ to be chosen later. Now equation (IV.9.34) reduces to

$$(E - e\phi - mc^2\sigma_3)P(r) = -ic\hbar \left(\left(\frac{d}{dr} + \frac{1}{r} - \varkappa\right)\sigma_1 + i\frac{(l+1)}{r}\sigma_2\right) P(r), \quad (IV.9.35)$$

or, equivalently,

$$(E - mc^2\sigma_3 - ic\hbar\varkappa\sigma_1)P(r) = -ic\hbar \left(\left(\frac{d}{dr} + \frac{1}{r}\right)\sigma_1 + i\frac{(l+1)}{r}\sigma_2 + \frac{ie\phi}{c\hbar}\right) P(r).$$
$$(IV.9.36)$$

Denote the matrix

$$M := \frac{i}{c\hbar}(E - mc^2\sigma_3 - ic\hbar\varkappa\sigma_1), \quad (IV.9.37)$$

and rewrite the Coulomb potential as $e\phi = -c\hbar\alpha/r$, where

$$\alpha := \frac{e^2}{c\hbar} \approx \frac{1}{137}$$

is the dimensionless Sommerfeld *fine-structure constant*. Let us seek (and find) the solution in the form

$$P(r) = r^\delta \sum_0^\infty P_k r^k, \quad (IV.9.38)$$

where we can assume that $P_0 \neq 0$ for a nontrivial solution (IV.9.30). Hence, we must choose $\varkappa > 0$ and $\delta > -3/2$ to ensure the finiteness of the charge (IV.9.4). Substituting (IV.9.38) into (IV.9.36), we obtain the equation

$$\sum_0^\infty r^{k+\delta}MP_k = \sum_0^\infty \left((k+\delta+1)\sigma_1 + i(l+1)\sigma_2 - i\alpha\right)r^{k+\delta-1}P_k. \quad (IV.9.39)$$

This gives the recurrence equations

$$MP_{k-1} = \left((k+\delta+1)\sigma_1 + i(l+1)\sigma_2 - i\alpha\right)P_k, \qquad k = 0, 1, \ldots, \quad (IV.9.40)$$

similar to (I.6.41). As in Section I.6.3, the charge (IV.9.4) is finite only if the series (IV.9.38) terminate; see the details in Section 202 of [32]. Hence,

$$P(r) = r^\delta \sum_0^n P_k r^k \qquad (IV.9.41)$$

with some $n < \infty$ and $P_n \neq 0$. To calculate δ, note that (IV.9.40) with $k = 0$ gives

$$\big((\delta + 1)\sigma_1 + i(l + 1)\sigma_2 - i\alpha\big)P_0 = 0. \qquad (IV.9.42)$$

However, $P_0 \neq 0$, and hence

$$\det\big((\delta + 1)\sigma_1 + i(l + 1)\sigma_2 - i\alpha\big) = 0. \qquad (IV.9.43)$$

This is equivalent to

$$(\delta + 1)^2 = (l + 1)^2 - \alpha^2. \qquad (IV.9.44)$$

Since α is small, one has $|\delta + 1| \approx l + 1$. Therefore, we must choose a positive value for $\delta + 1$, since for a negative value we would obtain $-\delta \approx l + 2 \geq 2$ while $\delta > -3/2$.

It remains to calculate the eigenvalue E from the recurrence equations (IV.9.40). First, taking $k - 1 = n$, we obtain

$$MP_n = 0. \qquad (IV.9.45)$$

However, $P_n \neq 0$, and hence

$$\det M = \det(E - mc^2\sigma_3 - ic\hbar\varkappa\sigma_1) = 0. \qquad (IV.9.46)$$

This is equivalent to

$$c^2\hbar^2\varkappa^2 = m^2c^4 - E^2 \qquad (IV.9.47)$$

and so, in particular, $E < mc^2$. This equation alone is not capable of determining the eigenvalues E since we have an additional unknown parameter $\varkappa$. Therefore, we need an additional equation. We derive it from the same recurrence equation (IV.9.40), taking $k = n$:

$$MP_{n-1} = \big((n + \delta + 1)\sigma_1 + i(l + 1)\sigma_2 - i\alpha\big)r^{n+\delta-1}P_n. \qquad (IV.9.48)$$

Indeed, the characteristic equation for the matrix M gives

$$(M - 2iE/c\hbar)M = 0. \qquad (IV.9.49)$$

Therefore, multiplying both sides of (IV.9.48) by $M - 2iE/c\hbar$, we see that

$$0 = (M - 2iE/c\hbar)\big((n + \delta + 1)\sigma_1 + i(l + 1)\sigma_2 - i\alpha\big)P_n. \qquad (IV.9.50)$$

Now equation (IV.9.45) suggests that in (IV.9.50) we move to the right the matrix $M - 2iE/c\hbar$. Hence, we need to calculate the commutation relations

$$(M - 2iE/c\hbar)\sigma_2 = -\sigma_2 M, \qquad (M - 2iE/c\hbar)\sigma_1 = -\sigma_1 M + 2\varkappa \quad \text{(IV.9.51)}$$

which follow from (III.3.6). Applying these relations in (IV.9.50) and using (IV.9.45), we obtain the new desired quantization condition

$$\alpha E = c\hbar\varkappa(n + \delta + 1). \tag{IV.9.52}$$

Hence $E > 0$, and so, substituting $c\hbar\varkappa = \alpha E/(n + \delta + 1)$ into (IV.9.47), we find the eigenvalues

$$E = E_{nj} = \frac{mc^2}{\sqrt{1 + \alpha^2/(n + \delta + 1)^2}}, \qquad n = 0, 1, \ldots, \tag{IV.9.53}$$

where $\delta = \delta(l) > -1$ with $l = j - 1/2$ by (IV.9.44) and (IV.9.32).

Comparison with the nonrelativistic case. Since α is small, we can approximate the eigenvalues by the binomial expansion:

$$E_{nj} \approx mc^2 - \frac{mc^2\alpha^2}{2(n + \delta + 1)^2}. \tag{IV.9.54}$$

This approximation with $\delta = 0$ coincides with the nonrelativistic spectrum (I.6.5) of the hydrogen atom up to the (unessential) additive constant mc^2. The relativistic formula depends on the angular momentum j through $\delta = \delta(l)$ by (IV.9.44) and (IV.9.32), while the nonrelativistic formula is independent of the angular momentum. This was another triumph of the Dirac theory: the agreement with the experimental observation of the *fine structure* of the spectrum.

Exercise IV.9.13. Prove (IV.9.49). **Hint:** $\det M = 0$ and $\operatorname{tr} M = 2iE/c\hbar$.

Exercise IV.9.14. Check (IV.9.51) and (IV.9.52).

Exercise IV.9.15. Calculate the eigenvalues corresponding to eigenfunctions of type (IV.9.29) with

$$R(r) := \begin{pmatrix} R_+^-(r) \\ R_-^-(r) \end{pmatrix} \not\equiv 0. \tag{IV.9.55}$$

Chapter V

Quantum Postulates and Attractors

The goal of this chapter is to discuss the problem of giving a mathematical interpretation of the basic postulates (or "principles") of quantum theory:

A. Transitions between quantum stationary orbits (N. Bohr, 1913);

B. Wave-Particle duality (L. de Broglie, 1923);

C. Probabilistic interpretation (M. Born, 1927).

The problem concerns the validity of these postulates in Schrödinger's quantum mechanics and it still remains an open problem. These and other questions have been frequently addressed in the 1920s and 1930s in heated discussions by N. Bohr, E. Schrödinger, A. Einstein, and others [10]. However, a satisfactory solution was not achieved, and a rigorous dynamical interpretation of these postulates is still unknown. This lack of theoretical clarity hinders a further progress in the theory (e.g., in superconductivity and in nuclear reactions) and in numerical simulation of many engineering processes (e.g., laser radiation and quantum amplifiers), since a computer can solve dynamical equations, but cannot take into account postulates.

In this chapter, we discuss possible relation of these postulates to the theory of attractors of Hamiltonian nonlinear PDEs in the context of semiclassical self-consistent Maxwell–Schrödinger equations, applying a novel general conjecture on global attractors of G-invariant nonlinear Hamiltonian partial differential equations with a Lie symmetry group G. This conjecture was inspired by a number of results on global attractors of nonlinear Hamiltonian PDEs obtained since 1990 for several model equations with three basic symmetry groups: the trivial group, the group of translations, and the unitary group $U(1)$. We briefly present these results. However, for the Maxwell–Schrödinger equations, the justification of global attraction still remains an open problem.

V.1   Quantum Jumps

In this section we discuss possible dynamical interpretation of Bohr's postulates.

V.1.1   Quantum jumps as global attraction

The perturbation arguments of Section I.4.2 suggest that Bohr's postulate (I.4.3) can be treated as the following long-time asymptotics *for all finite-energy solutions* of the Maxwell–Schrödinger equations (I.5.1) in the case of static external potentials (I.3.21):

$$(\mathbf{A}(\mathbf{x}, t),\ \mathbf{\Pi}(\mathbf{x}, t),\ \psi(\mathbf{x}, t)) \sim (\mathbf{A}_{\pm}(\mathbf{x}),\ 0,\ e^{-i\omega_{\pm}t}\varphi_{\pm}(\mathbf{x})), \qquad t \to \pm\infty. \quad (\text{V.1.1})$$

We conjecture that these long-time asymptotics hold in the H^1-norm on every bounded region of $\mathbb{R}^3$. However, a rigorous justification is still missing.

Experiments show that the transition time of quantum jumps (I.4.3) is of the order of 10^{-8} s, although the asymptotics (V.1.1) require infinite time. We suppose that this discrepancy can be explained by the following arguments:

i) 10^{-8} s is the transition time between very small neighborhoods of initial and final states;

ii) during this period of time, the atom emits an overwhelming part of the radiated energy.

The asymptotics (V.1.1) have not been proved for the Maxwell–Schrödinger system (I.5.1). On the other hand, similar asymptotics are now proved for a number of model Hamiltonian nonlinear PDEs with the symmetry group $U(1)$; see Section VI.4. In the next section, we state a general conjecture which reduces to the asymptotics (V.1.1) in the case of the Maxwell–Schrödinger system.

Definition V.1.1. Stationary orbits *of the Maxwell–Schrödinger nonlinear system (I.5.1) are finite energy solutions of equations (I.5.4) of the form*

$$(\mathbf{A}(\mathbf{x}, t),\ \psi(\mathbf{x}, t)) = (\mathbf{A}(\mathbf{x}),\ e^{-i\omega t}\varphi(\mathbf{x})). \qquad (\text{V.1.2})$$

The existence of stationary orbits for the system (I.5.1) was proved in [130] in the case

$$\mathbf{A}^{\text{ext}}(\mathbf{x}) \equiv 0, \qquad A_0^{\text{ext}}(\mathbf{x}) = -\frac{eZ}{|\mathbf{x}|}, \qquad \int |\varphi(\mathbf{x})|^2\, d\mathbf{x} \le Z. \qquad (\text{V.1.3})$$

The asymptotics (V.1.1) means that there is a *global attraction* to the set of stationary orbits. We suggest that a similar attraction takes place for the Maxwell–Dirac, Maxwell–Yang–Mills, and other coupled equations. In other words, we suggest the interpretation of quantum stationary states as the points that constitute the *global attractor* of the corresponding quantum dynamical equations.

V.1.2 Einstein–Ehrenfest's paradox. Bifurcation of attractors

The Stern–Gerlach experiment demonstrates an almost instantaneous reorientation of the atomic magnetic moment when the electron enters the inhomogeneous magnetic field, with the relaxation time $\sim 10^{-8}$ s. On the other hand, the Old Quantum Theory gives a much larger relaxation time [28]. In the linear Schrödinger theory, this phenomenon also did not find a satisfactory explanation.

On the other hand, this almost instantaneous orientation is exactly in line with the asymptotics of type (V.1.1) for solutions of the coupled nonlinear Maxwell–Pauli equations (III.3.8). First, let us consider the Pauli equation from (III.3.8) with

$$\mathbf{A}(\mathbf{x},t) \equiv 0, \quad A^0(\mathbf{x},t) \equiv 0, \quad \mathbf{A}^{\text{ext}}(\mathbf{x},t) \equiv \frac{1}{2}B(-x^2, x^1, 0), \quad A_0^{\text{ext}}(\mathbf{x},t) \equiv -\frac{e^2 Z}{|\mathbf{x}|},$$

where the external magnetic potential $\mathbf{A}^{\text{ext}}(\mathbf{x},t)$ corresponds to the uniform magnetic field $\mathbf{B}(\mathbf{x}) = (0,0,B)$. Then we obtain the linear Pauli equation (III.3.3):

$$i\hbar\dot\psi(\mathbf{x},t) = H_P\psi(\mathbf{x},t), \quad H_P = -\frac{\hbar^2}{2\mathrm{m}}\Delta - \frac{e^2 Z}{|\mathbf{x}|} - \Omega_\Lambda(\hat{L}_3 + 2\hat{s}_3). \quad (\text{V.1.4})$$

In the case $B = 0$, according to (III.3.4), each eigenvalue of H_P is of even multiplicity. In particular, the minimal eigenvalue E_1 is of multiplicity two, which implies that the set $\mathcal{G}$ of the corresponding ground states satisfying the normalization condition (I.3.8) is isomorphic to the three-dimensional sphere $\mathbb{S}^3 = \{(C_1, C_2) \in \mathbb{C}^2 : |C_1|^2 + |C_2|^2 = 1\}$:

$$\mathcal{G} = \{C_1\varphi_1(\mathbf{x}) + C_2\varphi_2(\mathbf{x}) : \ (C_1, C_2) \in \mathbb{S}^3\}, \quad (\text{V.1.5})$$

where φ_1 and φ_2 are some orthonormal eigenfunctions which are not uniquely defined. The corresponding solutions of minimal energy to equation (V.1.4) read as

$$\psi(\mathbf{x},t) = \varphi(\mathbf{x})e^{-i\omega t}, \qquad \varphi \in \mathcal{G}, \quad \omega = E_1/\hbar. \quad (\text{V.1.6})$$

The attractor of these solutions is the entire set $\mathcal{G}$.

On the other hand, when the magnetic field B is turned on, then, according to (III.3.4), the minimal eigenvalue E_1 splits into two different values

$$E_1^{\pm} = E_1 \pm \hbar\Omega_\Lambda.$$

Then the solutions (V.1.6) bifurcate into

$$\psi(\mathbf{x},t) = C^+\varphi^+(\mathbf{x})e^{-i\omega^+ t} + C^-\varphi^-(\mathbf{x})e^{-i\omega^- t}, \qquad (C^+, C^-) \in \mathbb{S}^3, \quad (\text{V.1.7})$$

where $\omega^{\pm} = E_1^{\pm}/\hbar$. Now the orthonormal eigenfunctions $\varphi^{\pm} \in \mathcal{G}$ are defined uniquely up to a unitary factor and correspond to the spin $\pm\frac{1}{2}$, respectively. We note that the two-frequency solutions (V.1.7) with $C^+ = 0$ or $C^- = 0$ reduce to single-frequency solutions. Obviously, these single-frequency solutions do not attract the two-frequency solutions with $C^{\pm} \neq 0$.

One can expect that a similar bifurcation also occurs for the coupled nonlinear Maxwell–Pauli system (III.3.8). Namely, in the case of zero external magnetic field, we expect that ground states form a three-dimensional manifold; when the magnetic field is turned on, this manifold bifurcates into two submanifolds of single-frequency solutions of type (V.1.1): one submanifold corresponding to the spin $1/2$ and other to the spin $-1/2$. The attraction to these submanifolds is suggested by the perturbation arguments relying on the nonlinear radiation mechanism; see Section I.4.2. Let us stress that this attraction is in contrast to the linear case discussed above.

Such a *bifurcation of attractor* for $B \neq 0$, if proved, would explain the splitting of the beam in the Stern–Gerlach experiments and also the Einstein–Ehrenfest paradox. Indeed, this bifurcation is not related to the *reorientation* of electrons, explaining much shorter relaxation time in the Stern–Gerlach experiment.

Remark V.1.2. Let us stress that we conjecture the asymptotics (V.1.1) for the Maxwell–Schrödinger equations (I.5.1) with *static external potentials* (I.3.21) which corresponds to the *zero absolute temperature $T = 0$*. The nonzero temperature provisionally corresponds to a random time-dependent external potentials which provide a long-time convergence to statistical equilibrium. This equilibrium corresponds to a random distribution over the stationary orbits instead of the global attraction to them.

V.2   Conjecture on Attractors

Let us consider general G-invariant *autonomous* Hamiltonian nonlinear dynamical systems of type

$$\dot{\Psi}(t) = F(\Psi(t)), \qquad t \in \mathbb{R}, \tag{V.2.1}$$

with a Lie symmetry group G acting on a suitable Hilbert or Banach phase space $\mathcal{E}$ via a linear representation T. The Hamiltonian structure means that

$$F(\Psi) = JD\mathcal{H}(\Psi), \qquad J^* = -J, \tag{V.2.2}$$

where $\mathcal{H}$ denotes the corresponding Hamiltonian functional. The G-invariance means that

$$F(T(g)\Psi) = T(g)F(\Psi), \qquad \Psi \in \mathcal{E} \tag{V.2.3}$$

for all $g \in G$. In that case, for any solution $\Psi(t)$ of equation (V.2.1), the trajectory $T(g)\Psi(t)$ is also a solution, so the representation commutes with the dynamical group $U(t) : \Psi(0) \mapsto \Psi(t)$,

$$T(g)U(t) = U(t)T(g). \tag{V.2.4}$$

Let us note that the theory of elementary particles deals systematically with the symmetry groups $SU(2)$, $SU(3)$, $SU(5)$, $SO(10)$ and others, such as e.g. the group

$$SU(4) \times SU(2) \times SU(2)$$

from the Pati–Salam model [123], one of the candidates for the "Grand Unified Theory".

Conjecture A (On attractors) *For* generic *G-invariant equations (V.2.1), any finite energy solution $\Psi(t)$ admits a long-time asymptotics*

$$\Psi(t) \sim e^{\hat{\lambda}_{\pm}t}\Psi_{\pm}, \qquad t \to \pm\infty, \tag{V.2.5}$$

in the appropriate topology of the phase space $\mathcal{E}$. Here $\hat{\lambda}_{\pm} = T'(e)\lambda_{\pm}$, where $\lambda_{\pm}$ belong to the corresponding Lie algebra $\mathfrak{g}$, while $\Psi_{\pm}$ are some limiting amplitudes depending on the trajectory $\Psi(t)$.

In other words, all solutions of the type $e^{\hat{\lambda}t}\Psi$ with $\hat{\lambda} = T'(e)\lambda$, where $\lambda \in \mathfrak{g}$, constitute a global attractor for *generic* G-invariant Hamiltonian nonlinear PDEs of type (V.2.1). This conjecture suggests defining *stationary G-orbits* for equation (V.2.1) as solutions of type

$$\Psi(t) = e^{\hat{\lambda}t}\Psi, \qquad t \in \mathbb{R}. \tag{V.2.6}$$

This definition leads to the corresponding *nonlinear eigenvalue problem*

$$\hat{\lambda}\Psi = F(\Psi). \tag{V.2.7}$$

In particular, in the case of the unitary symmetry group $U(1)$, the Lie algebra is $\mathfrak{g} = \mathbb{R}$, and λ is a real number. In the case of the symmetry group $G = SU(3)$, the generator ("eigenvalue") λ is a skew-Hermitian 3×3-matrix, and solutions (V.2.6) can be quasiperiodic in time.

Note that the conjecture (V.2.5) fails for linear equations, i.e., linear equations are exceptional, not "generic"; see Example V.2.5 below.

Example V.2.1. The Maxwell–Schrödinger system (I.5.1) with static external potentials (I.3.21) is invariant with respect to the symmetry group $U(1)$ in the representation (I.5.5), and the conjecture (V.2.5) reduces to the asymptotics (V.1.1). In this case,

$$\hat{\lambda}_\pm = \begin{pmatrix} 0 & 0 & 0 \\ 0 & 0 & 0 \\ 0 & 0 & -i\omega_\pm \end{pmatrix}, \qquad \omega_\pm \in \mathbb{R}.$$

Remark V.2.2. Let us emphasize that the conjecture (V.2.5) is formulated only for *finite energy solutions*. For infinite energy solutions, the asymptotics (V.2.5) is not expected. For example, the d'Alembert waves (V.2.11) satisfy the translation-invariant d'Alembert equation for any functions $f(x)$ and $g(x)$. In particular, $\psi(x, t) = x + t$ is a solution with infinite energy (VI.2.4), for which the asymptotics (V.2.5) does not hold, while it holds for finite energy solutions; see Section VI.2.1 below.

In Chapter VI, we give a brief survey of results obtained since 1990 that justify the conjecture (V.2.5) for a list of model equations of type (V.2.1). The details can be found in [145, 150, 151]. The results confirm the existence of finite-dimensional attractors in the Hilbert or Banach phase spaces, and demonstrate an explicit correspondence between the long-time asymptotics and the symmetry group G of equations. The results obtained so far concern equations (V.2.1) with the following four basic symmetry groups: the trivial group $G = \{e\}$, the group of translations $\mathbb{R}^n$ for translation-invariant equations, the unitary group $G = U(1)$ for phase-invariant equations, and the rotation group $SO(3)$ for "isotropic equations". Below we specify the asymptotics (V.2.5) in these cases.

V.2.1   Trivial symmetry group $G = \{e\}$

For such *generic* equations, the conjecture (V.2.5) means the attraction of *all finite energy solutions* to stationary states:

$$\Psi(t) \to S_\pm, \qquad t \to \pm\infty, \tag{V.2.8}$$

as illustrated on Fig. V.1. Here the states $S_\pm = S_\pm(\mathbf{x})$ depend on the trajectory $\Psi(t)$ under consideration, and the convergence holds in local seminorms of type $L^2(|\mathbf{x}| < R)$ with any $R > 0$. This convergence cannot hold in global norms (i.e., in norms corresponding to $R = \infty$) due to energy conservation. The asymptotics (V.2.8) can be symbolically written as the transitions

$$S_- \mapsto S_+, \tag{V.2.9}$$

which can be considered as the mathematical model of Bohr's "quantum jumps" (I.4.3).

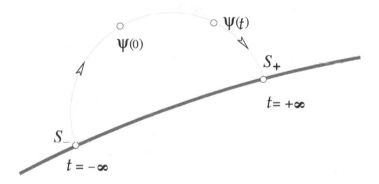

Figure V.1: Convergence to stationary states.

Such an attraction was established for a variety of model equations: i) for a string coupled to nonlinear oscillators; ii) for the three-dimensional wave equation coupled to a charged particle and for the Maxwell–Lorentz equations; iii) for the wave equation, the Dirac and Klein–Gordon equations with concentrated nonlinearities. The proofs rely on the analysis of radiation which irreversibly carries a portion of energy to infinity; see the surveys [150, 151].

In Section VI.2, we briefly describe the results [181]–[186] on global attraction (V.2.8) for the wave and Maxwell–Lorentz equations coupled to nonlinear oscillators and moving particles and for other PDEs.

In all the problems considered, the convergence (VI.2.1) implies, by the Fatou theorem, the inequality

$$\mathcal{H}(S_\pm) \leq \mathcal{H}(Y(t)) \equiv \text{const}, \ \ t \in \mathbb{R}, \tag{V.2.10}$$

where $\mathcal{H}$ is the corresponding conserved Hamiltonian (energy) functional. This inequality is an analog of the well-known property of weak convergence in Hilbert and Banach spaces. Simple examples show that the strict inequality in (V.2.10) is possible, which means that an irreversible scattering of energy to infinity occurs.

Example V.2.3. The d'Alembert waves. The asymptotics (V.2.8) with the strict inequality (V.2.10) can easily be demonstrated for the d'Alembert equation $\ddot{\psi}(x,t) = \psi''(x,t)$ with general solution

$$\psi(x,t) = f(x-t) + g(x+t). \qquad (V.2.11)$$

Indeed, the convergence $\psi(\cdot,t) \to 0$ in $L^2_{\mathrm{loc}}(\mathbb{R})$ obviously holds for any f, $g \in C_0^\infty(\mathbb{R})$. On the other hand, the convergence to zero in *global $L^2(\mathbb{R})$ norms* obviously fails if $f(x) \not\equiv 0$ or $g(x) \not\equiv 0$.

Example V.2.4. Nonlinear strong Huygens principle. Consider solutions of the 3D wave equation with a unit propagation velocity and initial data with support in a ball $|x| < R$. The corresponding solution is concentrated in spherical layers $|t| - R < |x| < |t| + R$. Therefore, the energy localized in any particular bounded region converges to zero as $t \to \pm\infty$, although the total energy remains constant. This also illustrates the strict inequality in (V.2.10).

This convergence to zero is known as *strong Huygens principle* in optics and acoustics (see [3]). Thus, global attraction to stationary states (VI.2.1) is a generalization of the strong Huygens principle to nonlinear equations. The difference is that for the linear wave equation, the limit is always zero, while for nonlinear equations the limit can be any stationary solution.

V.2.2   Symmetry group of translations $G = \mathbb{R}^n$

Let us consider, for example, the case of the simplest representation

$$[T(a)\Psi](x) := \Psi(x-a), \qquad x \in \mathbb{R}^n, \qquad (V.2.12)$$

with $a \in \mathbb{R}^n$. For *generic translation-invariant equations*, the conjecture (V.2.5) means the attraction of *all finite energy solutions* to *solitons* (traveling waves):

$$\Psi(x,t) \sim \Psi_\pm(x - v_\pm t) = e^{-v_\pm \nabla t}\Psi_\pm, \qquad t \to \pm\infty, \qquad (V.2.13)$$

where the asymptotics holds in local seminorms of type $L^2(|x - v_\pm t| < R)$ with any $R > 0$, i.e., *in the comoving frame of reference*. A trivial example is provided by the d'Alembert equation with general solution (V.2.11) corresponding to the asymptotics (V.2.13) with $v_+ = \pm 1$ and $v_- = \pm 1$.

Such global attraction to solitons was first proved for *integrable equations* (KdV, etc); see [192, 200]. Later on, this global attraction was extended in [193]–[196] to *nonintegrable* translation-invariant wave and Maxwell equations coupled to relativistic charged particles. We briefly describe these results in Section VI.3.1. In Section VI.5, we present the results of numerical simulations of the asymptotics (V.2.13) for relativistic nonlinear wave equations.

In Section VI.3.2, we describe the results on *adiabatic effective dynamics* of solitons in *slowly varying external potentials*; then, in Section VI.3.3, we discuss the related question of *mass-energy equivalence*.

V.2.3 Unitary symmetry group $G = U(1)$

Let us consider the simplest representation

$$[T(e^{i\theta})\Psi](x) := e^{i\theta}\Psi(x), \qquad x \in \mathbb{R}^n, \qquad (V.2.14)$$

with $\theta \in \mathbb{T} = [0, 2\pi)$. For *generic $U(1)$-invariant equations*, the conjecture (V.2.5) means the attraction of *all finite energy solution* to the *single-frequency trajectories*:

$$\Psi(x, t) \sim \Psi_{\pm}(x)e^{-i\omega_{\pm}t}, \qquad t \to \pm\infty, \qquad (V.2.15)$$

where $\omega_{\pm} \in \mathbb{R}$. Note that the interpretation of Bohr's transitions (V.1.1) is a particular case of asymptotics (V.2.15) for the coupled Maxwell–Schrödinger equations (I.5.1).

This asymptotics means that there is a global attraction of finite energy solutions to the solitary manifold formed by all *stationary orbits* (I.4.8). The asymptotics are considered in the local seminorms $L^2(|x| < R)$ with any $R > 0$. The global attractor is a smooth manifold formed by the circles that are the orbits of the action of the symmetry group $U(1)$, as illustrated on Fig. V.2.

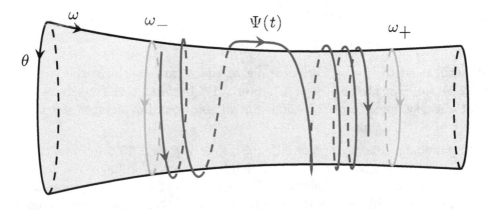

Figure V.2: Convergence to stationary orbits.

Such an attraction was proved for the first time i) in [209]–[217] for the Klein–Gordon and Dirac equations coupled to $U(1)$-invariant nonlinear oscillator, ii) in [210] for discrete approximations of such coupled systems (i.e., for the corresponding difference schemes), and iii) in [218]–[220] for the wave, Klein–Gordon, and Dirac equations with concentrated nonlinearities. More precisely, we have proved global attraction to the *solitary manifold* of all stationary orbits, though global attraction to particular stationary orbits, with fixed $\omega_{\pm}$, is still an open problem.

All these results were proved under the assumption that the equations are "strictly nonlinear". For linear equations, the global attraction obviously fails if the discrete spectrum consists of at least two different eigenvalues.

The proofs of all these results rely on i) a nonlinear analog of the Kato theorem on the absence of embedded eigenvalues, ii) a new theory of multipliers in the space of quasimeasures, and iii) a novel application of the Titchmarsh convolution theorem.

In Section VI.4, we present the streamlined detailed proofs of the results from [212]–[214].

V.2.4   Orthogonal symmetry group $G = SO(3)$

For *generic rotation-invariant equations*, the conjecture (V.2.5) means that *each finite energy solution* admits the long-time asymptotics

$$\psi(x,t) \sim e^{\hat{\Omega}_{\pm}t}\psi_{\pm}(x), \qquad t \to \pm\infty, \tag{V.2.16}$$

where $\hat{\Omega}_{\pm}$ are suitable representations of real skew-symmetric 3×3 matrices $\Omega_{\pm} \in \mathfrak{so}(3)$. This means that there is a global attraction to "stationary $SO(3)$-orbits". Such asymptotics are proved in [222] for the Maxwell–Lorentz equations with spinning particle.

V.2.5   Generic equations

We still need to specify the meaning of the term *generic* in the conjecture (V.2.5) (and in all results of Chapter VI). In fact, this conjecture means that the asymptotics (V.2.5) hold for all solutions for an *open dense set* of G-invariant equations.

i) In particular, the asymptotics (V.2.8), (V.2.13), (V.2.15), and (V.2.16) hold under appropriate conditions, which define the corresponding open dense subsets of G-invariant equations with the four types of symmetry group G. This asymptotic expression may break down if some of these conditions are not satisfied: this corresponds to some "exceptional" equations. For example, global attraction (VI.4.2) breaks down for the linear Schrödinger equations with at least two different eigenvalues. Thus, linear equations are exceptional, *not generic*; see Example V.2.5 below.

ii) The general situation is as follows. Let the Lie group G_1 be a (proper) subgroup of some larger Lie group G_2. Then G_2-invariant equations form an "exceptional subset" among all G_1-invariant equations and the corresponding asymptotics (V.2.5) may be completely different. For example, the trivial group $\{e\}$ is a proper subgroup in $U(1)$ and in $\mathbb{R}^n$, and the asymptotic expressions (VI.3.3) and (VI.4.2) may differ significantly from (VI.2.1).

Example V.2.5. The linear Schrödinger equations

$$i\dot{\psi}(x,t) = -\Delta\psi(x,t) + V(x)\psi(x,t), \qquad x \in \mathbb{R}^n,$$

are $U(1)$-invariant; however, the asymptotics (VI.4.2) for these equations generally fails. Namely, under suitable conditions on the potential $V(x)$, any finite energy solution admits the spectral representation

$$\psi(x,t) = \sum_k C_k\varphi_k(x)e^{-i\omega_k t} + \int_0^\infty C(\omega)\varphi(x,\omega)e^{-i\omega t}\,d\omega, \qquad (V.2.17)$$

where φ_k and $\Psi(\cdot,\omega)$ are the corresponding eigenfunctions of the discrete and continuous spectrum, respectively. The last integral is a dispersive wave, which decays to zero in the topology of $L^2_{\text{loc}}(\mathbb{R}^n)$ under additional conditions on the potential $V(x)$; see [133, 143]. Hence, (V.2.17) implies that the global attractor of all finite-energy solutions is the linear span of all the eigenfunctions φ_k. Thus, if the discrete spectrum contains at least two different points, then the long-time asymptotics does not reduce to a single term like (VI.4.2).

Thus, the linear equations are *non-generic* in the class of $U(1)$-invariant equations! Note that all the results [209]–[220] are established for a *strictly nonlinear case*, which does not include linear equations (see the condition (VI.4.16)).

V.2.6 Empirical evidence

The conjecture (V.2.5) agrees with the Gell-Mann–Ne'eman theory of baryons [119, 121, 122]. Indeed, in 1961, M. Gell-Mann and Y. Ne'eman suggested using the symmetry group $SU(3)$ for the strong interaction of baryons relying on the discovered parallelism between empirical data for the baryons and the "Dynkin diagram" of the Lie algebra $\mathfrak{g} = \mathfrak{su}(3)$ with 8 generators (the famous "eightfold way"). This theory resulted in the scheme of quarks in quantum chromodynamics [120], and in the prediction of a new baryon with prescribed values of its mass and decay products. This particle (the Ω^--hyperon) was promptly discovered experimentally [118].

On the other hand, the elementary particles seem to describe long-time asymptotics of quantum fields. Hence, the empirical correspondence between elementary particles and generators of the Lie algebras presumably gives evidence in favor of our general conjecture (V.2.5) for equations with Lie symmetry groups.

V.3 Wave-Particle Duality

The wave-particle duality (I.1.13), (I.1.15) was introduced by L. de Broglie in his 1923 PhD. Thesis. The identification of the beam of particles with the wave was suggested by de Broglie for *relativistic particles* as the antithesis to Einstein's corpuscular treatment of light as a beam of photons. For *nonrelativistic particles*, a similar identification holds only asymptotically, for sufficiently small momenta; this follows from (I.2.7) and (I.2.8).

This duality was the key source for the Schrödinger quantum mechanics; see Sections I.1.2 and I.2.1. In this section, we discuss a possible treatment of this nonrelativistic wave-particle duality in the framework of semiclassical Maxwell–Schrödinger equations for the following phenomena: i) reduction of wave packets; ii) the diffraction of electrons; iii) acceleration of electrons in the electron gun.

V.3.1 Reduction of wave packets

We suggest the appearance of the wave-particle duality relying on a generalization of conjecture (V.2.5) to the case of translation-invariant Maxwell–Schrödinger system (I.5.1) without external potentials, i.e., $\mathbf{A}^{\text{ext}}(\mathbf{x}, t) \equiv 0$, $A_0^{\text{ext}}(\mathbf{x}, t) \equiv 0$. In this case, the Schrödinger equation from (I.5.1) takes the form

$$i\hbar\dot{\psi}(\mathbf{x}, t) = \frac{1}{2\text{m}}\left[-i h\nabla - \frac{e}{c}\mathbf{A}(\mathbf{x}, t)\right]^2 \psi(\mathbf{x}, t) + eA_0(\mathbf{x}, t)\psi(\mathbf{x}, t), \quad \mathbf{x} \in \mathbb{R}^3.$$
(V.3.1)

Now the symmetry group of the system (I.5.1) is $G = \mathbb{R}^3 \times U(1)$, and our general conjecture (V.2.5) can be strengthened similarly to (VI.3.4) as the following long-time asymptotics of each finite energy solution:

$$\begin{cases} \mathbf{A}(\mathbf{x}, t) \sim \sum_n \mathbf{A}_n^{\pm}(\mathbf{x} - \mathbf{v}_n^{\pm}t) + \mathbf{A}^{\pm}(\mathbf{x}, t), \\ \psi(\mathbf{x}, t) \sim \sum_n \psi_n^{\pm}(\mathbf{x} - \mathbf{v}_n^{\pm}t)e^{i\Phi_n^{\pm}(\mathbf{x}, t)} + \psi^{\pm}(\mathbf{x}, t) \end{cases}, \quad t \to \pm\infty, \quad \text{(V.3.2)}$$

where $\mathbf{A}^{\pm}(\mathbf{x}, t)$ and $\psi^{\pm}(\mathbf{x}, t)$ stand for the corresponding *dispersive waves*. These asymptotics are considered in *global energy norms*, and suggest treating the solitons

$$\left(\mathbf{A}_n^{\pm}(\mathbf{x} - \mathbf{v}_n^{\pm}t), \psi_n^{\pm}(\mathbf{x} - \mathbf{v}_n^{\pm}t)e^{i\Phi_n^{\pm}(\mathbf{x}, t)}\right)$$
(V.3.3)

as electrons. This asymptotics is analogous to the *reduction (or collapse) of wave packets*, which was suggested by W. Heisenberg for photons [42] (see also [17]).

V.3.2   Diffraction of electrons

The most striking and direct manifestation of wave-particle duality is given by the diffraction of electrons that we considered in Section II.7. The traditional description of these experiments relies on the following steps:

I. The incident beam of electrons emitted by an electron gun is described by the plane wave

$$\psi^{\mathrm{in}}(\mathbf{x}, t) = C e^{i(\mathbf{k}\cdot\mathbf{x} - \omega t)} \qquad (V.3.4)$$

which satisfies the free Schrödinger equation.

II. The diffraction of a plane wave in reflection by a crystal or in scattering by the aperture in the scattering screen.

III. The formation of the diffraction pattern on the screen of registration.

We show in Section V.3.3 below that the formation of the incident plane wave (V.3.4) in the electron gun can be explained by the quasiclassical asymptotics for the *linear* Schrödinger equation.

The step II is traditionally described by the *linear* Schrödinger equation; see [108]–[117] and Section II.7. According to the *limiting amplitude principle*, the diffracted wave admits the long-time asymptotics (II.7.5), where ω is the frequency of the incident wave (V.3.4).

In Section II.7, we calculated the diffraction amplitudes $a_\infty(\mathbf{x})$ in the case of diffraction by the screen by using the *Kirchhoff approximation*; see the formulae (II.7.13) and (II.7.24). In particular, for the double-slit diffraction, the corresponding amplitude (II.7.28) is in a fine *quantitative* agreement with the results of recent diffraction experiments [108]. Indeed, the maxima of $|a_\infty(\mathbf{x})|$ on the screen agree very well with those of the diffraction pattern in experiments [108].

Thus, the arguments above rely on the linear Schrödinger theory. On the other hand, the formation of the diffraction pattern in step III is a genuinely nonlinear effect that we shall discuss in Section V.4.

V.3.3   Quasiclassical asymptotics for the electron gun

Here we use the quasiclassical asymptotics (I.2.19) to explain the formation of the plane wave (V.3.4) as a result of acceleration of electrons in an electron gun with voltage $V = -E/e$. A hot cathode emits electrons of small energy and after that the electrons are accelerated in the electrostatic field

$$\mathbf{E}(\mathbf{x}) = -\nabla\Phi(\mathbf{x})$$

in the region Ω_1 between the cathode and the anode. The electrons cross the region Ω_1 and pass through the aperture in the anode into the region Ω_2 behind the anode, where the external electric field $\mathbf{E}(\mathbf{x})$ vanishes. In this region, the electrons interfere, forming a diffraction pattern.

We will show that the corresponding quasiclassical solution of the Schrödinger equation admits the representation (V.3.4) at the points of the aperture, with E and $\mathbf{p}$ in (V.3.4) being the kinetic energy and momentum of the accelerated classical *nonrelativistic* electrons at the aperture.

First we note that the electron wave function in the region $\Omega_1 \cup \Omega_2$ is a solution of the corresponding Schrödinger equation

$$i\hbar\dot\psi(\mathbf{x}, t) = -\frac{\hbar^2}{2\mathrm{m}}\Delta\psi(\mathbf{x}, t) + e\Phi(\mathbf{x})\psi(\mathbf{x}, t), \quad \mathbf{x} \in \Omega_1 \cup \Omega_2, \qquad \text{(V.3.5)}$$

where $\Phi(\mathbf{x})$ is the external electrostatic potential which vanishes on the cathode. This potential is constant in Ω_2, where the electrostatic field vanishes, so equation (V.3.5) behind the anode is given by

$$i\hbar\dot\psi(\mathbf{x}, t) = -\frac{\hbar^2}{2\mathrm{m}}\Delta\psi(\mathbf{x}, t) + e\Phi_*\psi(\mathbf{x}, t), \qquad \mathbf{x} \in \Omega_2, \qquad \text{(V.3.6)}$$

where Φ_* is the value of the potential $\Phi(\mathbf{x})$ at the points of the aperture, so the voltage $V = \Phi*$.

Further, we suppose that the solution admits the quasiclassical asymptotics (I.2.19):

$$\psi(\mathbf{x}, t) \sim a(\mathbf{x}, t)e^{i\frac{S(\mathbf{x},t)}{\hbar}}, \qquad \hbar \to 0, \qquad \text{(V.3.7)}$$

where $a(\mathbf{x}, t)$ and $S(\mathbf{x}, t)$ are slowly varying functions. Substituting this asymptotics into (V.3.5), we obtain in the limit $\hbar \to 0$ the corresponding Hamilton–Jacobi equation (I.2.24) for the phase function,

$$-\partial_t S(\mathbf{x}, t) = \frac{1}{2\mathrm{m}}[\nabla S(\mathbf{x}, t)]^2 + e\Phi(\mathbf{x}), \qquad \mathbf{x} \in \Omega_1. \qquad \text{(V.3.8)}$$

The solution is given by the Lagrangian action integral over classical trajectories (I.2.27) starting at the time t_0 from the points $\mathbf{x}(t_0)$ of the cathode and satisfying the Hamiltonian equations

$$\dot{\mathbf{x}}(t) = \mathbf{p}(t)/\mathrm{m}, \qquad \dot{\mathbf{p}}(t) = -e\nabla\Phi(\mathbf{x}(t)) \qquad \text{(V.3.9)}$$

which correspond to the Hamiltonian function

$$\mathcal{H}(\mathbf{x}, \mathbf{p}) = \frac{1}{2\mathrm{m}}\mathbf{p}^2 + e\Phi(\mathbf{x}). \qquad \text{(V.3.10)}$$

The initial data $\mathbf{x}(t_0)$ and $\mathbf{p}(t_0)$ satisfy the relation

$$\mathbf{p}(t_0) = \nabla S(\mathbf{x}(t_0), t_0) \qquad \text{(V.3.11)}$$

as in (I.2.22). The corresponding solution of the system (V.3.9) can be treated as a trajectory of a classical electron emitted with the initial momentum $\mathbf{p}(t_0)$

at the time moment t_0 from the point $\mathbf{x}(t_0)$ of the cathode. According to (V.3.11) and (I.2.25), these trajectories satisfy

$$\mathbf{p}(t) = \nabla S(\mathbf{x}(t), t), \qquad t \in \mathbb{R}. \tag{V.3.12}$$

The initial kinetic energy of the *emitted electrons* is relatively small, so we can assume that

$$\mathcal{H}(\mathbf{x}(t_0), \mathbf{p}(t_0)) = 0. \tag{V.3.13}$$

Let us denote $\mathbf{x}_* = \mathbf{x}(t_*)$ and $\mathbf{p}_* := \mathbf{p}(t_*)$, where t_* is the time of arrival of the electron at the point $\mathbf{x}_*$ lying in the aperture. By the energy conservation, we also have

$$\mathcal{H}(\mathbf{x}_*, \mathbf{p}_*) = 0, \tag{V.3.14}$$

and hence

$$\frac{1}{2\mathrm{m}} \mathbf{p}_*^2 = -e\Phi_*, \qquad \partial_t S(\mathbf{x}_*, t_*) = 0, \tag{V.3.15}$$

where the last identity follows from (V.3.14) and the Hamilton–Jacobi equation (V.3.8) with $\mathbf{x} = \mathbf{x}_*$ and $t = t_*$. Now the Taylor expansion gives

$$S(\mathbf{x}, t) \approx S(\mathbf{x}_*, t_*) + \mathbf{p}_* \cdot (\mathbf{x} - \mathbf{x}_*), \qquad |\mathbf{x} - \mathbf{x}_*| + |t - t_*| \ll 1, \tag{V.3.16}$$

since $\partial_t S(\mathbf{x}_*, t_*) = 0$ by (V.3.15) and

$$\nabla S(\mathbf{x}_*, t_*) = \mathbf{p}_* \tag{V.3.17}$$

by (V.3.12). Therefore, at the points of the aperture, the wave function reads as

$$\psi(\mathbf{x}, t) = a(\mathbf{x}_*, t_*) e^{iS(\mathbf{x}_*, t_*)/\hbar} e^{i\mathbf{p}_* \cdot (\mathbf{x} - \mathbf{x}_*)/\hbar} \approx C e^{i\mathbf{p}_* \cdot \mathbf{x}/\hbar}, \tag{V.3.18}$$

since $a(\mathbf{x}_*, t_*)$ and $S(\mathbf{x}_*, t_*)$ are slowly varying functions.

Finally, the *constant electrostatic potential* Φ_* in the Schrödinger equation (V.3.6) behind the screen can be eliminated by the gauge transform

$$\psi(\mathbf{x}, t) \mapsto \psi_*(\mathbf{x}, t) := \psi(\mathbf{x}, t) e^{ie\Phi_* t/\hbar}. \tag{V.3.19}$$

Indeed, the transformed function $\psi_*(\mathbf{x}, t)$ behind the screen satisfies the free Schrödinger equation

$$i\hbar \dot{\psi}_*(\mathbf{x}, t) = -\frac{\hbar^2}{2\mathrm{m}} \Delta \psi_*(\mathbf{x}, t), \qquad \mathbf{x} \in \Omega_2. \tag{V.3.20}$$

The next key observation is that the transformed wave function admits the following asymptotics at the points of the small aperture:

$$\psi_*(\mathbf{x}, t) := \psi(\mathbf{x}, t) e^{ie\Phi_* t/\hbar} \approx C e^{i\mathbf{p}_* \cdot \mathbf{x}/\hbar} e^{ie\Phi_* t/\hbar} = C e^{i(\mathbf{p}_* \cdot \mathbf{x} - E_* t)/\hbar}, \tag{V.3.21}$$

as follows from (V.3.19) and (V.3.18); here

$$E_* = -e\Phi_* \tag{V.3.22}$$

is the energy of the classical electron at the aperture and $\mathbf{p}_*$ is its momentum. This justifies the asymptotics of type (V.3.4) for the *incident wave function* $\psi_*(x,t)$ at the points of the aperture, with

$$\mathbf{k} = \mathbf{p}_*/\hbar, \qquad \omega = E_*/\hbar, \tag{V.3.23}$$

exactly in agreement with de Broglie's relations (I.2.8).

Remark V.3.1. The diffraction pattern corresponding to the wave functions $\psi_*(\mathbf{x},t)$ and $\psi(\mathbf{x},t)$ behind the aperture coincide by (V.3.19) and by the *Born rule*, which we discuss in the next section.

V.4   Probabilistic Interpretation

In 1927, M. Born suggested the *Born rule*, which is the probabilistic interpretation of the wave function (see [24, 77]):

$$
\begin{array}{|c}
\textit{The probability of detecting an electron at a point } \mathbf{x} \textit{ at time } t \\
\textit{is proportional to } |\psi(\mathbf{x}, t)|^2.
\end{array} \qquad \text{(V.4.1)}
$$

V.4.1   Diffraction current

M. Born proposed the probabilistic interpretation to describe the diffraction experiments of C. Davisson and L. Germer of 1924–1927. In these experiments, the incident wave is diffracted by a *bounded aperture* Q lying on the scattering plane $x_3 = 0$, and the diffraction pattern is observed on a parallel plane of registration

$$
x_3 = \text{const} \gg \text{diam}(Q), \qquad \text{(V.4.2)}
$$

where $\text{diam}(Q)$ is the diameter of the aperture Q; see Section II.7.

Let us demonstrate that the rule (V.4.1) can be explained in the framework of these experiments by calculating the current (I.5.2) at the points of the screen of registration. Indeed, the diffraction pattern is registered either by atoms of photo emulsion or by registration counters located on the screen. In both cases the *rate of registration* (number of electrons per second and cm^2) is proportional to the current density. Thus, to explain the Born rule (V.4.1) in this situation, we have to show that the current density on the screen of registration is proportional to $|\psi(\mathbf{x}, t)|^2$,

$$
\mathbf{j}(\mathbf{x}, t) \sim |\psi(\mathbf{x}, t)|^2 (0, 0, 1). \qquad \text{(V.4.3)}
$$

The current density (I.5.2) is approximately given by

$$
\mathbf{j}(\mathbf{x}, t) \approx \frac{e}{m} \, \text{Re} \left(\overline{\psi}(\mathbf{x}, t) [-i\hbar \nabla \psi(\mathbf{x}, t)] \right). \qquad \text{(V.4.4)}
$$

Indeed, in the formula (I.5.2), we have $\mathbf{A}^{\text{ext}}(\mathbf{x}, t) = 0$ since there is no external fields between the scattering screen and the screen of observation. The term with $\mathbf{A}(\mathbf{x}, t)$ in (I.5.2) can be neglected, since its contribution contains an additional small factor e/c.

For large times, the wave function is given by the limiting amplitude principle (II.7.5). The corresponding limiting amplitude is given by formula (II.7.25) which holds in the framework of the *linear Schrödinger equation* (II.7.1). Finally, the long-range asymptotics (II.7.26) implies that near the screen of registration the current (V.4.4) has the asymptotics

$$
\mathbf{j}(\mathbf{x}, t) \approx \frac{e\hbar |\mathbf{k}|}{m} |a_\infty(\mathbf{x})|^2 (0, 0, 1), \qquad t \to \infty, \qquad \text{(V.4.5)}
$$

since the screen of registration is sufficiently far from the screen of scattering by (V.4.2). Finally, (II.7.5) implies that for large times we have

$$|a_\infty(\mathbf{x})|^2 \approx |\psi(\mathbf{x}, t)|^2,$$

so (V.4.5) implies (V.4.3).

Remark V.4.1. The key role in the proof of (V.4.3) is played by the asymptotics (II.7.26). This asymptotics holds with a high precision if the corresponding limiting amplitude is slowly varying function of transversal variables x^1 and x^2. This condition holds in the framework of the experiments of R. Bach *et al.* [108, 109] since the distance between maxima of the diffraction pattern $36, 4\,\mu m$ is much greater than the wavelength $\lambda = 50$ nm.

V.4.2 Discrete registration of electrons

In 1948, the probabilistic interpretation was given a new content and confirmation by the experiments of L. Biberman, N. Sushkin, and V. Fabrikant [110] with an electron beam of very low intensity. Later similar experiments were carried out by R. Chambers, A. Tonomura, S. Frabboni, R. Bach, and others [111, 117, 115, 108, 109]. In these experiments, the diffraction pattern is created as a time average of random discrete registration of individual electrons.

Below, we suggest two possible treatments of the probabilistic interpretation in these experiments. These treatments rely, respectively,

i) on a random interaction with the counters, and

ii) on the *soliton resolution conjecture* (V.3.2) in the framework of the coupled translation-invariant Maxwell–Schrödinger equations (I.5.1).

However, the corresponding rigorous justification is still an open problem.

I. Interaction with counters. One possible explanation of the discrete registration is a random triggering of a) registration counters located at the points of the screen of observation, or b) atoms of the photo emulsion. In both cases, the probability of triggering must be proportional to the electric current density near the screen of observation.

II. Reduction of wave packets. In the space between the scattering screen and the screen of observation, the *external fields vanish*. Hence, in this space, the Maxwell–Schrödinger system is translation-invariant, being covered by our conjecture (V.3.2) on the decay to solitons-electrons (V.3.3), see Fig. II.1. Such a decay should be regarded as a random process, since it is subject to microscopic fluctuations.

V.4.3   Superposition principle as a linear approximation

Treatment **II** of the discrete registration of electrons in the previous section is not self-consistent. Indeed, the justification of the formula (V.4.5) relies on the linear Schrödinger equation, while the reference to the soliton asymptotics (V.3.2) involves the nonlinear Maxwell–Schrödinger equations.

We suggest the following argument reconciling this formal contradiction: formula (V.4.5) holds in the *linear approximation*, while the soliton asymptotics (V.3.2) is suggested for the nonlinear Maxwell–Schrödinger equations.

Remark V.4.2. The formula (V.4.5) for the linear Schrödinger equation relies on the superposition principle which is the traditional argument in favor of the quantum mechanics being *perfectly linear*. On the other hand, the discrete registration of electrons cannot be described by the linear Schrödinger equation. We note that W. Heisenberg began developing a nonlinear theory of elementary particles in [43, 44].

Chapter VI

Attractors of Hamiltonian PDEs

In this chapter, we describe rigorous results which confirm our general conjecture (V.2.5) for a number of Hamiltonian nonlinear partial differential equations of type (V.2.1). This theory was initiated by the author in 1990 and was developed in collaboration with H. Spohn since 1995, and with V.S. Buslaev, A. Comech, V. Imaikin, E. Kopylova, D. Stuart and B. Vainberg since 2005; see [181]–[222]. We mention only sketchily many results in this direction; see the surveys [145, 150, 151, 191] for details.

The results obtained confirm the existence of finite-dimensional global attractors in the Hilbert phase space, and demonstrate a certain correspondence between global attractors and the symmetry group of equations in accordance with the conjecture (V.2.5).

So far, the results were obtained for model equations with three basic symmetry groups: the trivial symmetry group $G = \{e\}$, the translation group $G = \mathbb{R}^n$ for translation-invariant equations, the unitary group $G = U(1)$ for phase-invariant equations, and the orthogonal group $G = SO(3)$ for "isotropic equations".

VI.1   Global Attractors of Nonlinear PDEs

The theory of attractors for nonlinear PDEs began in Landau's 1944 seminal paper [165], where he proposed the first mathematical interpretation of the onset of turbulence as the growth of the dimension of attractors of the Navier–Stokes equations when the Reynolds number increases.

The foundation for the corresponding mathematical theory was laid in 1951 by E. Hopf, who first established the existence of global solutions of the 3D Navier–Stokes equations [140]. He introduced the *method of compactness*, which is a nonlinear version of Faedo–Galerkin approximations. This method is based on a priori estimates and Sobolev embedding theorems and has had an essential influence on the development of the theory of nonlinear PDEs (see [156]).

The modern development of the theory of attractors for general *dissipative systems*, i.e., systems with friction, originated in 1975–1985 in the publications of C. Foias, J.M. Ghidaglia, J.K. Hale, D. Henry, R. Temam, and was developed further by M.I. Vishik, A.V. Babin, V.V. Chepyzhov, A.A. Ilyin, V. Pata, E. Titi, S. Zelik, and others. A typical property of dissipative systems is global convergence to stationary states in the absence of external excitation: any finite energy solution of a dissipative autonomous equation in a region $\Omega \subset \mathbb{R}^n$ converges to a stationary state

$$\psi(x,t) \to S(x), \qquad t \to +\infty, \tag{VI.1.1}$$

where, as a rule, convergence holds in the $L^2(\Omega)$-metric. In particular, the relaxation to an equilibrium regime in chemical reactions is due to energy dissipation.

The results obtained concern a wide class of nonlinear dissipative PDEs, including fundamental equations of applied and mathematical physics: the Navier–Stokes equations, nonlinear parabolic equations, reaction-diffusion equations, damped wave equations, integro-differential equations, equations with delay, with memory, and so on. Very clever techniques of functional analysis of nonlinear PDEs were developed for the study of the structure of attractors, their smoothness and their fractal and Hausdorff dimensions, dependence on parameters, on averaging, and so on. An essential part of the theory up to 2000 was covered in the monographs [166]–[172].

The development of a similar theory for *Hamiltonian PDEs* seemed at first to be unmotivated and even impossible in view of energy conservation and time reversal for these equations. However, it turned out that such a theory is possible and its development was inspired by the problem of mathematical interpretation of the fundamental postulates of quantum theory which we discussed in Chapter V.

Rigorous results obtained in 1990–2020 suggest that such long-time asymptotics of solutions are in fact typical for nonlinear Hamiltonian PDEs. These results are presented briefly in this chapter. Presently this theory is only at the initial stage of its development and cannot compete with the theory of attractors of dissipative PDEs in the depth and diversity of results. For Hamiltonian PDEs, it differs significantly from the case of dissipative systems, where the attraction to stationary states is caused by energy dissipation caused by friction. For Hamiltonian equations, friction and energy dissipation are absent, and attraction is caused by radiation, which irreversibly carries energy to infinity.

The modern development of the theory of nonlinear Hamiltonian equations dates back to K. Jörgens [144], who established the existence of global solutions for nonlinear wave equations of the form

$$\ddot{\psi}(x,t) = \Delta\psi(x,t) + F(\psi(x,t)), \qquad x \in \mathbb{R}^n, \qquad \text{(VI.1.2)}$$

by developing the Hopf method of compactness. Subsequent studies in this direction were well presented by J.-L. Lions [156] and by T. Cazenave and A. Haraux [135, 136].

The first results on the long-time asymptotics of solutions of nonlinear Hamiltonian PDEs were obtained by I. Segal [175, 176] and C. Morawetz and W. Strauss [173, 174, 177]. In these papers, *local energy decay* was proved for solutions of equations (VI.1.2) with *defocusing type nonlinearities*

$$F(\psi) = -m^2\psi - \kappa|\psi|^{p-1}\psi,$$

where $m^2 \geq 0$, $\kappa > 0$, and $p > 1$. In this case, for any finite $R > 0$,

$$\int_{|x|<R} [|\dot{\psi}(x,t)|^2 + |\nabla\psi(x,t)|^2 + |\psi(x,t)|^2]\, dx \to 0, \qquad t \to \pm\infty, \quad \text{(VI.1.3)}$$

for sufficiently smooth and small initial data. Moreover, the corresponding nonlinear wave operators and scattering operators have been constructed. In [178, 179], Strauss established the completeness of nonlinear scattering for small solutions of more general equations. The decay (VI.1.3) means that the energy escapes any bounded region for large times.

For convenience, characteristic properties of all finite-energy solutions of an equation will be referred to as *global*, in order to distinguish them from the corresponding *local* properties of the solutions with initial data sufficiently close to an attractor. Note that global attraction to a (proper) attractor is impossible for all finite-dimensional Hamiltonian systems, because of energy conservation.

All the above-mentioned results on local energy decay (VI.1.3) mean that the corresponding *local attractor* of solutions with small initial states consists only of the zero point. *Global attraction* for Hamiltonian PDEs is obtained from the analysis of the irreversible energy radiation to infinity, which plays the role of energy dissipation. Such an analysis requires subtle methods of Harmonic Analysis: the Wiener Tauberian theorem, the Titchmarsh convolution theorem, the theory of quasimeasures, and the Paley–Wiener estimates.

Questions of asymptotic stability required the use of the stationary scattering theory of S. Agmon, A. Jensen and T. Kato [133, 143], and of the eigenfunction expansion for nonselfadjoint Hamiltonian operators [146, 147] based on M.G. Krein's theory of *J*-selfadjoint operators.

VI.2 Global Attraction to Stationary States

For generic equations with trivial symmetry group $G = \{e\}$, the conjecture (V.2.5) means the attraction of *all finite energy solutions* to stationary states (V.2.8) (see Fig. V.1 on p. 143)

$$\Psi(\mathbf{x}, t) \to S_\pm(\mathbf{x}), \qquad t \to \pm\infty. \qquad (VI.2.1)$$

Here the functions $S_\pm(\mathbf{x})$ are certain stationary states which depend on the considered trajectory $\Psi(\mathbf{x}, t)$, and the convergence holds in local seminorms of type $L^2(|x| < R)$ for any $R > 0$. The convergence (VI.2.1) in global norms (i.e., corresponding to $R = \infty$) *cannot hold* due to energy conservation. Such a global attraction was proved for a number of model equations in [181]–[186] and [190].

VI.2.1 The d'Alembert equation

The global attraction (VI.2.1) can easily be demonstrated using the trivial (but instructive) example of the d'Alembert equation

$$\ddot{\psi}(x, t) = \psi''(x, t), \qquad x \in \mathbb{R}. \qquad (VI.2.2)$$

All derivatives here and below are understood in the sense of distributions. This equation is formally equivalent to the Hamiltonian system

$$\dot{\psi}(t) = D_\pi \mathcal{H}, \quad \dot{\pi}(t) = -D_\psi \mathcal{H} \qquad (VI.2.3)$$

with the Hamiltonian

$$\mathcal{H}(\psi, \pi) = \frac{1}{2} \int [|\pi(x)|^2 + |\psi'(x)|^2] \, dx, \quad (\psi, \pi) \in \mathcal{E} := H_c^1(\mathbb{R}) \oplus [L^2(\mathbb{R}) \cap L^1(\mathbb{R})], \qquad (VI.2.4)$$

where $H_c^1(\mathbb{R})$ is the Hilbert space of complex-valued functions $\psi(x)$ with finite norm

$$\|\psi\|_{H_c^1(\mathbb{R})} := \|\psi'\|_{L^2(\mathbb{R})} + |\psi(0)|. \qquad (VI.2.5)$$

Let us consider solutions $(\psi(x, t), \pi(x, t))$ to (VI.2.3) with initial data

$$(\psi(x, 0), \pi(x, 0)) = (\psi_0(x), \pi_0(x)) \in \mathcal{E}.$$

Let us assume, moreover, that

$$\psi_0(x) \to C_\pm, \quad x \to \pm\infty. \qquad (VI.2.6)$$

For such initial data, the d'Alembert formula gives

$$\psi(x, t) \quad = \quad \frac{\psi_0(x + t) + \psi_0(x - t)}{2} + \frac{1}{2} \int_{x-t}^{x+t} \pi_0(y) \, dy$$

$$\xrightarrow[t \to \pm\infty]{} S_\pm(x) = \frac{C_+ + C_-}{2} \pm \frac{1}{2} \int_{-\infty}^{\infty} \pi_0(y) \, dy, \qquad (VI.2.7)$$

where the convergence is uniform on every finite interval $|x| < R$. Moreover,

$$\dot{\psi}(x,t) = \frac{\psi_0'(x+t) - \psi_0'(x-t)}{2} + \frac{\pi_0(x+t) + \pi_0(x-t)}{2} \to 0, \qquad t \to \pm\infty,$$
(VI.2.8)

where the convergence holds in $L^2(-R,R)$ for each $R > 0$. Thus, the set of stationary states $(\psi(x), \pi(x)) = (C, 0)$, where $C \in \mathbb{C}$ is an arbitrary constant, is the global attractor. Note that for positive and negative times the limits (VI.2.7) may be different.

Exercise VI.2.1. Prove that *for every two stationary solutions* $S_\pm(x) = (C_\pm, 0) \in \mathcal{E}$, *there exists a solution* $\Psi(x,t) = (\psi(x,t), \pi(x,t)) \in C(\mathbb{R}, \mathcal{E})$ *such that the convergence* (VI.2.1) *holds.*

VI.2.2   String coupled to a nonlinear oscillator

The first results on global attraction to stationary states (VI.2.1) for nonlinear Hamiltonian PDEs were established in [181, 182, 187] for the case of a point nonlinearity ("Lamb system"):

$$(1 + m\delta(x))\ddot{\psi}(x,t) = \psi''(x,t) + \delta(x)F(\psi(0,t)), \qquad x \in \mathbb{R}. \qquad \text{(VI.2.9)}$$

This equation describes transversal oscillations of a string with vector displacements $\psi(t) \in \mathbb{R}^d$, where $d \geq 1$, coupled to an oscillator attached at $x = 0$, and acting on the string with a force $F(\psi(0,t))$ orthogonal to the string; $m > 0$ is the mass of a particle attached to the string at the point $x = 0$; see Fig. VI.1. For a linear force function $F(\psi) = -k\psi$, such a system was first studied by H. Lamb [180].

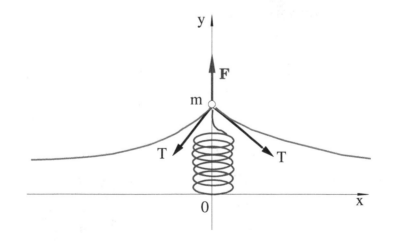

Figure VI.1: String coupled to an oscillator.

Equation (VI.2.9) can be written as the dynamical system

$$\dot{Y}(t) = \mathbf{F}(Y(t)), \qquad Y(t) := (\psi(x,t), \dot{\psi}(x,t), \dot{y}(t)), \ \ y(t) := \psi(0,t). \quad \text{(VI.2.10)}$$

We assume that

$$F \in C^1(\mathbb{R}^d, \mathbb{R}^d), \qquad F(y) = -\nabla U(y); \quad \text{(VI.2.11)}$$

$$U(y) \to +\infty, \qquad |u| \to \infty. \quad \text{(VI.2.12)}$$

In this case, equation (VI.2.10) can formally be written as a Hamiltonian system with the conserved Hamiltonian

$$\mathcal{H}(\psi, \pi, p) = \frac{1}{2} \int [|\pi(x)|^2 + |\psi'(x)|^2] \, dx + \frac{mp^2}{2} + U(\psi(0)). \quad \text{(VI.2.13)}$$

Let us introduce a suitable phase space $\mathcal{E}$ of finite energy states. Denote by $\| \cdot \|$ (resp. $\| \cdot \|_R$) the norm in the Hilbert space $L^2 := L^2(\mathbb{R}, \mathbb{R}^d)$ (resp. $L^2([-R, R], \mathbb{R}^d)$); denote $E_c := H_c^1(\mathbb{R}) \otimes \mathbb{R}^d$, where $H_c^1(\mathbb{R})$ is the Hilbert space with the norm (VI.2.5).

Definition VI.2.2. *i)* $\mathcal{E}$ *is the Hilbert phase space of triples* $(\psi(x), \pi(x), p) \in E_c \oplus L^2 \oplus \mathbb{R}^d$ *with finite energy norm*

$$\|(\psi, v, p)\|_{\mathcal{E}} = \|\psi\|_{E_c} + \|\pi\| + |p| = \|\psi'\|_R + |\psi(0)| + \|\pi\| + |p|. \quad \text{(VI.2.14)}$$

ii) $\mathcal{E}_F$ *is the space* $\mathcal{E}$ *endowed with the topology defined by the* local energy seminorms

$$\|(\psi, \pi, p)\|_{\mathcal{E},R} \equiv \|\psi'\|_R + |\psi(0)| + \|\pi\|_R + |p|, \quad R > 0. \quad \text{(VI.2.15)}$$

The space $\mathcal{E}_F$ is not complete, and the convergence in $\mathcal{E}_F$ is equivalent to the convergence in the metric

$$\text{dist}(Y_1, Y_2) = \sum_{R=1}^{\infty} 2^{-R} \frac{\|Y_1 - Y_2\|_{\mathcal{E},R}}{1 + \|Y_1 - Y_2\|_{\mathcal{E},R}}, \quad Y_1, Y_2 \in \mathcal{E}. \quad \text{(VI.2.16)}$$

The stationary states $S = (s(x), 0, 0) \in \mathcal{E}$ for equation (VI.2.10) are evidently determined: the set $\mathcal{S}$ of all stationary states $S \in \mathcal{E}$ is given by

$$\mathcal{S} = \{S_z = (z, 0, 0) : \ z \in Z\}, \qquad Z := \{z \in \mathbb{R}^d : \ F(z) = 0\}. \quad \text{(VI.2.17)}$$

The next theorem states that the set $\mathcal{S}$ is the global attractor of the system (VI.2.10) in the space $\mathcal{E}_F$.

Theorem VI.2.3. (cf. [181, 182]) *Let conditions* (VI.2.11), (VI.2.12) *hold. Then*

i) *For any initial state* $Y(0) \in \mathcal{E}$, *the equation* (VI.2.10) *admits a unique solution* $Y(t) \in C(\mathbb{R}, \mathcal{E})$, *and it is attracted to* $\mathcal{S}$:

$$Y(t) \xrightarrow{\mathcal{E}_F} \mathcal{S}, \qquad t \to \pm\infty, \tag{VI.2.18}$$

in the metric (VI.2.16). *This means that*

$$\mathrm{dist}(Y(t), \mathcal{S}) := \inf_{S \in \mathcal{S}} \mathrm{dist}(Y(t), S) \to 0, \qquad t \to \pm\infty. \tag{VI.2.19}$$

ii) *Suppose additionally that the set* Z *is a discrete subset in* $\mathbb{R}^d$. *Then for any solution* $Y(t) \in C(\mathbb{R}, \mathcal{E})$ *there exist stationary states* $S_\pm \in \mathcal{S}$ *such that*

$$Y(t) \xrightarrow{\mathcal{E}_F} S_\pm, \qquad t \to \pm\infty, \tag{VI.2.20}$$

as illustrated on Fig. V.1 on p. 143.

Remark VI.2.4. The discreteness of the set Z is essential for the attraction (VI.2.20). For example, let us consider a nonlinearity which vanishes on a C^1-submanifold of $\mathbb{R}^d$,

$$F(\psi) \equiv 0, \qquad \psi \in I. \tag{VI.2.21}$$

Then, in the case $m = 0$, the function $\psi(x,t) := f(x - t)$, with any smooth function f having values in I, is the solution of equation (VI.2.9). In particular, for $d = 1$ and $I = [-1, 1]$, we can take the function

$$\psi(x,t) = \sin \log(|x - t| + 2), \qquad (x,t) \in \mathbb{R}^2. \tag{VI.2.22}$$

In this case, the function $(\psi(x,t), \dot\psi(x,t), \psi(0,t)) \in C(\mathbb{R}, \mathcal{E})$ is the *finite energy solution* of system (VI.2.10) with $m = 0$, and the attraction (VI.2.20) obviously breaks down. On the other hand, the convergence (VI.2.18) for this solution holds. For $m > 0$, similar examples can also be easily constructed; see [182].

Further, let us denote $\mathcal{E}_0 = \{(\psi, v, 0) \in \mathcal{E}\}$, and let $\tilde{W}(t)(\psi, v, 0) := (W(t)(\psi, v), 0)$, where $W(t)$ is the dynamical group of the free wave equation corresponding to $F(y) \equiv 0$.

Theorem VI.2.5. ([187]) *Let conditions* (VI.2.11), (VI.2.12) *hold, and additionally, for the initial state* $Y(0) = (\psi_0(x), \pi_0(x), y_0)$, *let the following limits exist:*

$$C_\pm := \lim_{x \to \pm\infty} \psi_0(x), \qquad I_0 := \int_{-\infty}^{\infty} \pi_0(y)\, dy. \tag{VI.2.23}$$

Then the following scattering *asymptotics holds:*

$$Y(t) = S_\pm + \tilde{W}(t)\Psi_\pm + r_\pm(t), \tag{VI.2.24}$$

with $S_\pm \in \mathcal{S}$ and some asymptotic states $\Psi_\pm \in \mathcal{E}_0$; the remainder is small in the global energy norm (VI.2.14):

$$\|r_\pm(t)\|_\mathcal{E} \to 0, \quad t \to \pm\infty. \tag{VI.2.25}$$

The proofs of Theorems VI.2.3 and VI.2.5 for $m \geq 0$ rely on the reduced equation for the oscillator

$$m\ddot{y}(t) = F(y(t)) - 2\dot{y}(t) + 2\dot{w}_{\text{in}}(t), \qquad t > 0,$$

where the incident wave satisfies $\dot{w}_{\text{in}} \in L^2(0, \infty)$. This equation follows from the d'Alembert representation (V.2.11) for the solution $\psi(x, t)$ at $x > 0$ and $x < 0$ separately.

For equation (VI.2.9) with $m = 0$, the asymptotic completeness of the corresponding nonlinear scattering operator $S : \Psi_- \mapsto \Psi_+$ was proved in [188, 189].

Exercise VI.2.6. Prove that the attraction (VI.2.18) holds for the solution (VI.2.22), while (VI.2.20) breaks down.

VI.2.3 String coupled to several nonlinear oscillators

In [183], we have extended the results [181, 182] on global attraction to stationary states to the case of a string coupled to several nonlinear oscillators:

$$\ddot{\psi}(x, t) = \psi''(x, t) + \sum_{k=1}^{M} \delta(x - x_k) F_k(\psi(x_k, t)).$$

This equation reduces to a system of M ordinary differential equations with delay. Its study required a novel approach relying on a special analysis of *omega-limit points* of trajectories.

VI.2.4 Nonlinear string

The global attraction (VI.2.1) was established in [184] for general nonlinear 1D wave equations

$$\ddot{\psi}(x, t) = \psi''(x, t) + \chi(x) F(\psi(x, t)), \qquad x \in \mathbb{R}. \tag{VI.2.26}$$

Equation (VI.2.26) can be formally written as the Hamiltonian system (VI.2.3) with the Hamiltonian functional

$$\mathcal{H}(\psi, \pi) = \frac{1}{2} \int \left[|\pi(x)|^2 + |\psi'(x)|^2 + \chi(x) U(\psi(x, t)) \right] dx, \quad (\psi, \pi) \in \mathcal{E} := E_c \oplus L^2.$$

VI.2.5   Wave-particle system

The global attraction (VI.2.1) was established in [185] for *real-valued solutions* of the 3D wave equation coupled to a charged particle

$$\ddot{\psi}(\mathbf{x}, t) = \Delta\psi(\mathbf{x}, t) - \rho(\mathbf{x} - \mathbf{q}(t)), \qquad \mathbf{x} \in \mathbb{R}^3, \qquad \text{(VI.2.27)}$$

where $\rho \in C_0^\infty(\mathbb{R}^3)$ is a fixed function representing the charge density of the particle, and $\mathbf{q}(t) \in \mathbb{R}^3$ is the particle's position. The particle motion obeys the Hamiltonian equation with relativistic kinetic energy $\sqrt{1 + \mathbf{p}^2}$:

$$\dot{\mathbf{q}}(t) = \frac{\mathbf{p}(t)}{\sqrt{1 + \mathbf{p}^2(t)}}, \quad \dot{\mathbf{p}}(t) = -\nabla V(\mathbf{q}(t)) - \int \nabla\psi(\mathbf{x}, t)\rho(x - \mathbf{q}(t))\, dx. \quad \text{(VI.2.28)}$$

Here $-\nabla V(\mathbf{q})$ is the external force corresponding to a real potential $V(\mathbf{q})$, and the integral term is a self-force. Thus, the wave function ψ is generated by a charged particle, and plays the role of a potential, along with the external potential $V(\mathbf{q})$, in the second equation in (VI.2.28).

The system (VI.2.27), (VI.2.28) can formally be represented in Hamiltonian form

$$\dot{\psi} = D_\pi\mathcal{H}, \quad \dot{\pi} = -D_\psi\mathcal{H}, \quad \dot{\mathbf{q}}(t) = D_\mathbf{p}\mathcal{H}, \quad \dot{\mathbf{p}} = -D_\mathbf{q}\mathcal{H} \qquad \text{(VI.2.29)}$$

with the Hamiltonian (energy)

$$\mathcal{H}(\psi, \pi, \mathbf{q}, \mathbf{p}) = \frac{1}{2}\int [|\pi(\mathbf{x})|^2 + |\nabla\psi(\mathbf{x})|^2]\, d\mathbf{x} + \int \psi(\mathbf{x})\rho(\mathbf{x} - \mathbf{q})\, d\mathbf{x} + \sqrt{1 + \mathbf{p}^2} + V(\mathbf{q}).$$
$$\text{(VI.2.30)}$$

We assume that the external potential is confining:

$$V \in C^2(\mathbb{R}^3), \qquad V(\mathbf{q}) \to \infty, \qquad |\mathbf{q}| \to \infty. \qquad \text{(VI.2.31)}$$

In this case, the Hamiltonian (VI.2.30) is bounded from below:

$$\inf_{Y \in \mathcal{E}} \mathcal{H}(Y) = V_0 + \frac{1}{2}(\rho, \Delta^{-1}\rho), \qquad \text{(VI.2.32)}$$

where $\mathcal{E} := \dot{H}^1(\mathbb{R}^3) \oplus L^2(\mathbb{R}^3) \oplus \mathbb{R}^3 \oplus \mathbb{R}^3$ is the Hilbert phase space of finite energy states and

$$V_0 := \inf_{\mathbf{q} \in \mathbb{R}^3} V(\mathbf{q}) > -\infty. \qquad \text{(VI.2.33)}$$

Above, $\dot{H}^1(\mathbb{R}^3)$ is the completion of $C_0^\infty(\mathbb{R}^3)$ with respect to the norm

$$\|u\|_{\dot{H}^1} := \left(\int_{\mathbb{R}^3} |\nabla u|^2 dx\right)^{1/2}.$$

Under the conditions (VI.2.31), a solution

$$Y(t) = (\psi(\mathbf{x}, t), \dot{\psi}(\mathbf{x}, t), \mathbf{q}(t), \mathbf{p}(t)) \in C(\mathbb{R}, \mathcal{E})$$

of finite energy exists and is unique for any initial state $Y(0) \in \mathcal{E}$.

The global attraction of type (VI.2.1) for the system (VI.2.27), (VI.2.28) is proved in [185] under the following *Wiener condition* on the charge density ρ:

$$\hat{\rho}(\mathbf{k}) := \int e^{i\mathbf{k}\cdot\mathbf{x}} \rho(\mathbf{x}) \, d\mathbf{x} \neq 0, \qquad \mathbf{k} \in \mathbb{R}^3. \tag{VI.2.34}$$

This condition means that there is a strong coupling of the scalar wave field $\psi(\mathbf{x})$ to the particle. It is a suitable version of the "Fermi Golden Rule" for the system (VI.2.27), (VI.2.28): the perturbation $\rho(\mathbf{x} - \mathbf{q})$ is not orthogonal to eigenfunctions of the continuous spectrum of the Laplacian Δ. Such a condition was introduced by I.M. Sigal in the context of the nonlinear wave equation and the nonlinear Schrödinger equation [162].

The proof of global attraction (VI.2.1) for the system (VI.2.27), (VI.2.28) in [185] relies on

i) expressing the energy radiated to infinity via Liénard–Wiechert retarded potentials;

ii) its representation in the form of a convolution;

iii) a novel application of the Wiener Tauberian theorem.

The key point in the proof is the relaxation of acceleration

$$\ddot{\mathbf{q}}(t) \to 0, \qquad t \to \pm\infty. \tag{VI.2.35}$$

This relaxation was discovered about 100 years ago in classical electrodynamics and is known as the *radiation damping*; see [48, Chapter 16]. Traditional arguments are as follows:

1) the structure of Liénard–Wiechert formulae for the retarded potentials suggests that a particle with a nonzero acceleration radiates energy to infinity;

ii) the radiation has to decay in time, since the total energy of the solution is finite.

However, a rigorous proof of the relaxation (VI.2.35) is not so obvious. It first appeared in [185] under the Wiener condition (VI.2.34). This condition is necessary in the proof, which relies on an application of the Wiener Tauberian theorem. Note that this condition is sufficient but not necessary for the relaxation (VI.2.35): for example, the relaxation is also proved in [195, 196] for a weak coupling, i.e., in the case $\|\rho\|_{L^2(\mathbb{R}^3)} \ll 1$.

Remark VI.2.7. For a point charged particle which corresponds to the charge distribution $\rho(\mathbf{x}) = \delta(\mathbf{x})$, the system (VI.2.27)–(VI.2.28) is incorrect, since in this case any solution of the wave equation (VI.2.27) is singular at the point $x = q(t)$, and, accordingly, the integral in (VI.2.28) is not defined. The energy functional (VI.2.30) in this case is not bounded from below, because the last term in (VI.2.32) equals $-\infty$. Indeed, in the Fourier transform, this term has the form

$$(\rho, \Delta^{-1}\rho) = -\int \frac{|\hat{\rho}(\mathbf{k})|^2}{\mathbf{k}^2}\, d\mathbf{k}, \qquad (VI.2.36)$$

where $\hat{\rho}(\mathbf{k}) \equiv 1$. The divergence of the above integral at large $|\mathbf{k}|$ is the famous *ultraviolet divergence*. Thus, the self-energy of a point charge is infinite, which prompted M. Abraham to introduce the model of *extended electron* with a continuous charge density $\rho(\mathbf{x})$ [1], [2].

Exercise VI.2.8. Check (VI.2.29) and (VI.2.32).

VI.2.6   Coupled Maxwell–Lorentz equations

In [186], global attraction of the type (VI.2.1) was extended for the Maxwell–Lorentz equations with charged relativistic particle:

$$\begin{cases} \dot{\mathbf{E}}(\mathbf{x},t) = \operatorname{curl}\mathbf{B}(\mathbf{x},t) - \dot{\mathbf{q}}\rho(\mathbf{x}-\mathbf{q}), \qquad \dot{\mathbf{B}}(\mathbf{x},t) = -\operatorname{curl}\mathbf{E}(\mathbf{x},t) \\[2mm] \operatorname{div}\mathbf{E}(\mathbf{x},t) = \rho(\mathbf{x}-\mathbf{q}), \qquad \operatorname{div}\mathbf{B}(\mathbf{x},t) = 0 \\[2mm] \dot{\mathbf{q}} = \dfrac{\mathbf{p}}{\sqrt{1+\mathbf{p}^2}} \\[3mm] \dot{\mathbf{p}} = \displaystyle\int [\mathbf{E}(\mathbf{x},t) + \mathbf{E}^{\text{ext}}(\mathbf{x}) + \dot{\mathbf{q}}\times(\mathbf{B}(\mathbf{x},t) + \mathbf{B}^{\text{ext}}(\mathbf{x}))]\rho(\mathbf{x}-\mathbf{q})\, d\mathbf{x} \end{cases} . \quad (VI.2.37)$$

Here $\rho(\mathbf{x} - \mathbf{q})$ is the particle charge density, $\dot{\mathbf{q}}\rho(\mathbf{x} - \mathbf{q})$ is the corresponding current density, and

$$\mathbf{E}^{\text{ext}} = -\nabla\phi^{\text{ext}}(\mathbf{x}), \qquad \mathbf{B}^{\text{ext}} = -\operatorname{curl}\mathbf{A}^{\text{ext}}(\mathbf{x})$$

are external static Maxwell fields. Similarly to (VI.2.31), we assume that the *effective potential* is confining:

$$V(\mathbf{q}) := \int \phi^{\text{ext}}(\mathbf{x})\rho(\mathbf{x} - \mathbf{q})\, d\mathbf{x} \to \infty, \qquad |\mathbf{q}| \to \infty. \qquad (VI.2.38)$$

This system describes classical electrodynamics with *extended electron* introduced by M. Abraham [1], [2]. In the case of a point electron, when $\rho(\mathbf{x}) = \delta(\mathbf{x})$, such a system is not well-defined. Indeed, in this case, any solutions $\mathbf{E}(\mathbf{x},t)$ and $\mathbf{B}(\mathbf{x},t)$ of Maxwell's equations are singular at $\mathbf{x} = \mathbf{q}(t)$, and, accordingly, the integral in the last equation (VI.2.37) does not exist.

This system can be formally presented in Hamiltonian form if the fields are expressed in terms of the Maxwell potentials

$$\mathbf{E}(\mathbf{x}, t) = -\nabla \phi(\mathbf{x}, t) - \dot{\mathbf{A}}(\mathbf{x}, t), \qquad \mathbf{B}(\mathbf{x}, t) = -\operatorname{curl} \mathbf{A}(\mathbf{x}, t);$$

see [194]. The corresponding Hamiltonian functional reads

$$\mathcal{H}(\mathbf{E}, \mathbf{B}, \mathbf{q}, \mathbf{p}) = \frac{1}{2}[\langle \mathbf{E}, \mathbf{E} \rangle + \langle \mathbf{B}, \mathbf{B} \rangle] + V(\mathbf{q}) + \sqrt{1 + \mathbf{p}^2}. \qquad (VI.2.39)$$

The Hilbert phase space of finite energy states is now defined as

$$\mathcal{E} := [L^2(\mathbb{R}^3) \otimes \mathbb{R}^3] \oplus [L^2(\mathbb{R}^3) \otimes \mathbb{R}^3] \oplus \mathbb{R}^3 \oplus \mathbb{R}^3.$$

Under the condition (VI.2.38), the solution

$$Y(t) = (\mathbf{E}(\mathbf{x}, t), \mathbf{B}(\mathbf{x}, t), \mathbf{q}(t), \mathbf{p}(t)) \in C(\mathbb{R}, \mathcal{E})$$

of finite energy exists and is unique for any initial state $Y(0) \in \mathcal{E}$.

The key role in the proof of the global attraction (VI.2.1) in [186] is played again by the relaxation of acceleration (VI.2.35), which is derived by a suitable generalization of our methods [185]: the expression of energy radiated to infinity via Liénard–Wiechert retarded potentials, its representation in the form of a convolution, and the use of Wiener's Tauberian theorem.

In classical electrodynamics, the *radiation damping* (VI.2.35) is traditionally derived from the Larmor and Liénard formulae for radiation power of a point charged particle (formulae (14.22) and (14.24) of [48]). However, this approach ignores field feedback, although the latter plays the key role in the relaxation. The main problem is that this reverse field reaction is infinite for point charged particles. A rigorous treatment of these classical calculations was found first in [185], [186] for the Abraham model with *extended electron* under the Wiener condition (VI.2.34). The details can be found in [191] and [150, 151].

Problem VI.2.9. Check the conservation of energy (VI.2.39) for solutions of the system (VI.2.37).
Hint: differentiate w.r.t. time the energy $\mathcal{H}(t) := \mathcal{H}(\mathbf{E}(t), \mathbf{B}(t), \mathbf{q}(t), \mathbf{p}(t))$ and use equations (VI.2.37).

VI.3   Soliton-Like Asymptotics

In this section, we briefly describe known results on the global convergence to solitons (V.2.13) for translation-invariant equations. Moreover, we describe the results on adiabatic effective dynamics of solitons in slowly varying external potentials and discuss related questions on mass-energy equivalence.

VI.3.1   Global attraction to solitons

In [193], we proved the global attraction to solitons (V.2.13) for the system (VI.2.27), (VI.2.28) with zero external potential $V(\mathbf{q}) \equiv 0$. In this case, the system is translation-invariant, so the corresponding total momentum is conserved:

$$\mathbf{P}(t) := \mathbf{p}(t) - \int \nabla \psi(\mathbf{x}, t) \dot{\psi}(\mathbf{x}, t) \, d\mathbf{x} = \text{const}, \qquad t \in \mathbb{R}. \qquad \text{(VI.3.1)}$$

For any $\mathbf{v} \in \mathbb{R}^3$ with $|\mathbf{v}| < 1$, this system admits solitons, which are solutions of the form

$$S_{\mathbf{v}, \mathbf{q}}(t) \equiv (\varphi_{\mathbf{v}}(\mathbf{x} - \mathbf{v}t - \mathbf{q}), \mathbf{q} + \mathbf{v}t). \qquad \text{(VI.3.2)}$$

The global attraction to solitons means the long-time asymptotics

$$\psi(\mathbf{x}, t) \sim \varphi_{\mathbf{v}^{\pm}}(\mathbf{x} - q(t)), \quad \dot{\mathbf{q}}(t) \to \mathbf{v}^{\pm}, \qquad t \to \pm\infty \qquad \text{(VI.3.3)}$$

for any finite-energy solution, where the asymptotics for $\psi(\mathbf{x}, t)$ holds in local seminorms *in the comoving frame of reference*, i.e., in local seminorms of type $L^2(|\mathbf{x} - q(t)| < R)$ with any $R > 0$. Such soliton asymptotics were proved for the first time in [193] under the Wiener condition (VI.2.34).

In [194], similar asymptotics were established for the coupled Maxwell–Lorentz equations (VI.2.37) without external Maxwell fields, i.e., when

$$\mathbf{E}^{\text{ext}}(\mathbf{x}) \equiv 0 \equiv \mathbf{B}^{\text{ext}}(\mathbf{x}), \qquad \mathbf{x} \in \mathbb{R}^3.$$

This result gives the first rigorous proof of *radiation damping* for the translation-invariant system of classical electrodynamics. The proofs in [193] and [194] rely on

i) a *canonical transform* to the comoving frame;

ii) variational properties of solitons and their orbital stability;

iii) the relaxation of acceleration (VI.2.35).

More accurate soliton asymptotics *in global norms* with several solitons were discovered in 1965 by M. Kruskal and N.J. Zabusky in numerical simulation of the Korteweg–de Vries (KdV) equation: it is the decay into solitons,

$$\psi(\mathbf{x}, t) \sim \sum_n \varphi_n^{\pm}(\mathbf{x} - \mathbf{v}_n^{\pm} t) + w^{\pm}(\mathbf{x}, t), \qquad t \to \pm\infty, \qquad \text{(VI.3.4)}$$

where $w^{\pm}$ are certain dispersive waves. Later on, such asymptotics were proved by M.J. Ablowitz, H. Scgur, W. Eckhaus, A. van Harten and others for nonlinear *integrable* Hamiltonian translation-invariant equations (KdV, etc.), using the method of the *inverse scattering problem* [192].

In [197], results of numerical simulation were presented to confirm the soliton asymptotics (VI.3.4) also for general *nonintegrable* 1D *relativistic* nonlinear wave equations.

The *asymptotic stability* of solitons for *relativistic wave equations*

$$\ddot{\psi}(x,t) = \psi''(x,t) - m^2\psi(x,t) + F(\psi(x,t)), \qquad x \in \mathbb{R} \qquad (VI.3.5)$$

was first proved in [198, 199] for the nonlinearity of Ginzburg–Landau type.

Problem VI.3.1. Prove the total momentum conservation (VI.3.1) for solutions of the system (VI.2.27), (VI.2.28) with $V(\mathbf{q}) \equiv 0$.

VI.3.2 Adiabatic effective dynamics

The system (VI.2.27), (VI.2.28) admits soliton solutions (VI.3.2) in the case of identically zero external potential $V(\mathbf{q}) \equiv 0$. However, even in the case of nonzero external potential, *soliton-like solutions* of the form

$$\psi(\mathbf{x}, t) \approx \psi_{\mathbf{v}(t)}(\mathbf{x} - \mathbf{q}(t)) \qquad (VI.3.6)$$

may exist if the potential is slowly varying:

$$|\nabla V(\mathbf{q})| \leq \varepsilon \ll 1. \qquad (VI.3.7)$$

In this case, the total momentum (VI.3.1) is generally not conserved, but its slow evolution together with evolution of the parameters $\mathbf{q}(t)$, $\mathbf{v}(t)$ in (VI.3.6) can be described in terms of an appropriate finite-dimensional Hamiltonian dynamics.

Indeed, denote by $\mathbf{P} = \mathbf{P_v} \in \mathbb{R}^3$ the total momentum (VI.3.1) of the soliton $S_{\mathbf{v},\mathbf{q}}(t)$ in the notation (VI.3.2). It is important that the map $\mathcal{P} : \mathbf{v} \mapsto \mathbf{P_v}$ is an isomorphism of the ball $|\mathbf{v}| < 1$ on $\mathbb{R}^3$. Therefore, we can consider $\mathbf{q}$, $\mathbf{P}$ as global coordinates on the *solitary manifold*

$$\mathcal{S} := \{S_{\mathbf{v},\mathbf{q}}(0) : \mathbf{q}, \mathbf{v} \in \mathbb{R}^3, \ |\mathbf{v}| < 1\}.$$

The effective Hamiltonian functional is defined by

$$\mathcal{H}_{\text{eff}}(\mathbf{q}, \mathbf{P_v}) \equiv \mathcal{H}(S_{\mathbf{v},\mathbf{q}}), \qquad \mathbf{q}, \mathbf{P_v} \in \mathbb{R}^3, \qquad (VI.3.8)$$

where $\mathcal{H}$ is the Hamiltonian (VI.2.30) with $V(\mathbf{q}) \equiv 0$. This functional can be represented as

$$\mathcal{H}_{\text{eff}}(\mathbf{Q}, \mathbf{\Pi}) = E(\mathbf{\Pi}) + V(\mathbf{Q}),$$

since the first integral in (VI.2.30) does not depend on $\mathbf{Q}$, while the last integral vanishes on the solitons. Hence, the corresponding Hamiltonian equations have the form

$$\dot{\mathbf{Q}}(t) = \nabla E(\mathbf{\Pi}(t)), \qquad \dot{\mathbf{\Pi}}(t) = -\nabla V(\mathbf{Q}(t)). \tag{VI.3.9}$$

The main result in [205] is the following theorem. Let us denote by $\| \cdot \|_R$ the norm in $L^2(B_R)$, where B_R is the ball $\{\mathbf{x} \in \mathbb{R}^3 : |\mathbf{x}| < R\}$, and denote $\pi_\mathbf{v}(\mathbf{x}) := -\mathbf{v}\nabla\psi_\mathbf{v}(\mathbf{x})$.

Theorem VI.3.2. *Let condition* (VI.3.7) *hold, and suppose that the initial state*

$$(\psi(0), \pi(0), \mathbf{q}(0), \mathbf{p}(0)) \in \mathcal{S}$$

is a soliton with total momentum $\mathbf{P}(0)$. *Then the corresponding solution*

$$(\psi(\mathbf{x},t), \pi(\mathbf{x},t), \mathbf{q}(t), \mathbf{p}(t))$$

of the system (VI.2.27), (VI.2.28) *admits the "adiabatic asymptotics"*

$$|\mathbf{q}(t) - \mathbf{Q}(t)| \le C_0, \quad |\mathbf{P}(t) - \mathbf{\Pi}(t)| \le C_1\varepsilon \quad \text{for} \quad |t| \le C\varepsilon^{-1},$$

$$\sup_{t\in\mathbb{R}} \left[\|\nabla\psi(\mathbf{q}(t)+\mathbf{y}, t) - \nabla\psi_{\mathbf{v}(t)}(\mathbf{y})]\|_R + \|\dot{\psi}(\mathbf{q}(t)+\mathbf{y}, t) - \pi_{\mathbf{v}(t)}(\mathbf{y})\|_R \right] \le C\varepsilon,$$

where $\mathbf{P}(t)$ *denotes the total momentum* (VI.3.1), $\mathbf{v}(t) = \mathcal{P}^{-1}(\mathbf{\Pi}(t))$, *and* $(\mathbf{Q}(t), \mathbf{\Pi}(t))$ *is the solution of the effective Hamiltonian equations* (VI.3.9) *with initial conditions*

$$\mathbf{Q}(0) = \mathbf{q}(0), \qquad \mathbf{\Pi}(0) = \mathbf{P}(0).$$

We note that such relevance of the effective dynamics (VI.3.9) is due to the consistency of Hamiltonian structures:

I. The effective Hamiltonian (VI.3.8) is the restriction of the Hamiltonian functional (VI.2.30) with $V = 0$ to the solitary manifold $\mathcal{S}$.

II. As shown in [205], the canonical differential form of the Hamiltonian system (VI.3.9) is also the restriction to $\mathcal{S}$ of the canonical differential form of the system (VI.2.27), (VI.2.28): formally,

$$\mathbf{P} \cdot d\mathbf{Q} = \left[\mathbf{p} \cdot d\mathbf{q} + \int d\mathbf{x}\, \pi(\mathbf{x})\, d\psi(\mathbf{x}) \right]\Big|_{\mathcal{S}}.$$

Therefore, the total momentum $\mathbf{P}$ is canonically conjugate to the variable $\mathbf{Q}$ on the solitary manifold $\mathcal{S}$. This fact justifies the definition (VI.3.8) of the effective Hamiltonian as a function of the total momentum $\mathbf{P_v}$, and not of the particle momentum $\mathbf{p_v}$.

One of the important results in [205] is the following "effective dispersion relation":

$$E(\mathbf{\Pi}) \sim \frac{\mathbf{\Pi}^2}{2(1+m_e)} + \text{const}, \qquad |\mathbf{\Pi}| \ll 1. \qquad (\text{VI.3.10})$$

It means that the nonrelativistic mass of a slow soliton increases, due to the interaction with the field, by the amount

$$m_e = -\frac{1}{3}\langle \rho, \Delta^{-1}\rho \rangle. \qquad (\text{VI.3.11})$$

This increment is proportional to the field energy of the soliton at rest,

$$E_{\text{own}} = \mathcal{H}(\Delta^{-1}\rho, 0, 0, 0) = -\frac{1}{2}\langle \rho, \Delta^{-1}\rho \rangle, \qquad (\text{VI.3.12})$$

which agrees with the Einstein mass-energy equivalence principle (see Section VI.3.3 below).

Remark VI.3.3. The relation (VI.3.10) suggests that m_e is an increment of the effective mass. The true *dynamical justification* for such an interpretation is given by Theorem VI.3.2 which demonstrates the relevance of the effective dynamics (VI.3.9).

Generalizations. In [206], Theorem VI.3.2 was extended to solitons of the Maxwell–Lorentz equations (VI.2.37) with small external fields.

Following the articles [205, 206], suitable adiabatic effective dynamics was obtained in [203] and [204] for nonlinear Hartree and Schrödinger equations with slowly varying external potentials, and in [202, 207] and [208] for the nonlinear equations of Einstein–Dirac, Chern–Simon–Schrödinger and Klein–Gordon–Maxwell with small external fields. Similar adiabatic effective dynamics was established in [201] for an electron in the second-quantized Maxwell field in the presence of a slowly varying external potential.

The results of numerical simulation [197] (see the next chapter) confirm the adiabatic effective dynamics of solitons for relativistic 1D nonlinear wave equations.

VI.3.3 Mass-energy equivalence

M. Kunze and H. Spohn have established that in the case of the Maxwell–Lorentz equations (VI.2.37), the increment of nonrelativistic mass also turns out to be proportional to the energy of the static soliton's own field, [206].

Such an equivalence of the self-energy of a particle with its mass was first discovered in 1902 by M. Abraham: he showed by direct calculation that the energy E_{own} of electrostatic field of an electron at rest adds

$$m_e = \frac{4}{3}E_{\text{own}}/c^2$$

to its nonrelativistic mass (see [1, 2], and also [51, pp. 216–217]). By (VI.3.12) and (VI.2.36), this self-energy is infinite for a point electron with the charge density $\rho(x) = \delta(x)$. This means that the field mass for a point electron is infinite, which contradicts experiments. That is why M. Abraham introduced the model of electrodynamics with *extended electron* (VI.2.37), whose self-energy is finite.

At the same time, M. Abraham conjectured that the *entire mass* of an electron is due to its own electromagnetic energy, i.e., $m = m_e$: "*matter disappeared, only energy remains*", as philosophically-minded contemporaries wrote [46, pp. 63, 87, 88]. (smile :))

This conjecture was justified in 1905 by A. Einstein, who discovered the famous universal relation $E = m_0 c^2$ which follows from the Special Theory of Relativity [27]. The doubtful factor $\frac{4}{3}$ in Abraham's formula is due to the nonrelativistic character of the system (VI.2.37). According to the modern view, about 80% of the electron mass is of electromagnetic origin [29].

VI.4 Global Attraction to Stationary Orbits

The global attraction to stationary orbits (V.2.15) was proved i) in [209]–[217] for the Klein–Gordon and Dirac equations coupled to $U(1)$-invariant nonlinear oscillators, ii) in [210] for discrete in space and time difference approximations of such coupled systems, i.e., for the corresponding difference schemes, and iii) in [190] and [218]–[220] for the wave, Klein–Gordon, and Dirac equations with concentrated nonlinearities.

The attraction (V.2.15) was proved under the assumption that the equations are *strictly nonlinear*. For linear equations, the attraction can fail if the discrete spectrum consists of at least two points.

In this section, we present with detail the first results on global attraction (V.2.15) obtained in [212]–[214]. Besides the formal proof, in Subsection VI.4.10 we give an informal explanation of the *nonlinear radiative mechanism*, which causes the global attraction.

VI.4.1 Nonlinear Klein–Gordon equation

We consider the Klein–Gordon equation coupled to a nonlinear oscillator

$$\ddot{\psi}(x,t) = \psi''(x,t) - m^2\psi(x,t) + \delta(x)F(\psi(0,t)), \quad x \in \mathbb{R}, \quad t \in \mathbb{R}. \quad \text{(VI.4.1)}$$

The asymptotics (V.2.15) for this equation means that

$$\psi(x,t) \sim \psi_\pm(x)e^{-i\omega_\pm t}, \quad t \to \pm\infty. \quad \text{(VI.4.2)}$$

We consider complex solutions, identifying complex values $\psi \in \mathbb{C}$ with vectors $(\psi_1, \psi_2) \in \mathbb{R}^2$, where $\psi_1 = \operatorname{Re}\psi$ and $\psi_2 = \operatorname{Im}\psi$. Suppose that $F \in C^1(\mathbb{R}^2, \mathbb{R}^2)$ and that

$$F(\psi) = -\nabla_\psi U(\psi), \quad \psi \in \mathbb{C}, \quad \text{(VI.4.3)}$$

where U is a real-valued function and $\nabla_\psi := (\partial_{\psi_1}, \partial_{\psi_2})$. In this case, equation (VI.4.5) is formally equivalent to a Hamiltonian system of type (VI.2.3) in the Hilbert phase space $\mathcal{E} := H^1(\mathbb{R}) \oplus L^2(\mathbb{R})$. The Hamiltonian functional is

$$\mathcal{H}(\psi, \pi) = \frac{1}{2}\int \left[|\pi(x)|^2 + |\psi'(x)|^2 + m^2|\psi(x)|^2\right] dx + U(\psi(0)) \quad \text{(VI.4.4)}$$

for $(\psi, \pi) \in \mathcal{E}$. We can write (VI.4.1) in vector form,

$$\dot{Y}(t) = \mathcal{F}(Y(t)), \quad t \in \mathbb{R}; \quad Y(t) := (\psi(t), \dot{\psi}(t)). \quad \text{(VI.4.5)}$$

We assume that

$$\inf_{\psi \in \mathbb{C}} U(\psi) > -\infty; \quad \text{(VI.4.6)}$$

then a finite energy solution $Y(t) \in C(\mathbb{R}, \mathcal{E})$ to (VI.4.5) exists and is unique for any initial state $Y(0) \in \mathcal{E}$ (for details, see [214, Appendix C]). The a priori bound

$$\sup_{t \in \mathbb{R}} [\|\dot{\psi}(t)\|_{L^2(\mathbb{R})} + \|\psi(t)\|_{H^1(\mathbb{R})}] < \infty \qquad (VI.4.7)$$

holds due to the conservation of energy (VI.4.4). Note that the confining condition of type (VI.2.31) is no longer necessary, since the conservation of energy (VI.4.4) with $m > 0$ ensures the boundedness of solutions.

Further, we assume the $U(1)$-invariance of the potential:

$$U(\psi) = u(|\psi|), \qquad \psi \in \mathbb{C}. \qquad (VI.4.8)$$

Then the differentiation in (VI.4.3) gives us that

$$F(\psi) = a(|\psi|)\psi, \qquad \psi \in \mathbb{C}, \qquad (VI.4.9)$$

and therefore

$$F(e^{i\theta}\psi) = e^{i\theta}F(\psi), \qquad \theta \in \mathbb{R}. \qquad (VI.4.10)$$

By *stationary orbits* we mean solutions of the form

$$\psi(x, t) = \varphi_\omega(x)e^{-i\omega t} \qquad (VI.4.11)$$

with $\omega \in \mathbb{R}$ and $\varphi_\omega \in H^1(\mathbb{R})$. Each stationary orbit corresponds to some solution of the equation

$$-\omega^2\varphi_\omega(x) = \varphi_\omega''(x) - m^2\varphi_\omega(x) + \delta(x)F(\varphi_\omega(0)), \qquad x \in \mathbb{R}, \qquad (VI.4.12)$$

which is the *nonlinear eigenvalue problem*. Any solution $\varphi_\omega \in H^1(\mathbb{R})$ of this equation has the form

$$\varphi_\omega(x) = Ce^{-\kappa|x|}, \qquad \kappa := \sqrt{m^2 - \omega^2} > 0, \qquad (VI.4.13)$$

and the constant C satisfies the nonlinear algebraic equation

$$2\kappa C = F(C). \qquad (VI.4.14)$$

Hence, the solutions $\varphi_\omega \in H^1(\mathbb{R})$ exist for ω in some subset $\Omega \subset \mathbb{R}$ lying in the *spectral gap* $[-m, m]$. We denote the corresponding *solitary manifold* by $\mathcal{S}$:

$$\mathcal{S} = \{(e^{i\theta}\varphi_\omega, -i\omega e^{i\theta}\varphi_\omega) \in \mathcal{E} : \omega \in \Omega, \ \theta \in [0, 2\pi]\}. \qquad (VI.4.15)$$

Finally, we suppose that equation (VI.4.5) is *strictly nonlinear*:

$$U(\psi) = u(|\psi|^2) = \sum_0^N u_j|\psi|^{2j}, \qquad u_N > 0, \quad N \geq 2, \quad \psi \in \mathbb{C}. \qquad (VI.4.16)$$

For example, the well-known *Ginzburg–Landau potential* $U(\psi) = |\psi|^4/4 - |\psi|^2/2$ satisfies all three conditions (VI.4.6), (VI.4.8), and (VI.4.16).

Definition VI.4.1. *i)* $\mathcal{E}_F \subset H^1_{\text{loc}}(\mathbb{R}^3) \oplus L^2_{\text{loc}}(\mathbb{R}^3)$ *is the space* $\mathcal{E}$ *endowed with the seminorms*

$$\|Y\|_{\mathcal{E},R} := \|Y\|_{H^1(-R,R)} + \|Y\|_{L^2(-R,R)}, \qquad R = 1, 2, \ldots \qquad \text{(VI.4.17)}$$

ii) Convergence in $\mathcal{E}_F$ *is equivalent to convergence in each seminorm (VI.4.17).*

It is important to note that convergence in $\mathcal{E}_F$ is equivalent to convergence in a metric of type (VI.2.16),

$$\text{dist}(Y_1, Y_2) = \sum_{R=1}^{\infty} 2^{-R} \frac{\|Y_1 - Y_2\|_{\mathcal{E},R}}{1 + \|Y_1 - Y_2\|_{\mathcal{E},R}}, \qquad Y_1, Y_2 \in \mathcal{E}. \qquad \text{(VI.4.18)}$$

The main result of [212]–[214] is the following theorem.

Theorem VI.4.2. *Let conditions (VI.4.3), (VI.4.6), (VI.4.8) and (VI.4.16) hold. Then any finite energy solution* $Y(t) = (\psi(t), \dot{\psi}(t)) \in C(\mathbb{R}, \mathcal{E})$ *of (VI.4.5) is attracted to the solitary manifold (see Fig. V.2):*

$$Y(t) \xrightarrow{\mathcal{E}_F} \mathcal{S}, \qquad t \to \pm\infty, \qquad \text{(VI.4.19)}$$

where the attraction is in the sense of (VI.2.19).

Exercise VI.4.3. Check that equation (VI.4.1) is formally equivalent to the Hamiltonian system of type (VI.2.3) with the Hamiltonian (VI.4.4).

Problem VI.4.4. Prove the existence and uniqueness of the solution $Y(t) \in C_b(\mathbb{R}, \mathcal{E})$ for any initial state $Y(0) \in \mathcal{E}$, and the conservation of the energy (VI.4.4). **Hint:** apply the methods [156].

Exercise VI.4.5. Check (VI.4.9) and (VI.4.13), (VI.4.14).

VI.4.2 Generalizations and open questions

Generalizations: The global attraction (VI.4.19) was extended in [215] to the 1D Klein–Gordon equation with N nonlinear oscillators

$$\ddot{\psi}(x,t) = \psi''(x,t) - m^2\psi + \sum_{k=1}^{N} \delta(x - x_k) F_k(\psi(x_k, t)), \qquad x \in \mathbb{R}, \qquad \text{(VI.4.20)}$$

and in [209, 216, 217] it was extended to the Klein–Gordon and Dirac equations in $\mathbb{R}^n$ with $n \geq 3$ with a nonlocal interaction

$$\ddot{\psi}(x,t) = \Delta\psi(x,t) - m^2\psi + \sum_{k=1}^{N} \rho(x - x_k) F_k(\langle \rho(\cdot - x_k), \psi(\cdot, t)\rangle), \qquad \text{(VI.4.21)}$$

$$i\dot{\psi}(x,t) = \left(-i\boldsymbol{\alpha} \cdot \nabla + \beta m\right)\psi + \rho(x) F(\langle \rho, \psi(\cdot, t)\rangle), \qquad \text{(VI.4.22)}$$

under the Wiener condition (VI.2.34). Here $\boldsymbol{\alpha} = (\alpha^1, \ldots, \alpha^n)$ and β are the Dirac matrices.

Recently, the attraction (VI.4.19) was extended in [218, 190, 219] to the 3D wave and Klein–Gordon equations with concentrated nonlinearities, and in [220] it was extended to the 1D Dirac equation coupled to a nonlinear oscillator.

In addition, the attraction (VI.4.19) was extended in [210] to the nonlinear space-time discrete Hamiltonian equations that are discrete approximations of equations of type (VI.4.21), i.e., which are represented by the corresponding difference schemes. The proof relies on a new version of the Titchmarsh convolution theorem for distributions on a circle [221].

The analysis of solutions with compact spectrum to the (translation-invariant) Klein–Gordon and Schrödinger equations has been done in [211] with the aid of the version of the Titchmarsh theorem for partial convolutions. It is shown that, under rather general assumptions on the nonlinearity (which could be polynomial or, more generally, *algebraic*), the only solutions with compact spectrum are single-frequency solitary waves of the form $\varphi(x)e^{-i\omega t}$, $x \in \mathbb{R}^n$.

Open questions:

I. Global attraction (VI.4.2) to orbits with fixed frequencies $\omega_\pm$ has not yet been established.

II. Global attraction to stationary orbits for nonlinear Schrödinger equations has also not been established. In particular, such attraction is not proved for the 1D Schrödinger equation coupled to a nonlinear oscillator

$$i\dot\psi(x,t) = -\psi''(x,t) + \delta(x)F(\psi(0,t)), \qquad x \in \mathbb{R}. \tag{VI.4.23}$$

The main difficulty is the infinite "spectral gap" $(-\infty, 0)$ (see Remark VI.4.31).

III. Global attraction to solitons (VI.3.3) for the *translation-invariant* nonlinear Klein–Gordon equations, which are relativistically invariant, is an open problem. In particular, it is open for equation (VI.3.5) in one spatial dimension.

VI.4.3   Omega-limit trajectories

The proof of Theorem VI.4.2 is based on the general strategy of *omega-limit trajectories*, first introduced in [212] and developed further in [213]–[220].

Definition VI.4.6. *For a given function $Y(t) \in C(\mathbb{R}, \mathcal{E})$, its* omega-limit *trajectory is any limit function $Z(t)$ such that*

$$Y(t + s_j) \xrightarrow{\;\mathcal{E}_F\;} Z(t), \qquad t \in \mathbb{R}, \tag{VI.4.24}$$

for some sequence $s_j \to \infty$.

Definition VI.4.7. *A function $Y(t) \in C(\mathbb{R}, \mathcal{E})$ is* omega-compact *if for any sequence $s_j \to \infty$ there exists a subsequence $s_{j'} \to \infty$ such that (VI.4.24) holds with s_j replaced by $s_{j'}$.*

These concepts are useful in view of the following lemma, which lies at the basis of our approach.

Lemma VI.4.8. *Suppose that a given solution $Y(t) \in C(\mathbb{R}, \mathcal{E})$ of (VI.4.5) is omega-compact, and any its omega-limit trajectory $Z(t)$ corresponds to a stationary orbit (VI.4.11):*

$$Z(x, t) = (\varphi_\omega(x)e^{-i\omega t}, -i\omega\varphi_\omega(x)e^{-i\omega t}), \qquad (VI.4.25)$$

where $\omega \in \mathbb{R}$. Then the attraction (VI.4.19) holds for $t \to \infty$.

Proof. We need to show that

$$\lim_{t \to \infty} \operatorname{dist}(Y(t), \mathcal{S}) = 0.$$

Assume by contradiction that there exists a sequence $s_j \to \infty$ such that

$$\operatorname{dist}(Y(s_j), \mathcal{S}) \geq \delta > 0 \quad \forall j \in \mathbb{N}. \qquad (VI.4.26)$$

According to the omega-compactness of Y, the convergence (VI.4.24) holds for some subsequence $s_{j'} \to \infty$ and some stationary orbit (VI.4.25):

$$Y(t + s_{j'}) \xrightarrow{\mathcal{E}_F} Z(t), \qquad t \in \mathbb{R}, \qquad s_{j'} \to \infty. \qquad (VI.4.27)$$

But this convergence with $t = 0$ contradicts (VI.4.26), since $Z(0) \in \mathcal{S}$ by definition (VI.4.15). $\qquad \square$

For the proof of Theorem VI.4.2, it now suffices to check that

I. Each solution $Y(t) \in C(\mathbb{R}, \mathcal{E})$ of (VI.4.5) is omega-compact;

II. Any omega-limit trajectory is a stationary orbit (VI.4.25).

We check these conditions by analyzing the Fourier transform of solutions with respect to time. The main steps of the proof are as follows:

(1) Spectral representation for solutions of the nonlinear equation (VI.4.5):

$$\psi(t) = \frac{1}{2\pi} \int e^{-i\omega t} \tilde{\psi}(\omega) \, d\omega. \qquad (VI.4.28)$$

(2) The *absolute continuity* of the spectral density $\tilde{\psi}(\omega)$ on $(-\infty, -m) \cup (m, \infty)$, the bulk of the *continuous spectrum* of the free Klein–Gordon equation. This

is a nonlinear analog of the Kato theorem on the absence of embedded eigenvalues.

(3) The *omega-compactness* of each solution.

(4) The reduction of the spectrum *of each omega-limit trajectory* to a subset of the *spectral gap* $[-m, m]$. By the *spectrum* of a solution $\psi(t) := \psi(\cdot, t)$ we mean the support of its spectral density $\tilde{\psi}(\cdot)$, which is a tempered distribution of $\omega \in \mathbb{R}$ with values in H^1.

(5) Reduction of this spectrum to a *single point* using the *Titchmarsh convolution theorem*.

Below we follow this program, referring at some points to the articles [212] and [214] for technically important properties of quasimeasures.

VI.4.4   Limiting absorption principle

It suffices to prove the attraction (VI.4.19) only for positive times. We split the solution in two summands

$$\psi(x, t) = \psi_0(x, t) + \psi_1(x, t), \qquad (x, t) \in \mathbb{R}^2, \tag{VI.4.29}$$

where ψ_0 and ψ_1 satisfy

$$\left\{ \begin{array}{ll} \ddot{\psi}_0(x, t) = \psi_0''(x, t) - m^2 \psi_0(x, t), & \psi_0(x, 0) = \psi(x, 0), \quad \dot{\psi}_0(x, 0) = \dot{\psi}(x, 0) \\ \ddot{\psi}_1(x, t) = \psi_1''(x, t) - m^2 \psi_1(x, t) + \delta(x) F(\psi(0, t)), \ \psi_1(x, 0) = 0, \ \dot{\psi}_1(x, 0) = 0 \end{array} \right| . \tag{VI.4.30}$$

Both functions satisfy a priori bounds of type (VI.4.7):

$$\sup_{t \in \mathbb{R}} [\|\dot{\psi}_i(\cdot, t)\|_{L^2(\mathbb{R})} + \|\psi_i(\cdot, t)\|_{H^1(\mathbb{R})}] < \infty, \qquad i = 0, 1. \tag{VI.4.31}$$

Moreover, ψ_0 is a dispersive wave:

$$(\psi_0(\cdot, t), \dot{\psi}_0(\cdot, t)) \to 0, \qquad t \to \infty, \tag{VI.4.32}$$

where the convergence holds in the metric (VI.4.18). Accordingly, it suffices to prove the attraction (VI.4.19) only for $Y_1(t) = (\psi_1(\cdot, t), \dot{\psi}_1(\cdot, t))$, i.e.,

$$Y_1(t) \xrightarrow{\mathscr{E}_F} \mathcal{S}, \qquad t \to \pm\infty. \tag{VI.4.33}$$

Further, we extend $\psi_1(x, t)$ and $f(t) := F(\psi(0, t))$ by zero for $t < 0$:

$$\psi_+(x, t) := \left\{ \begin{array}{ll} \psi_1(x, t), & t > 0, \\ 0, & t < 0, \end{array} \right. \qquad f_+(t) := \left\{ \begin{array}{ll} f(t), & t > 0, \\ 0, & t < 0. \end{array} \right. \tag{VI.4.34}$$

From the second line of formula (VI.4.30) it follows that these functions satisfy the equation

$$\ddot{\psi}_+(x,t) = \psi''_+(x,t) - m^2\psi_+(x,t) + \delta(x)f_+(t), \qquad t > 0 \qquad (VI.4.35)$$

in the sense of distributions.

Remark VI.4.9. Let us stress that this equation is not equivalent to the original Klein–Gordon equation of type (VI.4.1) for $\psi_+(x,t)$ since generally $f_+(t) \not\equiv F(\psi_+(0,t))$.

The Fourier–Laplace transform. For tempered distributions $g(t)$ of $t \in \mathbb{R}$, we denote by $\tilde{g}(\omega)$ their Fourier transform, which is defined for $g \in C_0^\infty(\mathbb{R})$ by

$$\tilde{g}(\omega) = \int_\mathbb{R} e^{i\omega t}g(t)\,dt, \qquad \omega \in \mathbb{R}.$$

The a priori estimates (VI.4.31) imply that $\psi_+(x,t)$ and $f_+(t)$ are bounded functions of $t \in \mathbb{R}$ with values in the Sobolev space $H^1(\mathbb{R})$ and in $\mathbb{C}$, respectively. Therefore, their Fourier transforms are (by definition) *quasimeasures* of $\omega \in \mathbb{R}$ with values in $H^1(\mathbb{R})$ and in $\mathbb{C}$, respectively [137].

Let X be a Hilbert space.

Definition VI.4.10. *i) A quasimeasure $q(\omega)$ with values in X is the Fourier transform of a function $f(t) \in L^\infty(\mathbb{R}, X)$.*

ii) The convergence $\tilde{f}_n(\omega) \to \tilde{f}(\omega)$ as $n \to \infty$ in the space of quasimeasures is equivalent to the convergence $f_n(t) \to f(t)$ in $L^\infty_{loc}(\mathbb{R}, X)$ together with the bound

$$\sup_n \|f_n(t)\|_{L^\infty(\mathbb{R},X)} < \infty.$$

The Fourier transforms of $\psi_+(x,t)$ and $f_+(t)$ can be extended from the real axis to analytic functions in the upper complex half-plane $\mathbb{C}^+ := \{\omega \in \mathbb{C} : \operatorname{Im}\omega > 0\}$ with values in $H^1(\mathbb{R})$ and in $\mathbb{C}$, respectively:

$$\tilde{\psi}_+(x,\omega) = \int_0^\infty e^{i\omega t}\psi(x,t)\,dt, \qquad \tilde{f}_+(\omega) = \int_0^\infty e^{i\omega t}f(t)\,dt, \qquad \omega \in \mathbb{C}^+.$$

Indeed, by (VI.4.31), the functions $e^{-\varepsilon t}\psi_+(x,t)$ and $e^{-\varepsilon t}f_+(t)$ with $\varepsilon \geq 0$ are bounded in $L^\infty(\mathbb{R}, H^1)$ and in $L^\infty(\mathbb{R})$, and converge in the spaces $L^\infty_{loc}(\mathbb{R}, H^1)$ and $L^\infty_{loc}(\mathbb{R})$, respectively:

$$e^{-\varepsilon t}\psi_+(x,t) \to \psi_+(x,t), \qquad e^{-\varepsilon t}f_+(t) \to f_+(t), \qquad \varepsilon \to 0+.$$

Hence, by Definition VI.4.10, their Fourier transforms also converge in the sense of quasimeasures:

$$\tilde{\psi}_+(x,\omega+i\varepsilon) \to \tilde{\psi}_+(x,\omega), \qquad \tilde{f}_+(\omega+i\varepsilon) \to \tilde{f}_+(\omega), \qquad \varepsilon \to 0+. \quad (VI.4.36)$$

The limiting absorption principle. The Fourier transform of equation (VI.4.35) gives the stationary Helmholtz equation

$$-\omega^2 \tilde{\psi}_+(x,\omega) = \tilde{\psi}''_+(x,\omega) - m^2\tilde{\psi}_+(x,\omega) + \delta(x)\tilde{f}_+(\omega), \qquad x \in \mathbb{R}. \quad \text{(VI.4.37)}$$

This equation has two-parametric family of solutions, but only one of them admits analytic continuation to the upper complex half-plane $\operatorname{Im}\omega > 0$ with values in $H^1(\mathbb{R})$:

$$\tilde{\psi}_+(x,\omega) = -\tilde{f}_+(\omega)\frac{e^{ik(\omega)|x|}}{2ik(\omega)}, \qquad \operatorname{Im}\omega > 0. \quad \text{(VI.4.38)}$$

Here $k(\omega) := \sqrt{\omega^2 - m^2}$, where the branch has a positive imaginary part for $\operatorname{Im}\omega > 0$:

$$\operatorname{Im}k(\omega) > 0 \quad \text{for} \quad \operatorname{Im}\omega > 0.$$

If one uses in (VI.4.38) the other branch of the square root, one gets an expression which *grows exponentially* as $|x| \to \infty$, so it has to be discarded. Such an argument in the selection of solutions of stationary Helmholtz equations is known as the *limiting absorption principle* in the diffraction theory [142, 148, 149].

Spectral representation. We rewrite (VI.4.38) as

$$\tilde{\psi}_+(x,\omega) = \tilde{\alpha}(\omega)e^{ik(\omega)|x|}, \quad \operatorname{Im}\omega > 0, \quad \text{where } \alpha(t) := \psi_+(0,t). \quad \text{(VI.4.39)}$$

It is a nontrivial fact that the identity (VI.4.39) between analytic functions keeps its structure for their restrictions to the real axis, which are tempered distributions:

$$\tilde{\psi}_+(x,\omega + i0) = \tilde{\alpha}(\omega + i0)e^{ik(\omega+i0)|x|}, \qquad \omega \in \mathbb{R}, \quad \text{(VI.4.40)}$$

where $\tilde{\psi}_+(\cdot,\omega + i0)$ and $\tilde{\alpha}(\omega + i0)$ are the corresponding quasimeasures with values in $H^1(\mathbb{R})$ and $\mathbb{C}$, respectively. The problem is that the factor $M_x(\omega) := e^{ik(\omega+i0)|x|}$ is not smooth with respect to ω at the points $\omega = \pm m$. However, the identity (VI.4.40) can be justified by using the theory of quasimeasures [214] (see Problem VI.4.12 below).

Finally, applying the inverse Fourier transform to (VI.4.40), we obtain:

$$\psi_+(x,t) = \frac{1}{2\pi}\big\langle\!\big\langle e^{-i\omega t}, \tilde{\psi}_+(x,\omega+i0)\big\rangle\!\big\rangle = \frac{1}{2\pi}\big\langle\!\big\langle e^{-i\omega t}, \tilde{\alpha}(\omega+i0)e^{ik(\omega+i0)|x|}\big\rangle\!\big\rangle, \quad \text{(VI.4.41)}$$

where $\langle\!\langle \,\cdot\,, \cdot\, \rangle\!\rangle$ is the duality between distributions with compact support and smooth bounded functions, which we assume to be $\mathbb{C}$-linear with respect to each of the arguments. The right-hand side exists by Theorem VI.4.43; see the next section.

Problem VI.4.11. Prove (VI.4.32).
Hints: i) split the initial state $(\psi(x,0), \dot\psi(x,0))$ into two summands: choose the first one to be smooth and with compact support, while the second one has a small energy norm; ii) integrate by parts in the Fourier representation for the corresponding first solution and use the energy conservation for the second solution.

Problem VI.4.12. Prove (VI.4.40).
Hints: i) Prove that $M_x^\varepsilon(\omega) := e^{ik(\omega+i\varepsilon)|x|}$ is the *multiplier* in the space of quasimeasures for $\varepsilon \geq 0$, i.e., the multiplication by the function $M_x^\varepsilon(\omega)$ is a continuous operator with respect to the convergence of quasimeasures; deduce this by applying the Fourier transform to convert the product into the convolution and using the fact that $M_x^\varepsilon(\omega) = \tilde{K}_x^\varepsilon(\omega)$, where $K_x^\varepsilon \in L^1(\mathbb{R})$. ii) Prove that $\|K_x^\varepsilon - K_x^0\|_{L^1(\mathbb{R})} \to 0$ as $\varepsilon \to 0+$; see also [214, Lemma B.2].

VI.4.5 Nonlinear analog of the Kato theorem

It turns out that the properties of the quasimeasure $\tilde\alpha(\omega + i0)$ for $|\omega| < m$ and that for $|\omega| > m$ differ significantly. This is because the set $\{i\omega : |\omega| \geq m\}$ is the continuous spectrum of the operator

$$A = \begin{pmatrix} 0 & 1 \\ \frac{d^2}{dx^2} - m^2 & 0 \end{pmatrix},$$

which is the generator of the linear part of (VI.4.5). The following theorem plays a key role in the proof of Theorem VI.4.2. Denote

$$\Sigma := \{\omega \in \mathbb{R} : |\omega| > m\}. \tag{VI.4.42}$$

Below we will write $\tilde\alpha(\omega)$ and $k(\omega)$ instead of $\tilde\alpha(\omega + i0)$ and $k(\omega + i0)$ for $\omega \in \mathbb{R}$.

Theorem VI.4.13. ([214, Proposition 3.2]). *Let the conditions (VI.4.3), (VI.4.6), and (VI.4.8) hold, and let $\psi(t) \in C(\mathbb{R}, \mathcal{E})$ be any finite-energy solution of (VI.4.5). Then the corresponding tempered distribution $\tilde\alpha(\omega)$ is absolutely continuous on Σ. Moreover, $\alpha \in L^1(\Sigma)$ and*

$$\int_\Sigma |\tilde\alpha(\omega)|^2 |\omega\, k(\omega)|\, d\omega < \infty. \tag{VI.4.43}$$

Proof. We first explain the main idea of the proof. By (VI.4.41), the function $\psi_+(x,t)$ is formally a "linear combination" of the functions $e^{ik(\omega)|x|}$ with the amplitudes $\tilde\alpha(\omega)$:

$$\psi_+(x,t) = \frac{1}{2\pi} \int_\mathbb{R} \tilde\alpha(\omega) e^{ik(\omega)|x|} e^{-i\omega t}\, d\omega, \qquad x \in \mathbb{R}.$$

For $\omega \in \Sigma$, the functions $e^{ik(\omega)|x|}$ have an infinite $L^2(\mathbb{R})$-norm, whereas $\psi_+(\cdot, t)$ has a finite $L^2(\mathbb{R})$-norm. This is possible only if the amplitude $\tilde{\alpha}(\omega)$ is absolutely continuous in Σ. This idea is suggested by the Fourier integral

$$f(x) = \int_{\mathbb{R}} e^{-ikx} g(k)\, dk, \qquad (VI.4.44)$$

which belongs to $L^2(\mathbb{R})$ if and only if $g \in L^2(\mathbb{R})$. For example, if one takes $\tilde{\alpha}(\omega) = \delta(\omega - \omega_0)$ with $\omega_0 \in \Sigma$, then $\psi_+(\cdot, t)$ will be of infinite L^2-norm.

The rigorous proof relies on estimates of Paley–Wiener type. First note that the Parseval–Plancherel identity and (VI.4.31) imply that for $\varepsilon > 0$,

$$\int_{\mathbb{R}} \|\tilde{\psi}_+(\cdot, \omega + i\varepsilon)\|^2_{H^1(\mathbb{R})}\, d\omega = 2\pi \int_0^\infty e^{-2\varepsilon t} \|\psi_+(\cdot, t)\|^2_{H^1(\mathbb{R})}\, dt \le \frac{\text{const}}{\varepsilon}. \quad (VI.4.45)$$

On the other hand, we can exactly estimate the integral on the left-hand side of (VI.4.45). Indeed, according to (VI.4.41),

$$\tilde{\psi}_+(\cdot, \omega + i\varepsilon) = \tilde{\alpha}(\omega + i\varepsilon) e^{ik(\omega + i\varepsilon)|x|}.$$

Consequently, (VI.4.45) gives us

$$\varepsilon \int_{\mathbb{R}} |\tilde{\alpha}(\omega + i\varepsilon)|^2 \|e^{ik(\omega + i\varepsilon)|x|}\|^2_{H^1(\mathbb{R})}\, d\omega \le C < \infty, \qquad \varepsilon > 0. \quad (VI.4.46)$$

Here is a crucial observation about the asymptotics of the norm of $e^{ik(\omega + i\varepsilon)|x|}$ as $\varepsilon \to 0+$. Let us denote $\|\psi\|_{H^1(\mathbb{R})} := \left(\|\psi'\|^2_{L^2(\mathbb{R})} + m^2 \|\psi\|^2_{L^2(\mathbb{R})} \right)^{1/2}$.

Lemma VI.4.14. i) *For $\omega \in \mathbb{R}$,*

$$\lim_{\varepsilon \to 0+} \varepsilon \|e^{ik(\omega + i\varepsilon)|x|}\|^2_{H^1(\mathbb{R})} = n(\omega) := \begin{cases} \omega k(\omega), & |\omega| > m \\ 0, & |\omega| < m \end{cases}. \quad (VI.4.47)$$

ii) *For any $\delta > 0$, there exists an $\varepsilon_\delta > 0$ such that for $|\omega| > m + \delta$ and $\varepsilon \in (0, \varepsilon_\delta)$,*

$$\varepsilon \|e^{ik(\omega + i\varepsilon)|x|}\|^2_{H^1(\mathbb{R})} \ge \frac{n(\omega)}{2}. \quad (VI.4.48)$$

Proof. Let us compute the $H^1(\mathbb{R})$-norm using the Fourier representation. We set $k_\varepsilon = k(\omega + i\varepsilon)$ so that $\text{Im}\, k_\varepsilon > 0$; we then obtain

$$F_{x \to k} e^{ik_\varepsilon |x|} = 2ik_\varepsilon / (k_\varepsilon^2 - k^2), \qquad k \in \mathbb{R}.$$

Hence, by the Cauchy theorem on residues, we have

$$\|e^{ik_\varepsilon |x|}\|^2_{H^1(\mathbb{R})} = \frac{2|k_\varepsilon|^2}{\pi} \int_{\mathbb{R}} \frac{(k^2 + m^2)\, dk}{|k_\varepsilon^2 - k^2|^2} = -4\, \text{Im}\left[\frac{(k_\varepsilon^2 + m^2)\bar{k}_\varepsilon}{k_\varepsilon^2 - \bar{k}_\varepsilon^2} \right]. \quad (VI.4.49)$$

Substituting $k_\varepsilon^2 = (\omega + i\varepsilon)^2 - m^2$, we see that

$$\|e^{ik(\omega+i\varepsilon)|x|}\|^2_{H^1(\mathbb{R})} = \frac{1}{\varepsilon}\,\mathrm{Re}\left[\frac{(\omega+i\varepsilon)^2\overline{k(\omega+i\varepsilon)}}{\omega}\right], \quad \varepsilon > 0, \quad \omega \in \mathbb{R}, \quad \omega \neq 0.$$

The limits (VI.4.47) now follow, since the function $k(\omega)$ is real for $|\omega| > m$, but is purely imaginary for $|\omega| < m$. Therefore, the second assertion of the lemma also follows, since $n(\omega) > 0$ for $|\omega| > m$, and $n(\omega) \sim |\omega|^2$ for $|\omega| \to \infty$. $\square$

Remark VI.4.15. Clearly, $n(\omega) \equiv 0$ for $|\omega| < m$ without any calculations, since in that case the function $e^{ik(\omega)|x|}$ decays exponentially in x, and so the $H^1(\mathbb{R})$-norm of $e^{ik(\omega+i\varepsilon)|x|}$ remains finite as $\varepsilon \to 0+$.

Substituting (VI.4.48) into (VI.4.46), we obtain

$$\int_{\Sigma_\delta} |\tilde{\alpha}(\omega+i\varepsilon)|^2\omega k(\omega)\,d\omega \leq 2C, \quad 0 < \varepsilon < \varepsilon_\delta, \tag{VI.4.50}$$

with the same C as in (VI.4.46), and with the region $\Sigma_\delta := \{\omega \in \mathbb{R} : |\omega| > m + \delta\}$. We conclude that, for each $\delta > 0$, the set of functions

$$g_\varepsilon(\omega) = \tilde{\alpha}(\omega+i\varepsilon)|\omega k(\omega)|^{1/2}, \quad \varepsilon \in (0,\varepsilon_\delta),$$

is bounded in the Hilbert space $L^2(\Sigma_\delta)$, so it is weakly compact by the Banach Theorem. Hence, convergence of the distributions (VI.4.36) implies weak convergence in $L^2(\Sigma_\delta)$:

$$g_\varepsilon \rightharpoonup g, \quad \varepsilon \to 0+,$$

where the limit function $g(\omega)$ coincides with the distribution $\tilde{\alpha}(\omega)|\omega k(\omega)|^{1/2}$ restricted to Σ_δ. It remains to note that the norms of g in $L^2(\Sigma_\delta)$ for all $\delta > 0$ are bounded in view of (VI.4.50), and this implies (VI.4.43). Finally, $\tilde{\alpha}(\omega) \in L^1(\bar{\Sigma})$ by (VI.4.43) and by the Cauchy–Schwarz inequality. $\square$

Remark VI.4.16. Theorem VI.4.13 is a nonlinear analog of Kato's theorem on the absence of embedded eigenvalues in the continuous spectrum. Indeed, solutions of type $\psi_*(x)e^{-i\omega_* t}$ become

$$\psi_*(x)\left[\pi i\delta(\omega - \omega_*) + \mathrm{v.p.}\frac{1}{i(\omega - \omega_*)}\right]$$

in the Fourier–Laplace transform, and this is forbidden for $|\omega_*| > m$ by Theorem VI.4.13.

Exercise VI.4.17. Check (VI.4.49).

VI.4.6　Dispersive and bound components

Theorem VI.4.13 presupposes a splitting of the solutions (VI.4.41) into a "dispersive component" and a "bound component":

$$\psi_+(x,t) = \frac{1}{2\pi} \int_{|\omega|>m} (1-\zeta(\omega))\tilde{\alpha}(\omega)e^{ik(\omega)|x|}e^{-i\omega t}\,d\omega$$

$$+\frac{1}{2\pi}\langle\!\langle e^{-i\omega t}, \zeta(\omega)\tilde{\alpha}(\omega)e^{ik(\omega)|x|}\rangle\!\rangle = \psi_d(x,t) + \psi_b(x,t), \qquad (VI.4.51)$$

where

$$\zeta(\omega) \in C_0^\infty(\mathbb{R}), \quad \text{and} \quad \zeta(\omega)=1 \quad \text{for} \quad \omega \in [-m-1, m+1].$$

Note that $\psi_d(x,t)$ is a dispersive wave, because

$$\psi_d(x,t) := \frac{1}{2\pi}\int_{|\omega|>m}(1-\zeta(\omega))e^{-i\omega t}\tilde{\alpha}(\omega)e^{ik(\omega)|x|}\,d\omega \to 0, \quad t\to\infty \quad (VI.4.52)$$

according to the Riemann–Lebesgue theorem, since $\alpha \in L^1(\Sigma)$ by Theorem VI.4.13. Moreover, it is easy to show that

$$(\psi_d(\cdot,t), \dot{\psi}_d(\cdot,t)) \xrightarrow{\mathcal{E}_F} 0, \qquad t\to\infty. \qquad (VI.4.53)$$

Therefore, it remains to prove the attraction (VI.4.33) with the function $Y_b(t) := (\psi_b(\cdot,t), \dot{\psi}_b(\cdot,t))$ replacing $Y_1(t)$:

$$Y_b(t) \xrightarrow{\mathcal{E}_F} \mathcal{S}, \qquad t\to\infty. \qquad (VI.4.54)$$

Problem VI.4.18. Prove (VI.4.53). **Hint:** i) use (VI.4.43) to prove that

$$\psi_d(\cdot,t), \dot{\psi}_d(\cdot,t), \nabla\psi_d(\cdot,t) \in L^2(\mathbb{R}) \qquad (VI.4.55)$$

by splitting of (VI.4.52) into integrals over $(-\infty, -m-1]$ and $[m+1, \infty)$ and applying the Parseval–Plancherel theorem to each of the integrals; ii) split the function $\tilde{\alpha}(\omega)$ into two summands $\tilde{\alpha}(\omega) = \tilde{\alpha}_1(\omega) + \tilde{\alpha}_2(\omega)$, where $\tilde{\alpha}_1(\omega) \in C_0^\infty(\mathbb{R})$, while the integral of type (VI.4.43) with $\tilde{\alpha}_2(\omega)$ is small, and integrate by parts in the integral (VI.4.52) with $\tilde{\alpha}_1(\omega)$ in place of $\tilde{\alpha}(\omega)$.

VI.4.7　Omega-compactness

Here we establish the omega-compactness of the trajectory $Y_b(t)$, which is necessary for the application of Lemma VI.4.8. First, note that the bound component $\psi_b(x,t)$ is a smooth function for $x \neq 0$, and (VI.4.41) implies that

$$\partial_x^n\partial_t^l\psi_b(x,t) = \frac{1}{2\pi}\langle\!\langle(-i\omega)^l e^{-i\omega t}, \zeta(\omega)(ik(\omega)\,\text{sgn}\,x)^n\tilde{\alpha}(\omega)e^{ik(\omega)|x|}\rangle\!\rangle, \qquad (VI.4.56)$$

for $x \neq 0$ and any $n, l = 0, 1, \ldots$. These formulae must be justified, since the function $k(\omega)$ is not smooth at the points $\omega = \pm m$. The case $n = 0$ is trivial. For $n \geq 1$, the needed justification is done in [212, 214] by a suitable development of the theory of quasimeasures. These formulae imply the boundedness of each derivative.

Lemma VI.4.19. ([214, Proposition 4.1]). *For all $n, l = 0, 1, 2, \ldots$,*

$$\sup_{x \neq 0} \sup_{t \in \mathbb{R}} |\partial_x^n \partial_t^l \psi_b(x, t)| < \infty. \qquad (VI.4.57)$$

Proof. Note that in general the distribution $\tilde{\alpha}(\omega)$ is not a finite measure, since we only know that $\alpha(t) := \psi_+(0, t)$ is a bounded function by (VI.4.39) and (VI.4.7). To prove the lemma, it suffices to check that

$$\zeta(\omega)(ik(\omega) \operatorname{sgn} x)^n e^{ik(\omega)|x|}(-i\omega)^l = \tilde{g}_x(\omega), \qquad (VI.4.58)$$

where the function $g_x(\cdot)$ belongs to a bounded subset of $L^1(\mathbb{R})$ for $x \neq 0$ and $t \in \mathbb{R}$. This implies the lemma, since the right-hand side of (VI.4.56) is the convolution

$$\int_{\mathbb{R}} \alpha(t - s)g_x(s)\, ds,$$

where $\alpha(t)$ is a bounded function. $\square$

Remark VI.4.20. All properties of quasimeasures used above and below are justified in [212, 214] by similar arguments relying on the Parseval–Plancherel identity.

By the Ascoli–Arzelà theorem, the estimates (VI.4.57) imply that for any sequence $s_j \to \infty$ there is a subsequence $s_{j'} \to \infty$ such that

$$\partial_x^n \partial_t^l \psi_b(x, s_{j'} + t) \to \partial_x^n \partial_t^l \beta(x, t), \qquad x \neq 0, \ t \in \mathbb{R} \qquad (VI.4.59)$$

for any $n, l = 0, 1, \ldots$. Moreover, this convergence is uniform for $|x| + |t| \leq R$ with any $R > 0$, and

$$\sup_{x \neq 0} \sup_{t \in \mathbb{R}} |\partial_x^n \partial_t^l \beta(x, t)| < \infty. \qquad (VI.4.60)$$

Corollary VI.4.21. *Each solution $Y(t) \in C(\mathbb{R}, \mathcal{E})$ to (VI.4.5) is omega-compact.*

This follows from (VI.4.32), (VI.4.53), and (VI.4.59).

Problem VI.4.22. Prove (VI.4.56). **Hint:** i) it suffices to consider $n = 1$ and $l = 0$; ii) prove that for $x \neq 0$

$$\zeta(\omega)\frac{e^{ik(\omega)|x+\varepsilon|} - e^{ik(\omega)|x|}}{\varepsilon} = \tilde{f}_x^\varepsilon(\omega), \qquad (VI.4.61)$$

where $f_x^\varepsilon(\cdot) \in L^1(\mathbb{R})$ and $f_x^\varepsilon(\cdot) \to F_{\omega \to t}^{-1} \zeta(\omega) ik(\omega) \operatorname{sgn} x\, e^{ik(\omega)|x|}$ as $\varepsilon \to 0$, where the convergence holds in $L^1(\mathbb{R})$; iii) use the hints to Problem VI.4.12. See also [214, Proposition 4.1].

Problem VI.4.23. Check that the functions (VI.4.58) with $x \neq 0$ and $t \in \mathbb{R}$ belong to a bounded subset of $L^1(\mathbb{R})$. See also [214, Lemma B.3].

VI.4.8   Reduction of spectrum to spectral gap

For $n = l = 0$, the convergence (VI.4.59) and the representation (VI.4.56) imply the convergence of their Fourier transforms: for each $x \in \mathbb{R}$,

$$\zeta(\omega) \tilde{\alpha}(\omega) e^{ik(\omega)|x|} e^{-i\omega s_{j'}} \to \tilde{\beta}(x, \omega), \qquad j' \to \infty \tag{VI.4.62}$$

in the sense of quasimeasures as specified in Definition VI.4.10.

Lemma VI.4.24. *For any $x \in \mathbb{R}$,*

$$\tilde{\beta}(x, \omega) = 0, \qquad |\omega| > m. \tag{VI.4.63}$$

Proof. The convergence (VI.4.62) implies that for each $x \in \mathbb{R}$

$$\zeta(\omega) \tilde{\alpha}(\omega) e^{-i\omega s_{j'}} \to \tilde{\gamma}(\omega) := \tilde{\beta}(x, \omega) e^{-ik(\omega)|x|}, \qquad j' \to \infty, \tag{VI.4.64}$$

in the space of quasimeasures, since $e^{-ik(\omega)|x|}$ is a multiplier in this space. For the same reason, we obtain $\tilde{\beta}(x, \omega) = \tilde{\gamma}(\omega) e^{ik(\omega)|x|}$, and now (VI.4.56) implies that

$$\beta(x, t) = \frac{1}{2\pi} \langle\!\langle e^{-i\omega t}, \tilde{\gamma}(\omega) e^{ik(\omega)|x|} \rangle\!\rangle, \qquad (x, t) \in \mathbb{R}^2. \tag{VI.4.65}$$

Note that

$$\beta(0, t) = \gamma(t). \tag{VI.4.66}$$

Finally, the key observation is that (VI.4.64) and Theorem VI.4.13 imply that

$$\operatorname{supp} \tilde{\gamma} \subset [-m, m]. \tag{VI.4.67}$$

Now (VI.4.63) follows from (VI.4.66). $\square$

Remark VI.4.25. Note that the convergence (VI.4.62) in the sense of tempered distributions is not enough for the proof of Lemma VI.4.24. It is important that the function $e^{-ik(\omega)|x|}$ is a multiplier in the space of quasimeasures.

Exercise VI.4.26. Check the convergence (VI.4.62) in the sense of quasimeasures.

Exercise VI.4.27. Prove (VI.4.67). **Hint:** use Theorem VI.4.13, (VI.4.64) with $s_{j'} \to \infty$ and the Riemann–Lebesgue theorem.

VI.4.9 Reduction of spectrum to a single point

The question arises of the available means for verifying the representation (VI.4.25) for omega-limit trajectories. We have no formulae for solutions of the nonlinear equation (VI.4.1), and so the only hope is to use the nonlinear equation itself.

Equation for omega-limit trajectories and spectral inclusion

The key observation, albeit simple, is that $\beta(x,t)$ is a solution of equation (VI.4.5) for all $t \in \mathbb{R}$.

Lemma VI.4.28. *The function $\beta(x,t)$ satisfies the original equation (VI.4.1):*

$$\ddot{\beta}(x,t) = \beta''(x,t) - m^2\beta(x,t) + \delta(x)F(\beta(0,t)), \qquad (x,t) \in \mathbb{R}^2. \quad (VI.4.68)$$

Proof. This lemma follows from equation (VI.4.35), where we replace t by $s_{j'} + t$ and consider the limit $s_{j'} \to \infty$. Namely, we recall that for $t > 0$,

$$\psi_+(x,t) = \psi_d(x,t) + \psi_b(x,t), \quad f_+(t) = F(\psi_0(0,t) + \psi_d(0,t) + \psi_b(0,t)); \quad (VI.4.69)$$

see (VI.4.29), (VI.4.34), and (VI.4.51). Using (VI.4.32), (VI.4.53), (VI.4.59), respectively, we obtain that for $t \in \mathbb{R}$,

$$\psi_0(\cdot, s_{j'}+t) \to 0, \quad \psi_d(\cdot, s_{j'}+t) \to 0, \quad \psi_b(\cdot, s_{j'}+t) \to \beta(\cdot,t), \quad s_{j'} \to \infty, \quad (VI.4.70)$$

where each convergence holds in $H^1_{\mathrm{loc}}(\mathbb{R})$. Finally, substitute the representations (VI.4.69) into equation (VI.4.35) and replace t with $s_{j'}+t$. Now sending $s_{j'} \to \infty$ and using (VI.4.70), we obtain equation (VI.4.68) *in the sense of distributions.* $\qquad \square$

Remark VI.4.29. i) Let us recall that the function $\psi_+(x, s_{j'} + t)$ does not satisfy the original equation (VI.4.1) according to Remark VI.4.9.

ii) The limit equation (VI.4.68) holds for all $t \in \mathbb{R}$, whereas equation (VI.4.35) holds only for $t > 0$.

Applying the Fourier transform to equation (VI.4.68), we now obtain the corresponding *nonlinear stationary Helmholtz equation*:

$$-\omega^2 \tilde{\beta}(x,\omega) = \tilde{\beta}''(x,\omega) - m^2\tilde{\beta}(x,\omega) + \delta(x)\tilde{f}(\omega), \qquad (x,\omega) \in \mathbb{R}^2, \quad (VI.4.71)$$

where $f(t) := F(\beta(0,t)) = F(\gamma(t))$ in accordance with (VI.4.66). From (VI.4.9), we see that

$$f(t) = a(|\gamma(t)|)\gamma(t) = A(t)\gamma(t), \qquad A(t) := a(|\gamma(t)|), \qquad t \in \mathbb{R}.$$

Applying the Fourier transform, we obtain the convolution $\tilde{f} = \tilde{A} * \tilde{\gamma}$, which exists by (VI.4.67). Respectively, (VI.4.71) is now

$$-\omega^2 \tilde{\beta}(x, \omega) = \tilde{\beta}''(x, \omega) - m^2 \tilde{\beta}(x, \omega) + \delta(x)[\tilde{A} * \tilde{\gamma}](\omega), \qquad (x, \omega) \in \mathbb{R}^2.$$

This identity implies the key *spectral inclusion*

$$\operatorname{supp} \tilde{A} * \tilde{\gamma} \subset \operatorname{supp} \tilde{\gamma}, \qquad\qquad\qquad (VI.4.72)$$

because

$$\operatorname{supp} \tilde{\beta}(x, \cdot) \subset \operatorname{supp} \tilde{\gamma}, \qquad \operatorname{supp} \tilde{\beta}''(x, \cdot) \subset \operatorname{supp} \tilde{\gamma} \qquad (VI.4.73)$$

in view of the representation (VI.4.65). In the next section, we will derive (VI.4.25) from the inclusion (VI.4.72) using a fundamental result of Harmonic Analysis: the Titchmarsh convolution theorem.

Exercise VI.4.30. Prove (VI.4.73).

Titchmarsh Convolution Theorem

In 1926, E.C. Titchmarsh proved a theorem on the distribution of zeros of entire functions (see [163] and [154, p. 119]), which has, in particular, the following corollary (see [141, Theorem 4.3.3]):

Theorem. *Let $f(\omega)$ and $g(\omega)$ be distributions of $\omega \in \mathbb{R}$ with bounded supports. Then*

$$[\operatorname{supp} f * g] = [\operatorname{supp} f] + [\operatorname{supp} g],$$

where $[X]$ denotes the convex hull *of a set $X \subset \mathbb{R}$.*

Proof of Theorem VI.4.2. By Lemma VI.4.8 and Corollary VI.4.21, it suffices to check the representation (VI.4.25) for the above constructed omega-limit trajectory $\beta(x, t)$, since every omega-limit trajectory can be obtained in this way by (VI.4.32), (VI.4.53), and (VI.4.57).

First, we know that $\operatorname{supp} \tilde{\gamma}$ is bounded because of (VI.4.67). Consequently, $\operatorname{supp} \tilde{A}$ is also bounded, since $A(t) := a(|\gamma(t)|)$ is a polynomial in $|\gamma(t)|^2$ according to (VI.4.16). Now the spectral inclusion (VI.4.72) and the Titchmarsh theorem imply that

$$[\operatorname{supp} \tilde{A}] + [\operatorname{supp} \tilde{\gamma}] \subset [\operatorname{supp} \tilde{\gamma}],$$

hence it immediately follows that $[\operatorname{supp} \tilde{A}] \subset \{0\}$. Besides, $A(t) := a(|\gamma(t)|)$ is a bounded function due to (VI.4.60), because $\gamma(t) = \beta(0, t)$. Therefore, $\tilde{A}(\omega) = C\delta(\omega)$, and hence

$$a(|\gamma(t)|) = C_1, \qquad t \in \mathbb{R}. \qquad\qquad (VI.4.74)$$

Now, the strict nonlinearity condition (VI.4.16) implies

$$|\gamma(t)| = C_2, \qquad t \in \mathbb{R}. \tag{VI.4.75}$$

This immediately gives

$$\mathrm{supp}\, \tilde{\gamma} \subset \{\omega_+\} \tag{VI.4.76}$$

by the same Titchmarsh theorem for the convolution $\tilde{\gamma} * \overline{\tilde{\gamma}} = C_3 \delta(\omega)$. Therefore, we have $\tilde{\gamma}(\omega) = C_4 \delta(\omega - \omega_+)$, and now (VI.4.25) follows from (VI.4.65). Theorem VI.4.2 is proved.

Remark VI.4.31. In the case of the Schrödinger equation (VI.4.23), the Titchmarsh theorem does not work. The problem is that the continuous spectrum of the operator $-d^2/dx^2$ is the half-line $[0, \infty)$, so now the role of the spectral gap is played by the unbounded interval $(-\infty, 0)$. Accordingly, in this case the spectral inclusion (VI.4.77) only implies that $\mathrm{supp}\, \tilde{\beta}(x, \cdot) \subset (-\infty, 0)$, while the Titchmarsh theorem does not apply to distributions with unbounded supports.

Exercise VI.4.32. Deduce (VI.4.76) from (VI.4.75).

VI.4.10   Nonlinear radiative mechanism

Let us explain the informal arguments for the attraction to stationary orbits behind the formal proof of Theorem VI.4.2. The main part of the proof involves the study of the spectrum of omega-limit trajectories

$$\beta(x, t) = \lim_{s_{j'} \to \infty} \psi(x, s_{j'} + t).$$

Theorem VI.4.13 implies the spectral inclusion (VI.4.67), which leads to

$$\mathrm{supp}\, \tilde{\beta}(x, \cdot) \subset [-m, m], \qquad x \in \mathbb{R}. \tag{VI.4.77}$$

The Titchmarsh theorem then allows us to conclude that

$$\mathrm{supp}\, \tilde{\beta}(x, \cdot) \subset \{\omega_+\}. \tag{VI.4.78}$$

These two inclusions are suggested by the following two informal arguments:

A. *Dispersive radiation in the continuous spectrum;*

B. *Nonlinear spreading of spectrum and energy transfer from lower to higher harmonics.*

A. Dispersive radiation in the continuous spectrum. The inclusion (VI.4.77) is due to the dispersive mechanism, which can be illustrated by an example of energy radiation from a harmonic source with frequency lying in the continuous spectrum. Indeed, let us consider a one-dimensional linear Klein–Gordon equation with a harmonic source

$$\ddot{\psi}(x,t) = \psi''(x,t) - m^2\psi(x,t) + b(x)e^{-i\omega_0 t}, \qquad x \in \mathbb{R}, \qquad \text{(VI.4.79)}$$

where the amplitude $b \in L^2(\mathbb{R})$ and the real frequency ω_0 is different from $\pm m$. In this case, the *limiting amplitude principle* holds [148, 152, 157]:

$$\psi(x,t) \sim a(x)e^{-i\omega_0 t}, \qquad t \to \infty. \qquad \text{(VI.4.80)}$$

For equation (VI.4.79), this follows directly from the Fourier–Laplace transform in time

$$\tilde{\psi}(\omega,t) = \int_0^\infty e^{i\omega t}\psi(x,t)\,dt, \qquad x \in \mathbb{R}, \quad \text{Im}\,\omega > 0. \qquad \text{(VI.4.81)}$$

Indeed, applying this transform to equation (VI.4.79), we see that

$$-\omega^2\tilde{\psi}(x,\omega) = \tilde{\psi}''(x,\omega) - m^2\tilde{\psi}(x,\omega) + \frac{b(x)}{i(\omega - \omega_0)}, \qquad x \in \mathbb{R}, \quad \text{Im}\,\omega > 0,$$

where for the simplicity we assume zero initial data. Hence,

$$\tilde{\psi}(\cdot,\omega) = \frac{R(\omega)b}{i(\omega - \omega_0)} = \frac{R(\omega_0 + i0)b}{i(\omega - \omega_0)} + \frac{R(\omega)b - R(\omega_0 + i0)b}{i(\omega - \omega_0)} \qquad \text{(VI.4.82)}$$

for $\text{Im}\,\omega > 0$, where

$$R(\omega) := (H - \omega^2)^{-1}$$

is the resolvent of the Schrödinger operator $H := -d^2/dx^2 + m^2$. This resolvent is a convolution operator with the fundamental solution $-\frac{e^{ik(\omega)|x|}}{2ik(\omega)}$, where $k(\omega) = \sqrt{\omega^2 - m^2} \in \overline{\mathbb{C}^+}$ for $\omega \in \overline{\mathbb{C}^+}$, as in (VI.4.38). The last fraction of (VI.4.82) is a smooth function of ω at $\omega = \omega_0$, and therefore its contribution decays like (VI.4.53) in local energy seminorms. Consequently, the long-time asymptotics of $\psi(x,t)$ is determined by the middle fraction in (VI.4.82). Therefore, (VI.4.80) holds with the limiting amplitude $a(x) = R(\omega_0 + i0)b$. The Fourier transform of this limiting amplitude is equal to

$$\hat{a}(k) = -\frac{\hat{b}(k)}{k^2 + m^2 - (\omega_0 + i0)^2}, \qquad k \in \mathbb{R}.$$

This formula shows that the properties of the limiting amplitudes *differ significantly* in the cases $|\omega_0| < m$ and $|\omega_0| \geq m$: $a(x) \in H^2(\mathbb{R})$ for $|\omega_0| < m$, however,

$$a(x) \notin L^2(\mathbb{R}) \quad \text{for} \quad |\omega_0| \geq m, \qquad \text{(VI.4.83)}$$

if $|\hat{b}(k)| > 0$ in a neighborhood of the sphere $|k|^2 + m^2 = \omega_0^2$ (which in the 1D case consists of two points $k = \pm\sqrt{\omega_0^2 - m^2}$). This means the following:

I. In the case $|\omega_0| \geq m$, the energy of the solution $\psi(x,t)$ tends to infinity for large times according to (VI.4.80) and (VI.4.83). This means that energy is transmitted from the harmonic source to the wave field!

II. In the opposite case $|\omega_0| < m$, the energy of the solution remains bounded, so there is no energy radiation.

It is this energy radiation in the case of $|\omega_0| \geq m$ that prohibits the occurrence of harmonics with such frequencies in omega-limit trajectories. Indeed, any omega-limit trajectory cannot radiate at all, since the total energy is finite and bounded from below, and hence the radiation has to decay with time. These physical arguments make the inclusion (VI.4.77) plausible, although a rigorous proof of it, as was seen above, requires special arguments.

Recall that the set $i\Sigma := \{i\omega, \ \omega \in \mathbb{R}, \ |\omega| \geq m\}$ coincides with the continuous spectrum of the generator of the free Klein–Gordon equation. Radiation in the continuous spectrum is well known in the theory of waveguides. Namely, waveguides can transmit only signals with a frequency $|\omega_0| > \mu$, where μ is a *threshold frequency*, which is an edge point of the continuous spectrum [155]. In our case, the waveguide occupies the "entire space" $x \in \mathbb{R}$ and is described by the nonlinear Klein–Gordon equation (VI.4.1) with the threshold frequency m.

B. Nonlinear inflation of the spectrum and energy transfer from lower to higher harmonics. Let us show that the single-frequency spectrum (VI.4.78) is due to the inflation of the spectrum by nonlinear functions. For example, consider the potential $U(\psi) = |\psi|^4$. Accordingly, $F(\psi) = -\nabla_{\overline{\psi}}U(\psi) = -4|\psi|^2\psi$. We consider the sum $\psi(t) = e^{i\omega_1 t} + e^{i\omega_2 t}$ of two harmonics, whose spectrum is shown on Fig. VI.2:

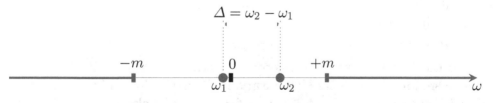

Figure VI.2: Two-point spectrum.

Let us substitute this sum into the nonlinearity:

$$F(\psi(t)) \sim \psi(t)\overline{\psi(t)}\psi(t) = e^{i\omega_2 t}e^{-i\omega_1 t}e^{i\omega_2 t} + \ldots = e^{i(\omega_2 + \Delta)t} + \ldots, \quad \Delta := \omega_2 - \omega_1.$$

The spectrum of this expression contains harmonics with the new frequencies $\omega_1 - \Delta$ and $\omega_2 + \Delta$. As a result, all the frequencies $\omega_1 - \Delta$, $\omega_1 - 2\Delta$, $\ldots$ and

$\omega_2 + \Delta$, $\omega_2 + 2\Delta$, ... also will appear in the nonlinear dynamics described by (VI.4.1) (see Fig. VI.3). Consequently, these frequencies will also appear in the nonlinear δ-function term which plays the role of the source.

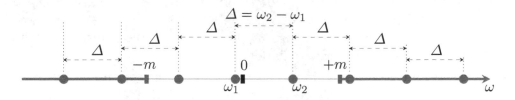

Figure VI.3: Nonlinear inflation of the spectrum.

As we already know, these frequencies, which lie in the continuous spectrum $|\omega| > m$, will surely cause energy radiation. This radiation will continue until the spectrum of the solution contains at least two different frequencies. It is this fact that prohibits the presence of two different frequencies in omega-limit trajectories, because the total energy is finite, and thus the radiation cannot continue forever.

However, we stress that

i) the precise meaning of the arguments "... until the spectrum of the solution contains at least two different frequencies" is given by the concept of omega-limit trajectory;

ii) the inflation of the spectrum by a nonlinearity is justified by the Titchmarsh convolution theorem.

Nonlinear radiative mechanism. The above arguments physically mean that the following binary nonlinear radiative mechanism takes place:

I. a nonlinearity inflates the spectrum, which means energy transfer from lower to higher harmonics;

II. the dispersive radiation carries energy to infinity.

For the first time we have rigorously justified such a nonlinear radiative mechanism for the nonlinear $U(1)$-invariant Klein–Gordon and Dirac equations (VI.4.1) and (VI.4.20)–(VI.4.22), and for other models; see [190, 210, 209, 215, 216, 217, 218, 219]. Our numerical experiments demonstrate a similar nonlinear radiative mechanism for *relativistic nonlinear wave equations* (see Remark VI.5.1 below). However, a rigorous proof is still missing.

VI.5 Numerical Simulation of Soliton Asymptotics

Here we describe the results of joint work with Arkadii Vinnichenko (1945–2009) on numerical simulation of i) global attraction to solitons (VI.3.3) and (VI.3.4), and ii) adiabatic effective dynamics of solitons (VI.3.9) for relativistic 1D nonlinear wave equations. Additional information can be found in [197].

VI.5.1 Kinks of relativistic Ginzburg–Landau equations

First let us describe numerical simulations of solutions of relativistic 1D nonlinear wave equations with the polynomial nonlinearity

$$\ddot\psi(x,t) = \psi''(x,t) + F(\psi(x,t)), \quad x \in \mathbb{R}, \tag{VI.5.1}$$

where $F(\psi) := -\psi^3 + \psi$ for $\psi \in \mathbb{R}$. Since $F(\psi) = 0$ for $\psi = 0, \pm 1$, there are three equilibrium states: $S(x) \equiv -1, 0, 1$. Equation (VI.5.1) is formally equivalent to the Hamiltonian system (VI.2.3) with the Hamiltonian

$$\mathcal{H}(\psi, \pi) = \int \left(\frac{1}{2}|\pi(x)|^2 + \frac{1}{2}|\psi'(x)|^2 + U(\psi(x)) \right) dx. \tag{VI.5.2}$$

where $U(\psi) = \frac{\psi^4}{4} - \frac{\psi^2}{2} + \frac{1}{4}$. This Hamiltonian is finite for functions $(\psi, \pi) \in \mathcal{E}$, where $\mathcal{E} = H_c^1 \oplus L^2$ (see (VI.2.5)), for which the convergence

$$\psi(x) \to \pm 1, \quad |x| \to \infty,$$

is sufficiently fast. The potential $U(\psi)$ has minima at $\psi = \pm 1$ and a maximum at $\psi = 0$. Accordingly, the two finite energy solutions $\psi = \pm 1$ are stable, and the solution $\psi = 0$ with infinite energy is unstable. Such potentials with two wells are called potentials of Ginzburg–Landau type.

In addition to the constant stationary solutions $S(x) \equiv -1, 0, +1$, there is also a nonconstant solution $S(x) = \tanh x/\sqrt{2}$, called a *kink*. Its shifts and its reflections, $\pm S(\pm x - a)$, are also stationary solutions, as well as their Lorentz transformations

$$\pm S(\gamma(\pm x - a - vt)), \quad \gamma = 1/\sqrt{1 - v^2}, \quad |v| < 1.$$

These are uniformly moving "traveling waves" (i.e., solitons). The kink is strongly compressed when the velocity v is close to ± 1. This compression is known as the *Lorentz contraction*.

VI.5.2 Numerical simulation

Our numerical experiments show a decay of finite energy solutions to a finite set of kinks and dispersive waves outside the kinks; this corresponds to the asymptotics of (VI.3.4). The result of one of the experiments is shown on Fig. VI.4: a finite energy solution of equation (VI.5.1) decays into three kinks. Here the vertical line is the time axis, and the horizontal line is the space axis. The spatial scale redoubles at $t = 20$ and $t = 60$. The red color corresponds to the values $\psi > 1 + \varepsilon$, the blue color to the values $\psi < -1 - \varepsilon$, and the yellow color to the intermediate values $-1 + \varepsilon < \psi < 1 - \varepsilon$, where $\varepsilon > 0$ is sufficiently small. Thus, the yellow stripes represent kinks, while the blue and red zones outside the yellow stripes are filled with dispersive waves.

For $t = 0$, the solution begins with a rather chaotic behavior, when there are no visible kinks. After 20 seconds, three separate kinks appear, which subsequently move almost uniformly.

The validity of the numerical simulations is confirmed by the appearance of the Lorentz contraction and of the Einstein time delay, and also by the observed group velocities of the dispersive waves. In the remainder of this section, we provide some related comments.

The Lorentz contraction

The left kink moves to the left at a low velocity $v_1 \approx 0.24$, the central kink is almost still, with its velocity $v_2 \approx 0.02$ being very small, and the right kink moves very rapidly with velocity $v_3 \approx 0.88$. The Lorentz spatial contraction with the factor $\sqrt{1 - v_k^2}$ is clearly visible in this picture: the central kink is the widest, the left is a bit narrower, and the right one the narrowest.

The Einstein time delay

The Einstein time delay is also very pronounced. Namely, all three kinks pulsate because of the presence of a nonzero eigenvalue in the equation linearized on the kink. Indeed, substituting $\psi(x, t) = S(x) + \varepsilon\varphi(x, t)$ in (VI.5.1), we obtain in the first-order approximation the linearized equation

$$\ddot{\varphi}(x, t) = \varphi''(x, t) - 2\varphi(x, t) - V(x)\varphi(x, t), \qquad \text{(VI.5.3)}$$

where the potential

$$V(x) = 3S^2(x) - 3 = -\frac{3}{\cosh^2 x/\sqrt{2}} \qquad \text{(VI.5.4)}$$

decays exponentially for large $|x|$. It is very fortunate that for this potential the spectrum of the corresponding Schrödinger operator

$$H := -\frac{d^2}{dx^2} + 2 + V(x)$$

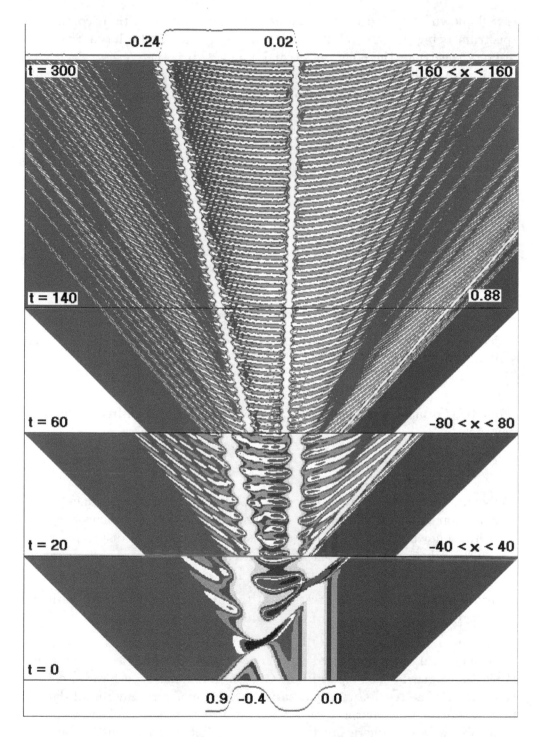

Figure VI.4: Decay to three kinks.

is well known [200]. Indeed, the operator H is nonnegative, with its continuous spectrum being the interval $[2, \infty)$. It turns out that H also has a two-point discrete spectrum: the points $\lambda = 0$ and $\lambda = 3/2$. It is the nonzero eigenvalue that is responsible for the pulsations that we observe for the central slow kink, with frequency $\omega_2 \approx \sqrt{3/2}$ and period $T_2 \approx 2\pi/\sqrt{3/2} \approx 5$. On the other hand, for the fast kinks, the ripples are much slower, i.e., the corresponding period is longer. This time delay agrees numerically with the Lorentz formulae, confirming the relevance of the numerical simulation results.

The dispersive waves

An analysis of dispersive waves provides an additional confirmation. Indeed, the space outside the kinks on Fig. VI.4 is filled with dispersive waves whose values are very close to ± 1, with accuracy of 0.01. These waves satisfy with high accuracy the linear Klein–Gordon equation obtained by linearization of the Ginzburg–Landau equation (VI.5.1) on the stationary solutions $\psi_{\pm} \equiv \pm 1$:

$$\ddot{\varphi}(x, t) = \varphi''(x, t) + 2\varphi(x, t), \qquad (VI.5.5)$$

which is quite different from (VI.5.3). The corresponding dispersion relation $-\omega^2 = -k^2 + 2$ yields

$$\omega(k) = \pm\sqrt{k^2 - 2}, \qquad |k| > \sqrt{2}.$$

From this formula, one derives the group velocities of high-frequency wave packets:

$$\omega'(k) = \pm\frac{k}{\sqrt{k^2 - 2}} = \pm\frac{\sqrt{\omega^2 + 2}}{\omega}, \qquad |k| > \sqrt{2}. \qquad (VI.5.6)$$

These wave packets are clearly visible on Fig. VI.4 as straight lines whose propagation velocities converge to ± 1. This convergence is explained by the high-frequency limit $|\omega'(k)| \to 1$ as $|\omega| \to \infty$. For example, for dispersive waves emitted by the central kink, the frequencies $\omega = \pm n\omega_2 \to \pm\infty$ are generated by the polynomial nonlinearity in (VI.5.1) in accordance with Fig. VI.3.

Remark VI.5.1. These observations of the dispersive waves agree with the nonlinear radiative mechanism from Section VI.4.10.

The nonlinearity in (VI.5.1) is chosen exactly because of the known discrete spectrum of the linearized equation (VI.5.3). In numerical experiments [197], more general nonlinearities of Ginzburg–Landau type have also been considered. The results were qualitatively the same: for "any initial data of finite energy", the solution decays for large times to a sum of kinks and of dispersive waves. Numerically, this is clearly visible, but rigorous justification remains an open problem.

Exercise VI.5.2. Check (VI.5.3)–(VI.5.4), and (VI.5.5).

VI.5.3 Soliton asymptotics

Besides the kinks, the numerical experiments [197] also reveal soliton-like asymptotics of type (VI.3.4) and adiabatic effective dynamics (VI.3.9) for complex solutions of the 1D relativistic nonlinear wave equations (VI.3.5). Polynomial potentials of the form

$$U(\psi) = a|\psi|^{2m} - b|\psi|^{2n}, \qquad \psi \in \mathbb{C}, \tag{VI.5.7}$$

were considered with a, $b > 0$ and $m > n = 2, 3, \dots$. Accordingly,

$$F(\psi) = 2am|\psi|^{2m-2}\psi - 2bn|\psi|^{2n-2}\psi. \tag{VI.5.8}$$

The parameters a, b, m, n were taken as follows:

N	a	m	b	n
1	1	3	0.61	2
2	10	4	2.1	2
3	10	6	8.75	5

Various "smooth initial data" $\psi(x,0), \dot{\psi}(x,0)$ with supports on the interval $[-20, 20]$ were considered. The second-order difference scheme with $\Delta x \sim 0.01$ and $\Delta t \sim 0.001$ was employed. In all cases, asymptotics of type (VI.3.4) were observed with the number of solitons 0, 1, 3, and 5 for $t > 100$.

VI.5.4 Adiabatic effective dynamics

In the numerical experiments [197], the adiabatic effective dynamics (VI.3.9) was also observed for soliton-like solutions of type (VI.3.6) of the 1D equations (VI.3.5) with a slowly varying external potential (VI.3.7):

$$\ddot{\psi}(x,t) = \psi''(x,t) - \psi(x,t) + F(\psi(x,t)) - V(x)\psi(x,t), \qquad x \in \mathbb{R}. \tag{VI.5.9}$$

This equation is formally equivalent to the Hamiltonian system of type (VI.2.3) with the Hamiltonian

$$\mathcal{H}_V(\psi, \pi) = \int \left(\frac{1}{2}|\pi(x)|^2 + \frac{1}{2}|\psi'(x)|^2 + U(\psi(x)) + \frac{1}{2}V(x)|\psi(x)|^2 \right) dx. \tag{VI.5.10}$$

The soliton-like solutions are of the form (cf. (VI.3.6))

$$\psi(x,t) \approx e^{i\Theta(t)}\varphi_{\omega(t)}\big(\gamma_{v(t)}(x - q(t))\big). \tag{VI.5.11}$$

The numerical experiments [197] qualitatively confirm the Hamiltonian adiabatic effective dynamics for the parameters Θ, ω, q, and v, but it has not yet been rigorously justified.

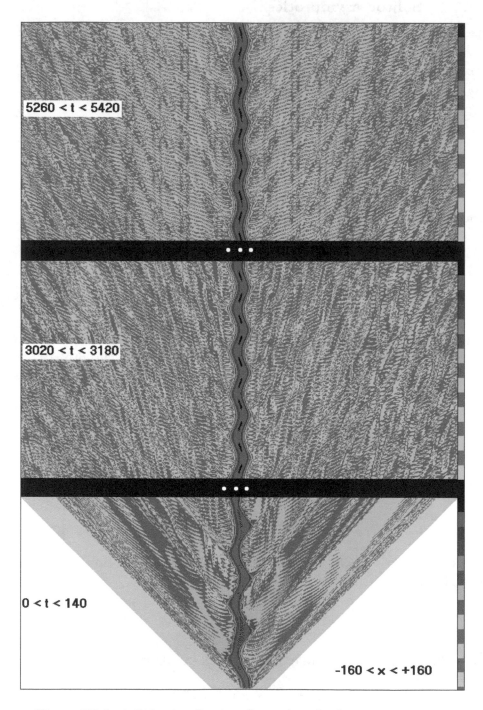

5260 < t < 5420

3020 < t < 3180

0 < t < 140

-160 < x < +160

Figure VI.5: Adiabatic effective dynamics of relativistic solitons.

For example, Fig. VI.5 represents a solution of equation (VI.5.9) with the potential (VI.5.7), where $a = 10$, $m = 6$, $b = 8.75$, and $n = 5$. The potential is $V(x) = -0.2\cos(0.31x)$ and the initial conditions are

$$\psi(x, 0) = \varphi_{\omega_0}(\gamma_{v_0}(x - q_0)), \qquad \dot{\psi}(x, 0) = 0, \qquad \text{(VI.5.12)}$$

where $v_0 = 0$, $\omega_0 = 0.6$ and $q_0 = 5.0$. We note that the initial state does not belong to the solitary manifold. The effective width (half-amplitude) of the solitons is in the range $[4.4, 5.6]$. It is quite small when compared to the spatial period of the potential $2\pi/0.31 \approx 20$. The results of the numerical simulations are shown on Fig. VI.5:

• The blue and green colors represent a dispersive wave with values

$$|\psi(x, t)| < 0.01,$$

while the red color represents the top of a soliton with values

$$|\psi(x, t)| \in [0.4, 0.8].$$

• The soliton trajectory (the "red snake" in the middle of Fig. VI.5) corresponds to oscillations of a classical particle in the potential $V(x)$.

• For $0 < t < 140$, the solution is far from the solitary manifold, and the radiation is rather intense.

• For $3020 < t < 3180$, the solution approaches the solitary manifold, and the radiation weakens. The amplitude of oscillations of the soliton is almost unchanged over a long time, confirming the Hamiltonian type of the effective dynamics.

• However, for $5260 < t < 5420$, the amplitude of the soliton oscillations is halved. This suggests that on a large time scale the deviation from Hamiltonian effective dynamics becomes essential. Consequently, the effective dynamics gives a good approximation only on the adiabatic time scale of type $t \sim \varepsilon^{-1}$.

• The deviation of the effective dynamics from being Hamiltonian is due to radiation, which plays the role of dissipation.

• The radiation is realized as dispersive waves, which carry the excess energy to infinity. The dispersive waves combine into uniformly moving wave packets with a discrete set of group velocities, as on Fig. VI.4. The magnitude of the solution is of order ~ 1 on the trajectory of the soliton, while the values of the dispersive waves is less than 0.01 for $t > 200$, so that their energy density does not exceed 0.0001. The amplitude of the dispersive waves decays at large times.

• In the limit as $t \to \pm\infty$, the soliton should converge to a static position corresponding to a local minimum of the potential $V(x)$. However, the numerical observation of this "ultimate stage" is hopeless, since the rate of the convergence slows down with the decrease of radiation.

Exercise VI.5.3. Check that equation (VI.5.9) formally reads as the Hamiltonian system (VI.2.3) with Hamiltonian (VI.5.10).

Appendix A

Old Quantum Theory

In this appendix, we recall the main achievements of the classical atomic Lorentz–Thomson theory (1897–1900) and of the Old Quantum Theory (1900–1925), as developed by M. Planck, A. Einstein, N. Bohr, P. Debye, A. Sommerfeld, H. Moseley, W. Pauli, and many others [11, 26, 86, 82, 90]. In particular, we recall the discovery of the black-body radiation law, the introduction of the Bohr–Sommerfeld quantization rules, selection rules, and the theory of the normal Zeeman effect. This reminiscence is necessary for the introduction of the Schrödinger quantum mechanics and of the Pauli theory of electron spin.

A.1 Black-body radiation law

The key role in the emergence of quantum theory was played by the investigation of the *spectral intensity* of radiation which is in thermodynamic equilibrium with matter at some fixed temperature $T > 0$.

The spectral intensity depends on the temperature T, which is well known from our everyday life as the temperature dependence of color of burning carbon, or of an electric stove, or an incandescent lamp. The key peculiarity of this spectral intensity is that it *does not depend on the substance*. This phenomenon was well known from observations on melting metals as early as in the sixth millennium BC: it is impossible to distinguish visually the heated iron ore from the burning carbon in thermodynamic equilibrium during fusion in the blast furnace. This coincidence of colors was used as an indication of the beginning of the melting. It means that at a fixed temperature, there is the *same radiation field* in the thermodynamic equilibrium with different substances.

The total radiation from a substance is a sum of reflected waves and the "own radiation" of the substance. In particular, the reflected waves are negligible in the case of black substance (e.g., carbon), and then the spectral intensity of the own radiation is approximately the universal spectral intensity. However, at a fixed temperature, the *total* spectral intensity of the own radiation and of the absorbed light are the same for any substance. This is why the universal spectral intensity is called the *black body radiation law*; see details in [50]. Note that our ability to distinguish the colors of bodies is due to the fact that we observe them in the *reflected light* of a source (sun or lamp), which is not in thermodynamic equilibrium with the bodies.

The great attention to the absorption and radiation of different substances was due to R. Bunsen and G. Kirchhoff's observations during 1859–1862 of spectra of different substances in a gas burner. In particular, they discovered the coincidence of spectral lines of the substance radiation and absorption, relying on J. Fraunhofer's observations of 1814 who mapped the wavelength over 570 *dark lines* in the solar spectrum. Some of the dark lines were first observed in 1802 by W.H. Wollaston.

The universal spectral intensity was measured experimentally in 1865–1899 by brilliant experimental physicists of that time — J. Tyndall, J. Stefan, W. Weber, F. Paschen, A. Pringsheim, H. Rubens, and others — for a broad variety of temperatures $T > 0$ and frequencies ω.

The results of these empirical observations were summarized in 1896 by W. Wien in the formula known as the *Wien law* [51, (1.19)]. This formula was in a perfect agreement with the experimental observation for large frequencies ω, but was unsatisfactory for small frequencies.

On the other hand, in 1894, Lord Rayleigh and J. Jeans obtained the formula which agreed perfectly with the experimental observation for small ω but was unsatisfactory for large frequencies. The formula was obtained on the basis of Boltzmann's *equipartition principle*.

The key progress was achieved by M. Planck in 1900 who refined the Wien law and discovered the famous formula (that earned him the 1918 Nobel Prize), now known as the *Kirchhoff–Planck law* (I.1.1). This formula is in perfect agreement with the experimental observations for large and small values of ω.

A.2   The Thomson electron and the Lorentz theory

The existence of electric currents in atoms was predicted by A. Ampère back in 1820. In 1838, M. Faraday discovered *cathode rays* in a vacuum tube. In 1895, J. Perrin experimentally proved that these rays carry a negative charge. During 1893–1897, J.J. Thomson conducted a series of ingenious experiments on the deflection of these rays in electric and magnetic fields, and came to the conclusion that the cathode rays consist of negatively charged particles (electrons) with trajectories that obey the Lorentz equation (I.2.11):

$$\mathrm{m}\ddot{\mathbf{x}}(t) = e\left[\mathbf{E}^{\mathrm{ext}}(\mathbf{x}(t),t) + \frac{1}{c}\dot{\mathbf{x}}(t) \times \mathbf{B}^{\mathrm{ext}}(\mathbf{x}(t),t)\right], \qquad \text{(A.2.1)}$$

where $e < 0$ is the electron charge, m is its mass, and $\mathbf{E}^{\text{ext}}(\mathbf{x}, t)$ and $\mathbf{B}^{\text{ext}}(\mathbf{x}, t)$ are the external Maxwell fields. In these experiments, J.J. Thomson for the first time measured with great accuracy the ratio e/m. It is for these experiments that J.J. Thomson was awarded the Nobel Prize in 1906.

During 1892–1900, H.A. Lorentz applied equation (A.2.1) to atomic electrons in a constant uniform magnetic field $\mathbf{B}$, replacing the first term on the right by an elastic force:

$$\text{m}[\ddot{\mathbf{x}}(t) + \omega_0^2 \mathbf{x}(t)] = \frac{e}{c}\dot{\mathbf{x}}(t) \times \mathbf{B}. \tag{A.2.2}$$

This classical equation led to a remarkable explanation of basic atomic phenomena (polarization, dispersion and others), and served as a basis for P. Drude's classical theory of metals (1900).

A.3   The Zeeman effect

In 1897, P. Zeeman discovered a splitting of spectral lines of atoms in magnetic fields. A part of these experimental results (*normal Zeeman effect*) was explained in 1897 by H.A. Lorentz in the framework of equation (A.2.2) with a uniform magnetic field. Namely, in this case the linear equation (A.2.2) is easy to solve: choosing $\mathbf{B} = (0, 0, B)$, we obtain:

$$\ddot{x}^1 + \omega_0^2 x^1 = 2\Omega_\Lambda \dot{x}^2, \qquad \ddot{x}^2 + \omega_0^2 x^2 = -2\Omega_\Lambda \dot{x}^1, \qquad \ddot{x}^3 + \omega_0^2 x^3 = 0, \quad (A.3.1)$$

where Ω_Λ is the *Larmor frequency* (III.1.3). Therefore, we have the solution $x^3(t) = \sin(\omega_0 t - \phi_0)$, and we arrive at the following equation for the complex function $z(t) := x^1(t) + ix^2(t)$:

$$\ddot{z} + \omega_0^2 z = -2i\Omega_\Lambda \dot{z}. \tag{A.3.2}$$

The corresponding characteristic equation is $-\omega^2 + \omega_0^2 = 2\Omega_\Lambda \omega$, and hence

$$\omega_\pm = -\Omega_\Lambda \pm \sqrt{\omega_0^2 + \Omega_\Lambda^2} \sim -\Omega_\Lambda \pm \omega_0, \qquad B \to 0, \tag{A.3.3}$$

where $\omega_0 > 0$ and $0 < \Omega_\Lambda \ll \omega_0$. Therefore, for small B, we have two solutions $z(t) = e^{i\omega_\pm t}$ with frequencies $|\omega_\pm| \sim \omega_0 \pm \Omega_\Lambda > 0$.

So, the spectrum of the solution contains three different frequencies that make up the *Zeeman triplet*:

$$\omega_0, \qquad \omega_0 \pm \Omega_\Lambda, \tag{A.3.4}$$

which correspond to the splitting of the "atomic spectral line" ω_0 into three components. Such a splitting was observed experimentally by P. Zeeman in 1895.

This *classical* interpretation of atomic spectra turned out to be only asymptotically correct for highly excited atoms, as it was shown later by the Bohr *correspondence principle* (1913); see Section A.5 below.

Moreover, H. Lorentz calculated the corresponding angular distribution of atomic radiation and its polarization [51, Section 14.5]. All these calculations were in a perfect agreement with P. Zeeman's experimental observations. In 1902, H. Lorentz, together with P. Zeeman, were awarded the Nobel Prize.

A.4   Debye's quantization rule

In 1913, N. Bohr formulated the postulates (I.4.3) and (I.4.4) on transitions between quantum stationary orbits. The question of determining the energies of the stationary orbits $E_n = \hbar \omega_n$ immediately came up. For the hydrogen atom, the Balmer empirical formula (1885) gives the values (I.6.5) with high accuracy.

The greatest surprise was caused by the discreteness of the set of possible energies. In 1911–1913, P. Ehrenfest conjectured the relationship of this discreteness with adiabatic invariance of the energy under slow variation of parameters [3, Section 52]. In particular, all basic integrals of the dynamical equations are adiabatic invariants: energy, momentum, and angular momentum.

In 1913, P. Debye calculated the Lagrangian action $S := \oint \mathbf{p} \cdot d\mathbf{q}$ for periodic orbits of the classical electron in the hydrogen atom and discovered that the Balmer formula (I.6.5) is equivalent to the "quantization rule"

$$S = \oint \mathbf{p} \cdot d\mathbf{q} = hn, \quad n = 1, 2, \ldots, \qquad h = 2\pi\hbar \qquad (A.4.1)$$

(see [51, p. 20]), where the integral is taken over the period of the classical electron orbit with energy E_n. The calculations of Debye rely on the Rutherford classical model of the atom as a point particle (electron) in the Coulomb field of the nucleus.

Let us recall these calculations assuming that the position of the nucleus is fixed since its mass is about 1840 times the mass of the electron. However, all the calculations admit a suitable generalization to the two-body problem, which takes into account the motion of the nucleus.

The dynamics of the electron in the Rutherford model of atom with nucleus of charge $|e|Z$ is described by the Lorentz equation (A.2.1) with the Coulomb field $\mathbf{E}^{\text{ext}}(\mathbf{x}) = -eZ\mathbf{x}/|\mathbf{x}|^3$ and $\mathbf{B}^{\text{ext}}(\mathbf{x}) \equiv 0$. In the Gaussian units [48, p. 781], [49], the equation reads as

$$\mathrm{m}\ddot{\mathbf{x}}(t) = -\frac{e^2 Z \mathbf{x}(t)}{|\mathbf{x}(t)|^3} = -\nabla U(\mathbf{x}(t)), \qquad U(\mathbf{x}) = -\frac{e^2 Z}{|\mathbf{x}|}. \qquad (A.4.2)$$

Multiplying by $\dot{x}$, we obtain the energy integral $E = m\dot{x}^2/2 + U(\mathbf{x})$. Any trajectory lies in a certain plane. We can choose the coordinates depending on the solution so that the trajectory lies in the plane $x^3 = 0$. In particular, for a circular orbit $\mathbf{x}(t) = x^1(t) + ix^2(t) = re^{-i\omega t}$, equation (A.4.2) gives

$$m\omega^2 r = e^2 Z/r^2, \qquad (A.4.3)$$

and the energy integral turns into $E = m\omega^2 r^2/2 - e^2 Z/r$. Eliminating ω, we obtain:

$$r = -\frac{e^2 Z}{2E}, \qquad |\omega|r = \sqrt{e^2 Z/(mr)} = \sqrt{-2E/m}, \qquad (A.4.4)$$

which implies that $E < 0$. Hence, for a circular orbit with period $T = 2\pi/|\omega|$, the Lagrangian action reads:

$$S = \oint \mathbf{p} \cdot d\mathbf{q} = m\int_0^T \dot{\mathbf{x}}^2(t)\, dt = 2\pi m|\omega|r^2 = -2\pi m\sqrt{-2E/m}\,\frac{e^2 Z}{2E}, \qquad (A.4.5)$$

since $\mathbf{p} := \nabla_{\dot{\mathbf{x}}}\Lambda = m\dot{\mathbf{x}} = -i\omega mre^{-i\omega t}$, where $\Lambda = m\dot{\mathbf{x}}^2/2 + e^2 Z\mathbf{x}/|\mathbf{x}|$ is the Lagrangian of the Lorentz equation (A.4.2). Therefore, (A.4.1) is equivalent to

$$e^2 Z\sqrt{-\frac{m}{2E_n}} = \hbar n. \qquad (A.4.6)$$

This result coincides with formulae (I.6.5), (I.6.6) obtained in the later Schrödinger theory.

Exercise A.4.1. Check (A.4.4).

A.5   Correspondence principle and selection rules

In a general form, the correspondence principle was formulated by N. Bohr in 1920, although he applied it in various formulations since 1913. For example,

The frequency of atomic radiation (I.4.4) for large quantum numbers n, n' asymptotically equals the frequency of revolution of the classical electron.

Until 1913, this frequency of revolution was identified with the frequency of the atomic radiation as in the Lorentz theory of the normal Zeeman effect in Section A.2. The revolution frequency in circular orbits is calculated from the Lorentz equation (A.4.2). Indeed, eliminating r and $E = E_n$ from (A.4.4) and (A.4.6), we obtain the *classical frequency*

$$\omega_{\mathrm{cl}} = \frac{me^4 Z^2}{\hbar^3 n^3}. \qquad (A.5.1)$$

On the other hand, by (I.6.5) and (I.6.6), quantum radiation frequencies (I.4.4) for large n, n' and bounded differences $n' - n$ admit the asymptotics

$$\omega_{nn'} = -\frac{2\pi cR}{n'^2} + \frac{2\pi cR}{n^2} \sim \frac{\mathrm{m}e^4 Z^2}{\hbar^3 n^3}(n' - n). \qquad (A.5.2)$$

Therefore, the correspondence principle dictates the *selection rule*

$$n' - n = \pm 1 \qquad (A.5.3)$$

for large quantum numbers n.

Exercise A.5.1. Check (A.5.1).

A.6   The Bohr–Sommerfeld quantization

In 1915–1916, A. Sommerfeld and W. Wilson suggested extending the Debye quantization rule (A.4.1) to solutions of Hamiltonian systems with several "periodic degrees of freedom" q^k and the corresponding canonically conjugate momenta p_k. In this case, they suggested the following quantization rules: for each k,

$$\oint p_k \, dq^k = hn_k, \quad n_k = 0, \pm 1, \ldots, \qquad h = 2\pi\hbar. \qquad (A.6.1)$$

In particular, for the Rutherford model of atom (A.4.2) with the Coulomb potential $U = -e^2 Z/|x|$, the corresponding Lagrangian, momentum and Hamiltonian are as follows:

$$\begin{cases} \Lambda(\mathbf{x}, \dot{\mathbf{x}}) &= \frac{\mathrm{m}\dot{\mathbf{x}}^2}{2} + \frac{e^2 Z}{|\mathbf{x}|}, \quad \mathbf{p} := \nabla_{\dot{\mathbf{x}}}\Lambda = \mathrm{m}\dot{\mathbf{x}} \\[2mm] \mathcal{H}(\mathbf{p}, \mathbf{x}) &= \mathbf{p} \cdot \dot{\mathbf{x}} - \Lambda = \frac{\mathbf{p}^2}{2\mathrm{m}} - \frac{e^2 Z}{|\mathbf{x}|} \end{cases} \Bigg|. \qquad (A.6.2)$$

Let us consider a fixed bounded trajectory $\mathbf{x}(t)$ of equation (A.4.2). The trajectory lies in some plane P passing through the nucleus located at the origin $\mathbf{x} = 0$. Therefore, we can assume that this plane is given by $\theta = 0$ in the appropriate spherical coordinates r (radius), φ (longitude), θ (latitude). Each bounded trajectory is periodic, and the quantization rules (A.6.1) take the form of integrals over a period

$$\oint p_r \, dr = hn_r, \qquad \oint p_\varphi \, d\varphi = hn_\varphi, \qquad \oint p_\theta \, d\theta = hn_\theta = 0. \qquad (A.6.3)$$

Let us show that these quantization rules imply the Debye condition (A.4.1), and, respectively, the Balmer formula (I.6.5) for the energies. First, in the polar coordinates r, φ in the plane $\theta = 0$, the Lagrangian reads as

$$\Lambda = \mathrm{m}(\dot{r}^2 + r^2\dot{\varphi}^2)/2 + e^2 Z/r,$$

which implies

$$p_r := \partial_{\dot{r}} \Lambda = m\dot{r}, \qquad p_\varphi := \partial_{\dot{\varphi}} \Lambda = mr^2 \dot{\varphi} = \text{const},$$

since $\dot{p}_\varphi = \partial_\varphi \Lambda = 0$. Therefore,

$$\oint p_\varphi \, d\varphi = 2\pi p_\varphi = h n_\varphi, \qquad p_\varphi = \hbar n_\varphi, \quad n_\varphi = 0, \pm 1, \ldots . \qquad (A.6.4)$$

The Euler–Lagrange equation for p_r gives

$$m\ddot{r} = mr\dot{\varphi}^2 - \frac{e^2 Z}{r^2} = \frac{p_\varphi^2}{mr^3} - \frac{e^2 Z}{r^2}; \qquad (A.6.5)$$

after the integration, we obtain:

$$\frac{m\dot{r}^2}{2} = -\frac{p_\varphi^2}{2mr^2} + \frac{e^2 Z}{r} + E. \qquad (A.6.6)$$

Hence,

$$p_r = m\dot{r} = \pm \frac{\sqrt{2m[Er^2 + e^2 Zr] - p_\varphi^2}}{r}. \qquad (A.6.7)$$

The minimal and maximal values of the radius $r_\pm$ on the trajectory are obtained from the condition $\dot{r} = 0$ (that is, $r_\pm$ correspond to zeros of the right-hand side of (A.6.7)). It is easy to verify that $E = \mathcal{H}$, i.e., E is the energy of the trajectory. The energy E must be negative, since otherwise the expression under the square root would remain positive for arbitrarily large r and the trajectory would be unbounded (corresponding to a parabolic or hyperbolic motion). Therefore, evaluating the integral in the first quantization condition (A.6.3), we arrive at

$$0 \le \oint p_r \, dr = 2 \int_{r_-}^{r_+} \sqrt{2m[Er^2 + e^2 Zr] - p_\varphi^2} \, \frac{dr}{r} = 2\pi \left[\sqrt{\frac{me^4 Z^2}{2|E|}} - |p_\varphi| \right] = h n_r.$$

$$(A.6.8)$$

Substituting here $|p_\varphi| = \hbar |n_\varphi|$ from (A.6.4), we obtain

$$\sqrt{\frac{me^4 Z^2}{2|E|}} = \hbar n, \qquad n = n_r + |n_\varphi|.$$

Finally, the quantized energy levels are

$$E_n = -|E_n| = -\frac{me^4 Z^2}{2\hbar^2 n^2}, \qquad n = 1, \ldots, \qquad (A.6.9)$$

which coincides with $\hbar\omega_n$ given by (I.6.5), (I.6.6). Obviously, $|p_\varphi| = mr^2|\dot{\varphi}|$ coincides with the magnitude of the electron's angular momentum $\mathbf{L} := \mathbf{x} \times \mathbf{p}$, and so

$$|\mathbf{L}| = p_\varphi = \hbar|n_\varphi|, \qquad 0 \le |n_\varphi| \le n \qquad (\text{A.6.10})$$

since $n_r \ge 0$ by (A.6.8). Finally, note that the Debye quantization condition (A.4.1) also holds, since $\mathbf{p} \cdot d\mathbf{q}$ is the invariant *canonical differential form*

$$\mathbf{p} \cdot d\mathbf{q} := \sum_{k=1}^{3} p_k \, dq^k = p_r \, dr + p_\varphi \, d\varphi + p_\theta \, d\theta,$$

so $\oint \mathbf{p} \cdot d\mathbf{q} = \hbar(n_r + n_\varphi)$ by (A.6.3).

Exercise A.6.1. Check that the constant E in (A.6.6) coincides with the Hamiltonian $\mathcal{H}$.

Exercise A.6.2. Justify the inequality in the formula (A.6.8). **Hint:**

$$\oint p_r \, dr := m \int_0^T \dot{r}^2(t) \, dt, \qquad (\text{A.6.11})$$

where $T > 0$ is the period.

Exercise A.6.3. Evaluate the integral (A.6.8). **Hint:** see [51, p. 262].

Remark A.6.4. The positivity property (A.6.8) is invariant with respect to the time reversal $t \mapsto s = -t$, since

$$\int_0^T \left(\frac{dr}{dt}\right)^2 dt = -\int_0^{-T} \left(\frac{d\rho}{ds}\right)^2 ds = \int_{-T}^0 \left(\frac{d\rho}{ds}\right)^2 ds, \qquad (\text{A.6.12})$$

where $\rho(s) := r(-s)$.

A.7   Atom in magnetic field

The Bohr–Sommerfeld quantization (A.6.1) can also be applied to the classical three-dimensional model of an atom with the Coulomb potential of the nucleus in a uniform magnetic field $\mathbf{B}$. Indeed, in this case, the Lorentz equation (A.2.1) takes the form

$$m\ddot{\mathbf{x}}(t) = -\frac{e^2 Z \mathbf{x}(t)}{|\mathbf{x}(t)|^3} + \frac{e}{c}\dot{\mathbf{x}}(t) \times \mathbf{B} = -\frac{e^2 Z \mathbf{x}(t)}{|\mathbf{x}(t)|^3} + m\tilde{\mathbf{B}}\dot{\mathbf{x}}(t), \qquad (\text{A.7.1})$$

where $\tilde{\mathbf{B}}$ is a skew-symmetric matrix such that $\tilde{\mathbf{B}}\mathbf{v} = \frac{e}{mc}\mathbf{v} \times \mathbf{B}$ for $\mathbf{v} \in \mathbb{R}^3$. The corresponding Lagrangian and Hamiltonian read as

$$\left\{ \begin{array}{rcl} \Lambda_{\mathbf{B}}(\mathbf{x}, \dot{\mathbf{x}}) & = & \dfrac{m\dot{\mathbf{x}}^2}{2} + \dfrac{e^2 Z}{|\mathbf{x}|} + \dfrac{e}{c}\mathbf{x} \cdot \mathbf{A}(\mathbf{x}) \\[4mm] \mathcal{H}_{\mathbf{B}}(\mathbf{x}, \mathbf{p}) & = & \mathbf{p} \cdot \dot{\mathbf{x}} - \Lambda_{\mathbf{B}} = \dfrac{1}{2m}\left[\mathbf{p} - \dfrac{e}{c}\mathbf{A}(\mathbf{x})\right]^2 - \dfrac{e^2 Z}{|\mathbf{x}|} \end{array} \right|, \qquad (A.7.2)$$

where the vector potential of the uniform magnetic field is

$$\mathbf{A}(\mathbf{x}) = \frac{1}{2}\mathbf{B} \times \mathbf{x}, \qquad (A.7.3)$$

so that $\mathbf{B} = \operatorname{curl}\mathbf{A}(\mathbf{x})$, and the electron's momentum is [51, (12.82)]

$$\mathbf{p} := \nabla_{\dot{\mathbf{x}}}\Lambda_{\mathbf{B}} = m\dot{\mathbf{x}} + \frac{e}{c}\mathbf{A}(\mathbf{x}). \qquad (A.7.4)$$

A.7.1 Rotating frame of reference. Larmor's theorem

The homogeneous and *sufficiently small* magnetic field can be eliminated by the transition to the frame rotating around the direction of the magnetic field $\mathbf{B}$ with the Larmor angular velocity $\Omega_L = \dfrac{e|\mathbf{B}|}{2mc}$ introduced in (III.1.3). Indeed, in this frame, the trajectory $\mathbf{x}(t)$ becomes $\mathbf{x}'(t) = e^{-\frac{\tilde{\mathbf{B}}}{2}t}\mathbf{x}(t)$, and, substituting $\mathbf{x}(t) = e^{\frac{\tilde{\mathbf{B}}}{2}t}\mathbf{x}'(t)$ into (A.7.1), we obtain:

$$m\ddot{\mathbf{x}}'(t) = -\frac{e^2 Z\mathbf{x}'(t)}{|\mathbf{x}'(t)|^3} + \mathcal{O}(|\mathbf{B}|^2), \qquad |\mathbf{B}| \to 0. \qquad (A.7.5)$$

Thus, for a small magnetic field, the trajectory $\mathbf{x}'(t)$ approximately satisfies equation (A.7.1) with the magnetic field $\mathbf{B}' = 0$ ("Larmor's theorem"). Hence, the Lagrangian in the rotating frame reads (approximately) as

$$\Lambda'(\mathbf{x}', \dot{\mathbf{x}}') = \frac{m|\dot{\mathbf{x}}'|^2}{2} + \frac{e^2 Z}{|\mathbf{x}'|}, \qquad (A.7.6)$$

according to (A.6.2). Respectively, any trajectory $\mathbf{x}'(t)$ lies in some plane P' passing through the nucleus and given by the equation $\theta' = 0$ in the appropriate spherical coordinates r' (radius), φ' (longitude), θ' (latitude) with origin at the nucleus. Due to the spherical symmetry of the Coulomb potential in (A.7.6), the corresponding vector of the electron angular momentum is conserved:

$$\mathbf{L}' := \mathbf{x}'(t) \times \mathbf{p}'(t), \qquad \mathbf{p}'(t) = \nabla_{\dot{\mathbf{x}}'}\Lambda' = m\dot{\mathbf{x}}'(t). \qquad (A.7.7)$$

Note that $\mathbf{L}' \perp P'$. Each bounded trajectory is periodic, and the quantization conditions (A.6.1) take the form of integrals over the period:

$$\oint p'_{r'}\, dr' = hn'_{r'}, \qquad \oint p'_{\varphi'}\, d\varphi' = hn'_{\varphi'}, \qquad \oint p'_{\theta'}\, d\theta' = hn'_{\theta'} = 0. \quad \text{(A.7.8)}$$

Here $n'_{r'}$, $n'_{\varphi'}$, $n'_{\theta'} = 0, \pm 1, \ldots$. As was shown above, these quantization conditions imply formulae (A.6.9) and (A.6.10). Hence,

$$E'_n = -\frac{me^4}{2\hbar^2 n^2}, \quad n = 1, \ldots; \qquad |\mathbf{L}'| = |p'_{\varphi'}| = \hbar|n'_{\varphi'}|, \quad |n'_{\varphi'}| \le n. \quad \text{(A.7.9)}$$

Respectively, the trajectory $\mathbf{x}(t) = e^{-\tilde{\mathbf{B}}t}\mathbf{x}'(t)$ lies in the rotating plane, and generally is not periodic, since the conservation of the angular momentum $\mathbf{L}'$ implies the *precession of the angular momentum* $\mathbf{L}(t)$ around the direction of the magnetic field $\mathbf{B}$:

$$\mathbf{L}(t) := \mathbf{x}(t) \times \mathbf{p}(t) = e^{-\tilde{\mathbf{B}}t}\mathbf{L}'. \qquad\qquad \text{(A.7.10)}$$

On the other hand, the orbit $\mathbf{x}(t)$ is *approximately periodic* for small magnetic field $|\mathbf{B}| \ll 1$. This is why Sommerfeld suggested extending to this case the quantization rules (A.6.3). The extension was suggested by the fact that these conditions hold for $\mathbf{B} = 0$, and hence should also hold by continuity for a small magnetic fields due to the *discreteness of the quantized observables*.

Exercise A.7.1. Prove "Larmor's theorem" (A.7.5).

Problem A.7.2. Check (A.7.2) and (A.7.4). **Hint:** see [51, (12.81), (12.90)].

A.7.2   Spatial quantization

Sommerfeld obtained the following identity [82, Vol. I, (II.8.4)]:

$$n_r + n_\varphi + n_\theta = n_{r'} + n_{\varphi'} + n_{\theta'}. \qquad\qquad \text{(A.7.11)}$$

It is derived from the expression of the kinetic energy [82, Vol. I, (II.7.22)]:

$$E_{\text{kin}}(t) = \frac{1}{2}\sum_{k=1}^{3} p_k(t)\,\dot{q}_k(t), \qquad\qquad \text{(A.7.12)}$$

which holds in any choice of canonical variables. Indeed, integration of (A.7.12) over the period yields [82, Vol. I, (II.7.23)]:

$$\oint E_{\text{kin}}(t)\, dt = \frac{1}{2}\sum_{k=1}^{3} \oint p_k(t)\, dq_k(t), \qquad\qquad \text{(A.7.13)}$$

which implies (A.7.11) according to the quantization rules (A.6.3) and (A.7.8). Finally, (A.7.11) implies the relation [82, (II.8.3)]:

$$n_\varphi + n_\theta = n'_{\varphi'}, \tag{A.7.14}$$

since $n'_{\theta'} = 0$, while $n_r = n'_{r'}$ due to $p_r = p'_{r'}$.

Further, let us show that the quantization rules (A.6.3) and (A.7.8) imply the fundamental relation [82, (II.8.5)]:

$$n_\varphi = n'_{\varphi'} \cos \alpha, \tag{A.7.15}$$

where α is the angle between the vectors $\mathbf{L}'$ and $\mathbf{B}$, which is also equal to the angle between the vectors $\mathbf{L}$ and $\mathbf{B}$ because of the precession (A.7.10). We will deduce this relation from the formula

$$p_\varphi = p'_{\varphi'} \cos \alpha, \tag{A.7.16}$$

which is obtained by the following computation. First,

$$\dot{x}^2 = \dot{r}^2 + r^2 \cos^2\theta \, \dot{\varphi}^2 + r^2 \dot{\theta}^2, \qquad p_\varphi := \partial_{\dot\varphi} \Lambda_{\mathbf{B}} = mr^2 \cos^2\theta \, \dot\varphi. \tag{A.7.17}$$

Second, $p'_{\varphi'} := \partial_{\dot\varphi'} \Lambda' = mr^2 \cos^2\theta' \, \dot\varphi' = mr^2 \dot\varphi'$ because $\theta' \equiv 0$. Both momenta p_φ and $p'_{\varphi'}$ are conserved, and $\dot\varphi = \dot\varphi' / \cos\theta$ at the maximal latitude $\theta = \alpha$, which implies (A.7.16). Finally, (A.7.16), together with (A.6.3) and (A.7.8), implies (A.7.15) due to the conservation of p_φ and $p'_{\varphi'}$.

The relation (A.7.15) means that the angle α is quantized. The precession (A.7.10) implies that $|\mathbf{L}| = |\mathbf{L}'|$, and the projection of the vector $\mathbf{L}$ onto the direction of the magnetic field equals

$$L_{\mathbf{B}} = |\mathbf{L}| \cos \alpha.$$

Moreover, $|\mathbf{L}'| = mr^2|\dot\varphi'| = |p'_{\varphi'}| = \hbar|n'_{\varphi'}|$ according to (A.7.9). Therefore, (A.7.15) implies that

$$L_{\mathbf{B}} = \pm \hbar n_\varphi, \qquad |n_\varphi| \le |n'_{\varphi'}|. \tag{A.7.18}$$

The formulae (A.7.9) can be rewritten as

$$\left.\begin{cases} E_n &= -\dfrac{me^4}{2\hbar^2 n^2}, \quad n = 1, \ldots, \\[2mm] |\mathbf{L}| &= \hbar|n'_{\varphi'}|, \quad |n'_{\varphi'}| \le n = 1, \ldots \end{cases}\right|. \tag{A.7.19}$$

In the notation $n'_{\varphi'} = l$ and $\pm n_\varphi = m$, the relations (A.7.19), (A.7.18) imply that

$$|\mathbf{L}| = \hbar l, \quad L_{\mathbf{B}} = \hbar m, \quad l = 0, 1, \ldots, n, \quad m = -l, -l+1, \ldots, l. \tag{A.7.20}$$

Thus, the quantum stationary states of the electron in an atom are numbered by three quantum numbers n, l, m (*principal, azimuthal,* and *magnetic* quantum number, respectively). Their quantum energy, the angular momentum, and the latter's projection onto the direction of the magnetic field are given by formulae (A.7.19) and (A.7.20).

Remark A.7.3. The quantized values (A.7.19) and (A.7.20) were obtained for an "infinitely small magnetic field" $\mathbf{B} \approx 0$. However, the same expressions should hold also for a *finitely small* magnetic field, since the values of the quantum numbers n, l, m are discrete, while the action integrals (A.6.3) are *adiabatic invariants* [95].

Exercise A.7.4. Check (A.7.12). **Hint:** the kinetic energy is the quadratic form of $\dot{q}_k$.

A.8   Normal Zeeman effect

In 1916, A. Sommerfeld obtained from the quantization rules (A.7.20) the same results for the normal Zeeman effect (A.3.4) as obtained by Lorentz from equation (A.2.2). Indeed, the classical angular and magnetic moments of the electron are determined by formulae

$$\mathbf{L} := \mathbf{x} \times \mathbf{p}, \qquad \mathbf{M} := \frac{1}{2c}\mathbf{x} \times \mathbf{j}, \tag{A.8.1}$$

where $\mathbf{j} = e\mathbf{p}/m$ is the current. Therefore,

$$\mathbf{M} = \frac{e}{2mc}\mathbf{L}. \tag{A.8.2}$$

The classical energy (A.7.2) of the electron in a small magnetic field (A.7.3) can be expanded as

$$\mathcal{H}_{\mathbf{B}}(\mathbf{p}, \mathbf{x}) = \frac{\mathbf{p}^2}{2m} - \frac{e^2 Z}{|\mathbf{x}|} - \frac{e}{mc}\mathbf{p} \cdot \mathbf{A}(\mathbf{x}) + \frac{e^2}{c^2}\mathbf{A}^2(\mathbf{x}) \approx \mathcal{H}_0(\mathbf{x}, \mathbf{p}) - \frac{e}{mc}\mathbf{p} \cdot \mathbf{A}(\mathbf{x}), \tag{A.8.3}$$

where $\mathcal{H}_0$ corresponds to the case $\mathbf{B} = 0$. Thus, in a small magnetic field, the classical energy acquires an additional portion

$$-\frac{e}{mc}\mathbf{p} \cdot \mathbf{A}(\mathbf{x}) = -\frac{e}{mc}\mathbf{p} \cdot \left[\frac{1}{2}\mathbf{B} \times \mathbf{x}\right] = -\frac{e}{2mc}\mathbf{B} \cdot [\mathbf{x} \times \mathbf{p}] = -\mathbf{B} \cdot \mathbf{M} = -\Omega_\Lambda L_{\mathbf{B}}, \tag{A.8.4}$$

where Ω_Λ is the Larmor frequency (III.1.3). Hence, the classical energy (A.8.3) reads as follows:

$$\mathcal{H}_{\mathbf{B}}(\mathbf{p}, \mathbf{x}) \approx \mathcal{H}_0(\mathbf{p}, \mathbf{x}) - \mathbf{B} \cdot \mathbf{M} = \mathcal{H}_0(\mathbf{p}, \mathbf{x}) - \Omega_\Lambda L_{\mathbf{B}}. \tag{A.8.5}$$

The corresponding quantum energy levels are obtained by replacing i) $\mathcal{H}_0(\mathbf{p}, \mathbf{x})$ by $E_n - \hbar\omega_n$, where ω_n are given by (I.6.5), and ii) $L_\mathbf{B}$ by (Λ.7.20). As a result, we find the following possible values of the energy, of the angular momentum, and of its projection onto the direction of the magnetic field

$$\left\{ \begin{array}{ll} E = E_{nm} = \hbar\omega_n - \Omega_\Lambda \hbar m & \\ |\mathbf{L}| = \hbar l, \qquad L_\mathbf{B} = \hbar m & \end{array} \right. \left| \begin{array}{l} n = 1, \ldots \\ l = 0, 1, \ldots, n \\ m = -l, \ldots, l \end{array} \right| ; \qquad (A.8.6)$$

this coincides with the results of the Schrödinger theory (III.1.4) (except for the range of values for l). Hence, the spectral lines (I.4.4) acquire the additional term $-\Omega_\Lambda(m' - m)$. So, the spectral lines in the magnetic field are given by

$$\omega_{nn'} - \Omega_\Lambda(m' - m), \qquad (A.8.7)$$

which coincides with the result of the Schrödinger theory (III.1.5). Finally, Bohr's correspondence principle dictates *selection rules*, similar to (A.5.3):

$$m' - m = 0, \pm 1; \qquad (A.8.8)$$

see [82, Supplements 7 and 8]). As a result, each spectral line (I.4.4) becomes the *Zeeman triplet* (A.3.4).

Remark A.8.1. The selection rules (A.8.8) also hold in the Schrödinger theory; see the proof in Section III.1.2.

A.9   The Bohr–Pauli theory of periodic table

The Bohr and Pauli theory of the Mendeleev periodic table of elements (1921–1923) was the highest achievement of the Old Quantum Theory. This theory relies on the following postulates:

1) Electrons in multiple-electron atoms are weakly coupled, so that their stationary states can be described separately by the Bohr–Sommerfeld quantization rules.

2) Stationary orbits of each electron are numbered by four quantum numbers n, l, m, s, where $s = \pm\frac{1}{2}$, and the quantum energy, angular momentum and its projection onto the direction of the magnetic field are determined by formulae (A.6.9), (A.7.20). However, now *the value $l = n$ is excluded*:

$$n = 1, 2, \ldots; \qquad l = 1, \ldots, n - 1; \qquad m = -l, \ldots, l; \qquad s = \pm\frac{1}{2}. \ (A.9.1)$$

3) The sets of quantum numbers n, l, m, s are different for different electrons in accordance with the *Pauli exclusion principle*.

4) Electrons with the same energy belong to the same electron shell: the shell K corresponds to $n = 1$, L corresponds to $n = 2$, M – to $n = 3$, N – to $n = 4$, and so on. The electron shells were introduced by H. Moseley in his interpretation of the scattering of X-rays by atoms (1913) [26].

The bound $l \leq n - 1$ was found empirically in the Old Quantum Theory, and it holds automatically in the Schrödinger theory by (I.6.9). This bound, together with the *Pauli exclusion principle*, leads to a wonderful explanation of the periods in the table of elements.

For example, the ground state of the atom corresponds to the minimal energy of the electron configuration. Hence, in the ground state, the electrons must belong to electron shells with possible minimal values of n.

The Pauli exclusion principle implies that for $Z \geq 1$ the electron shell K can contain at most two electrons with the smallest energy corresponding to quantum numbers $n = 1$, $l = m = 0$, and $s = \pm\frac{1}{2}$.

For $Z \geq 3$, the shell L can contain at most 8 electrons with

$$n = 2; \qquad l = 0, \ldots, n - 1 = 1; \qquad m = -l, \ldots, l; \qquad s = \pm\frac{1}{2}.$$

Similarly, for a shell with any number $n \geq 1$, the maximal number of electrons is

$$N(n) = 2 \sum_{l=0}^{n-1} (2l + 1) = 2n^2.$$

These *occupation numbers* coincide with the famous "kabbalistic sequence" 2, 8, 18, 32, 50, ... of lengths of periods in the table of chemical elements. It is exactly for this reason that W. Pauli introduced in 1923 the fourth two-valued quantum number $s = \pm 1$. Otherwise, all periods of the table would turn out to be twice shorter than necessary.

A.10   The Hamilton–Jacobi and eikonal equations

After 1915, the quantization rules (A.4.1) and (A.6.1) attracted everyone's attention to the study of the *Lagrangian action* and to its calculation via the Hamilton–Jacobi equation (I.2.24). Many attempts were made to derive new electron dynamics from this equation, using Hamilton's optical-mechanical analogy (1840), which is based on the parallelism between the P. Fermat and P.L. Maupertuis variation principles [3, p. 246].

E. Schrödinger was the first to recognize the role of the Hamilton–Jacobi equation as the equation for the phase function of short-wave packets, which is similar to the role played by the *eikonal equation* in optics, see [79]. This idea allowed him to find the *corresponding wave equation* which is the Schrödinger equation.

Appendix B

The Noether Theory of Invariants

In this appendix, on a *formal level*, we present the Noether theory of conservation laws for general Hamiltonian systems with symmetries, [71]. We follow the exposition of [138].

Let $\mathcal{E}$ be the *real Hilbert phase space* with real inner product $\langle \cdot, \cdot \rangle$ and let the Hamiltonian functional $\mathcal{H}$ be defined on a dense subset of $\mathcal{E}$. The Hamiltonian structure is determined by a skew-adjoint linear operator J in $\mathcal{E}$. The corresponding Hamiltonian equation reads as (V.2.1)–(V.2.2):

$$\dot{\Psi}(t) = J\mathcal{H}'(\Psi(t)), \qquad t \in \mathbb{R}, \tag{B.0.1}$$

where $\mathcal{H}'$ is the variational derivative of the Hamiltonian. For such equations, energy conservation holds, since we have *formally*

$$\partial_t \mathcal{H}(\Psi(t)) = \langle \mathcal{H}'(\Psi(t)), \dot{\Psi}(t) \rangle = \langle \mathcal{H}'(\Psi(t)), J\mathcal{H}'(\Psi(t)) \rangle = 0, \quad t \in \mathbb{R}. \tag{B.0.2}$$

Similarly, a functional $\mathcal{F}(\Psi)$ is conserved along trajectories of (B.0.1) if and only if the corresponding *Poisson bracket* vanishes:

$$[\mathcal{F}, \mathcal{H}](\Psi) := \langle \mathcal{F}'(\Psi), J\mathcal{H}'(\Psi) \rangle = 0 \tag{B.0.3}$$

for a dense subset of $\Psi \in \mathcal{E}$.

The Noether theory [138, 158] provides explicit construction of conserved functionals for G-invariant equations (B.0.1). Namely, let the Lie group G act on the space $\mathcal{E}$ according to a *unitary representation* T, and assume that the Hamiltonian functional is invariant, i.e.,

$$\mathcal{H}(T(g)\Psi) = \mathcal{H}(\Psi), \qquad \Psi \in \mathcal{E}, \quad g \in G. \tag{B.0.4}$$

217

Moreover, we assume the commutation

$$JT(g) = T(g)J, \qquad g \in G. \tag{B.0.5}$$

Then the symmetry condition (V.2.3) holds for $F(\Psi) = J\mathcal{H}'(\Psi)$. Indeed, differentiating (B.0.4) in Ψ, we obtain the identity

$$T^*(g)\mathcal{H}'(T(g)\Psi) = \mathcal{H}'(\Psi).$$

However, $T^*(g) = T^{-1}(g)$ since the operators $T(g)$ are unitary. Therefore,

$$\mathcal{H}'(T(g)\Psi) = T(g)\mathcal{H}'(\Psi). \tag{B.0.6}$$

This identity and (B.0.5) imply (V.2.3).

Further, the differential of the representation T acts on the Lie algebra $\mathfrak{g}$, which is the tangent space to G at the unit element e. Denote the operators $T_\omega := dT(e)\omega$ for $\omega \in \mathfrak{g}$ which are formally skew-symmetric and commute with J by (B.0.5). Differentiating (B.0.4) w.r.t. g, we obtain the identity

$$\langle \mathcal{H}'(\Psi), T_\omega \Psi \rangle = 0, \qquad \omega \in \mathfrak{g}. \tag{B.0.7}$$

Finally, assume that the operator J is invertible in $\mathcal{E}$, and denote the quadratic form

$$Q_\omega(\Psi) := \frac{1}{2}\langle \Psi, B_\omega \Psi \rangle, \qquad \omega \in \mathfrak{g}, \tag{B.0.8}$$

where the operators $B_\omega := J^{-1}T_\omega$ are formally symmetric because of the commutation (B.0.5). Hence,

$$Q_\omega'(\Psi) = B_\omega \Psi,$$

and for "smooth solutions" of (B.0.1) we can write

$$\partial_t Q_\omega(\Psi(t)) = \langle B_\omega \Psi, \dot{\Psi}(t) \rangle = -\langle T_\omega \Psi, \mathcal{H}'(\Psi(t)) \rangle = 0 \tag{B.0.9}$$

by (B.0.7). This means that the values of the quadratic forms Q_ω are conserved along the trajectories of (B.0.1).

Example B.0.1. The conservations of the total momentum (I.3.28) and of the angular momenta (I.3.31), (III.3.25), (IV.4.2) are particular cases of developed formalism corresponding to suitable symmetry groups and their unitary representations.

Exercise B.0.2. Check that the operators B_ω are formally symmetric.
Hints: i) Use (B.0.5) to prove the commutation

$$J^{-1}T_\omega = T_\omega J^{-1}. \tag{B.0.10}$$

ii) Use that both operators, J^{-1} and T_ω, are formally skew-symmetric.

Exercise B.0.3. Apply the Noether theory to construct the conserved momentum for the wave equation

$$\ddot{\psi}(\mathbf{x}, t) = \Delta\psi(\mathbf{x}, t), \qquad \mathbf{x} \in \mathbb{R}^3, \tag{B.0.11}$$

where $\psi(\mathbf{x}, t)$ is real-valued. **Hints:** i) Check the Hamiltonian structure (B.0.1) for this equation with

$$\Psi(\mathbf{x}, t) := \begin{pmatrix} \psi(\mathbf{x}, t) \\ \pi(\mathbf{x}, t) \end{pmatrix}, \quad J = \begin{pmatrix} 0 & 1 \\ -1 & 0 \end{pmatrix}, \quad H(\Psi) = \frac{1}{2}\int_{\mathbb{R}^3} [|\pi(\mathbf{x})|^2 + |\nabla\psi(\mathbf{x})|^2]\, d\mathbf{x}.$$
$$\tag{B.0.12}$$

ii) Check (B.0.4)–(B.0.5) for the representation of the translation group $G = \mathbb{R}^3$:

$$T(\mathbf{g})\Psi(\mathbf{x}, t) = \Psi(\mathbf{x} - \mathbf{g}, t), \qquad \mathbf{x}, \mathbf{g} \in \mathbb{R}^3. \tag{B.0.13}$$

iii) Calculate T_ω and obtain

$$T_\omega = \begin{pmatrix} -\omega \cdot \nabla & 0 \\ 0 & -\omega \cdot \nabla \end{pmatrix}, \qquad \omega \in \mathfrak{g} = \mathbb{R}^3. \tag{B.0.14}$$

iv) Check (B.0.7) with the Lie algebra $\mathfrak{g} = \mathbb{R}^3$.

v) Calculate the quadratic form (B.0.8) and obtain

$$Q_\omega(\Psi) = -\omega \cdot \int_{\mathbb{R}^3} \pi(\mathbf{x})\nabla\psi(\mathbf{x})\, d\mathbf{x}. \tag{B.0.15}$$

Thus,

$$P(\Psi) := -\int_{\mathbb{R}^3} \pi(\mathbf{x})\nabla\psi(\mathbf{x})\, d\mathbf{x}$$

is the conserved momentum for equation (B.0.11).

Appendix C

Perturbation Theory

In practice it is very important to have a simple formula for the correction ΔE to the eigenvalue E of the perturbed Schrödinger operator $H + \varepsilon H'$ with $|\varepsilon| \ll 1$; i.e.,

$$(E + \Delta E)\psi'(\mathbf{x}) = (H + \varepsilon H')\psi'(\mathbf{x}), \qquad \mathbf{x} \in \mathbb{R}^3. \tag{C.0.1}$$

Let us consider the case of a *simple eigenvalue* E. In this case, the formula reads (see [7, 77])

$$\Delta E = \langle \psi, \varepsilon H' \psi \rangle + \mathcal{O}(\varepsilon^2), \qquad \varepsilon \to 0, \tag{C.0.2}$$

where ψ is the unperturbed eigenfunction corresponding to the eigenvalue E and satisfying the normalization condition $\langle \psi, \psi \rangle = 1$, with $\langle \cdot, \cdot \rangle$ the Hermitian inner product (I.3.3) in $L^2(\mathbb{R}^3)$.

Let us prove this formula for the case of Hermitian operators H and H' in a finite-dimensional complex space. The corresponding eigenvalues and eigenfunctions are analytic functions of ε for small $|\varepsilon|$. So, for $|\varepsilon| < r$ with some small $r > 0$, we have:

$$\Delta E \sim \mathcal{O}(\varepsilon) \quad \text{and} \quad \psi' - \psi + \varepsilon \phi + \mathcal{O}(\varepsilon^2), \qquad \varepsilon \to 0. \tag{C.0.3}$$

Substituting this into (C.0.1), we obtain:

$$(E + \Delta E)(\psi + \varepsilon\phi + \mathcal{O}(\varepsilon^2)) = (H + \varepsilon H')(\psi + \varepsilon\phi + \mathcal{O}(\varepsilon^2)). \tag{C.0.4}$$

Taking into account that $H\psi = E\psi$, we can write

$$E\varepsilon\phi + \Delta E\psi = H\varepsilon\phi + \varepsilon H'\psi + \mathcal{O}(\varepsilon^2). \tag{C.0.5}$$

We can assume the normalization condition $\|\psi'\| = 1$; then one can show that

$$\langle \psi, \phi \rangle = 0. \tag{C.0.6}$$

Therefore, taking the inner product of (C.0.5) with ψ, we arrive at

$$\Delta E \langle \psi, \psi \rangle = \langle \psi, \varepsilon H' \psi + \mathcal{O}(\varepsilon^2) \rangle, \qquad (\text{C.0.7})$$

since $\langle \psi, H \varepsilon \phi \rangle = \langle H \psi, \varepsilon \phi \rangle = \varepsilon E \langle \psi, \phi \rangle = 0$. Formula (C.0.2) follows.

Exercise C.0.1. Indicate the step in the proof where we used i) the condition that E is a simple eigenvalue, and ii) the condition that the operator H is selfadjoint.

Exercise C.0.2. Prove (C.0.6). **Hints:** i) square the identity (C.0.3) and obtain

$$1 = 1 + \varepsilon \langle \psi, \phi \rangle + \bar{\varepsilon} \langle \phi, \psi \rangle + \mathcal{O}(|\varepsilon|^2), \qquad \varepsilon \to 0;$$

ii) Use this identity in the cases when ε is real and when it is purely imaginary.

Bibliography

Quantum Theory and Classical Electrodynamics

[1] M. Abraham, Prinzipien der Dynamik des Elektrons, *Physikal. Zeitschr.* **4** (1902), 57–63.

[2] M. Abraham, Theorie der Elektrizität, Bd.2: Elektromagnetische Theorie der Strahlung, Teubner, Leipzig, 1905.

[3] V. I. Arnold, Mathematical Methods of Classical Mechanics, Springer, New York, 1989.

[4] R. Becker, Théorie des Électrons, Librairie Félix Alcan, Paris, 1938.

[5] R. Becker, Electromagnetic Fields and Interactions, Vol. I, II: Quantum Theory of Atoms and Radiation, Blaisdell, 1964.

[6] F. A. Berezin, M. A. Shubin, The Schrödinger Equation, Kluwer Academic Publishers, Dordrecht, 1991.

[7] H. Bethe, Intermediate Quantum Mechanics, Benjamin, NY, 1964.

[8] H. Bethe, E. Salpeter, Quantum Mechanics of One- and Two-Electron Atoms, Berlin, Springer, 1957.

[9] J. D. Bjorken, S. D. Drell, Relativistic Quantum Mechanics, McGraw-Hill, NY, 1964; Relativistic Quantum Fields, McGraw-Hill, NY, 1965.

[10] N. Bohr, Discussion with Einstein on epistemological problems in atomic physics, pp. 201–241 in: Schilpp, P.A., Ed., *Albert Einstein: Philosopher-Scientist*, Vol. 7, Library of Living Philosophers, Evanston Illinois, 1949.

[11] M. Born, Atomic Physics, Blackie & Son, London–Glasgow, 1951.

[12] M. Born, E. Wolf, Principles of Optics, Cambridge University Press, Cambridge, 1966.

[13] https://physics.nist.gov/cgi-bin/cuu/Value?ryd

[14] J. J. Brehm, W. J. Mullins, Introduction to the Structure of Matter: A Course in Modern Physics, Wiley, 1989.

[15] L. Catto, C. Le Bris, P. L. Lions, The Mathematical Theory of Thermodynamic Limits: Thomas–Fermi Type Models, Clarendon Press, Oxford, 1998.

[16] S. J. Chang, Introduction to Quantum Field Theory, World Scientific, Singapore, 1990.

[17] R. Chiao, P. Kwiat, Heisenberg's Introduction of the 'Collapse of the Wavepacket' into Quantum Mechanics, arXiv:quant-ph/0201036.

[18] E. U. Condon, G. H. Shortley, The Theory of Atomic Spectra, Cambridge University Press, Cambridge, 1963.

[19] H.L. Cycon, R.G. Froese, W. Kirsch, B. Simon, Schrödinger Operators with Applications to Quantum Mechanics and Global Geometry, Springer, Berlin, Springer, 1987.

[20] C. G. Darwin, The wave equation of the electron, *Proc. Roy. Soc. London A* **118** (1928), 654–680.

[21] M. Defranceschi, C. Le Bris, Mathematical Models and Methods for ab Initio Quantum Chemistry, Lecture Notes in Chemistry 74, Springer, Berlin, 2000.

[22] P. A. M. Dirac, The quantum theory of the electron, I, *Proc. Roy. Soc. London A* **117** (1928), 610–624.

[23] P. A. M. Dirac, The quantum theory of the emission and absorption of radiation, *Proc. Roy. Soc. A* **114** (1927), 243–265.

[24] P. A. M. Dirac, The Principles of Quantum Mechanics, Oxford Univ. Press, Oxford, 1999.

[25] P. A. M. Dirac, Lectures on Quantum Field Theory, Yeshiva University, New York, 1967.

[26] R.G. Egdell, E. Bruton, Henry Moseley, X-ray spectroscopy and the periodic table, *Phil. Trans. R. Soc. A* **378** (2020), 20190302.

[27] A. Einstein, Ist die Trägheit eines Körpers von seinem Energieinhalt abhängig?, *Annalen der Physik*, **18** (1905), 639–643.

[28] A. Einstein, P. Ehrenfest, Quantentheoretische Bemerkungen zum Experiment von Stern und Gerlach, *Zeitschrift für Physik* **11** (1922), 31–34.

[29] R. P. Feynman, R. B. Leighton and M. Sands, The Feynman Lectures on Physics. Vol. 2: Mainly Electromagnetism and Matter, Addison-Wesley Publishing Co., Inc., Reading, Mass.-London, 1964.

[30] R. Feynman, Quantum Electrodynamics, Addison-Wesley, Reading, Massachusetts, 1998.

[31] R. P. Feynman, R. B. Leighton, M. Sands, The Feynman Lectures on Physics III, Addison–Wesley, Reading, Mass, 1965.

[32] S. Flügge, Practical Quantum Mechanics II, Springer, New York, 2011.

[33] Ya. Frenkel, *Zs. Phys.* **37** (1926), 43.

[34] I. M. Gelfand, R. A. Minlos, Z. Ya. Shapiro, Representations of the Rotation and Lorentz Groups and their Applications, Pergamon Press, Oxford, 1963.

[35] H. Goldstein, C.P. Poole, J. Safko, Classical Mechanics, Pearson, New York, 2001.

[36] W. Greiner, Quantum Mechanics: An Introduction, Berlin, Springer, 2001.

[37] W. Greiner, B. Müller, Quantum Mechanics: Symmetries, Springer, New York, 1994.

[38] F. Gross, Relativistic Quantum Mechanics and Field Theory, Wiley, New York, 1999.

[39] W. Gordon, Die Energieniveaus der Wassenstoffatoms nach der Diracschen Quantentheorie des Elektrons, *Zeitschrift f. Phys.* **48** (1928), 11–14.

[40] S. J. Gustafson, I. M. Sigal, Mathematical Concepts of Quantum Mechanics, Springer, Berlin, 2011.

[41] K. Hannabuss, An Introduction to Quantum Theory, Clarendon Press, Oxford, 1997.

[42] W. Heisenberg, The Physical Principles of the Quantum Theory, University of Chicago Press, Chicago, 1930 (reprinted by Dover Publications).

[43] W. Heisenberg, Der derzeitige Stand der nichtlinearen Spinortheorie der Elemen- tarteilchen, *Acta Phys. Austriaca* **14** (1961), 328–339.

[44] W. Heisenberg, Introduction to the Unified Field Theory of Elementary Particles, Interscience, London, 1966.

[45] P.D. Hislop, I.M. Sigal, Introduction to Spectral Theory. With applications to Schrödinger operators, Springer, NY, 1996.

[46] L. Houllevigue, L'Évolution des Sciences, A. Collin, Paris, 1908.

[47] C. Itzykson, J.B. Zuber, Quantum Field Theory, McGraw-Hill, NY, 1980.

[48] R. D. Jackson, Classical Electrodynamics, Wiley, New York, 1999.

[49] https://en.wikipedia.org/wiki/Lorentz–Heaviside_units

[50] F.A. Jenkins, H.E. White, Fundamentals of Optics, McGrow–Hill, New York, 2001.

[51] A. Komech, Quantum Mechanics: Genesis and Achievements, Springer, Dordrecht, 2013.

[52] H. A. Kramers, The quantum theory of dispersion, *Nature* **113** (1924), 673–676. [pp. 177–180 in: Sources in Quantum Mechanics, ed. B. L. van der Waerden, North-Holland, Amsterdam, 1967.]

[53] H. A. Kramers, The law of dispersion and Bohr theory of spectra, *Nature* **114** (1924), 310–311. [pp. 199–202 in: Sources in Quantum Mechanics, ed. B. L. van der Waerden, North-Holland, Amsterdam, 1967.]

[54] H. A. Kramers, W. Heisenberg, Über die Streuung von Strahlen durch Atome, em Z. f. Physik bf 31 (1925), 681–708. [English translation: On the dispersion of radiation by atoms, pp. 223–252 in: Sources in Quantum Mechanics, ed. B.L. van der Waerden, North-Holland, Amsterdam, 1967.]

[55] H.A. Kramers, pp. 545–557 in: La diffusion de la lumière par les atomes, Atti Cong. Intern. Fisica (Transactions of Volta Centenary Congress) Como 2, 1927.

[56] R. de L. Kronig, On the theory of the dispersion of X-rays, *J. Opt. Soc. Am.* **12** (1926), 547–557. DOI:10.1364/JOSA.12.000547.

[57] http://en.wikipedia.org/wiki/Kramers-Kronig_relation

[58] L. D. Landau, E. M. Lifshitz, Quantum Mechanics. Nonrelativistic Theory, Pergamon Press, London–New York, 1965.

[59] A. Landé, Über den anomalen Zeemaneffekt (Teil I) *Zs. f. Phys.* **5** (1921), no. 4, 231–241.

[60] D. F. Lawden, The Mathematical Principles of Quantum Mechanics, Methuen & Co LTD, London, 1967.

[61] V. G. Levich, Yu. A. Vdovin, V. A. Myamlin, Course of Theoretical Physics, v. II, Nauka, Moscow, 1971 [in Russian].

[62] E. H. Lieb, R. Seiringer, The Stability of Matter, Cambridge University Press, Cambridge, 2010.

[63] K. Lüders, R. O. Pohl (Editors), Pohl's Introduction to Physics Volume 2: Electrodynamics and Optics, Springer Nature, Cham (Switzerland), 2018.

[64] M. Maggiore, A Modern Introduction in Quantum Field Theory, Oxford University Press, Oxford, 2005.

[65] F. Mandl, G. Shaw, Quantum Field Theory, Wiley, Chichester, 1999.

[66] A. Messiah, Quantum Mechanics, North Holland, Wiley, 1966.

[67] E. Merzbacher, Quantum Mechanics, Wiley, New York, 1998.

[68] L. Navarro, E. Pérez, Paul Ehrenfest: The genesis of the adiabatic hypothesis, 1911–1914, *Archive for History of Exact Sciences* **60** (2006), 209–267.

[69] J. von Neumann, Mathematical Foundations of Quantum Mechanics, Princeton University Press Princeton, 1955.

[70] R. Newton, Quantum Physics, Springer, New York, 2002.

[71] E. Noether, Gesammelte Abhandlungen. Collected papers, Springer, Berlin, 1983.

[72] W. Pauli, Diracs Wellengleichungen des Elektrons und geometrische Optik, *Helv. Phys. Acta* **5** (1932), 179–199.

[73] https://en.wikipedia.org/wiki/Rayleigh_scattering

[74] J.J. Sakurai, Modern Quantum Mechanics, Cambridge University Press, Cambridge, 2017.

[75] J.J. Sakurai, Advanced Quantum Mechanics, Addison-Wesley, Reading, Massachusets, 1967.

[76] G. Scharf, Finite Quantum Electrodynamics. The Causal Approach, Springer, Berlin, 1995.

[77] L.I. Schiff, Quantum Mechanics, McGraw-Hill, New York, 1955.

[78] E. Schrödinger, Quantisierung als Eigenwertproblem, *Ann. d. Phys.* I, II **79** (1926) 361, 489; III **80** (1926) 437; IV **81** (1926) 109. (English translation in: E. Schrödinger, Collected Papers on Wave Mechanics, Blackie & Sohn, London, 1928.)

[79] E. Schrödinger, The fundamental idea of wave mechanics, Nobel Lecture, December 12, 1933.

[80] F. Schwabl, Advanced Quantum Mechanics, Springer, Berlin, 1997.

[81] A. Sommerfeld, Optics, Acad. Press, New York, 1954.

[82] A. Sommerfeld, Atombau und Spektrallinien, Vol. I and II, Friedr. Vieweg & Sohn, Braunschweig, 1951.

[83] A. Sommerfeld, Thermodynamics and Statistical Mechanics, Academic Press, New York, 1956.

[84] A. Szabo, N.S. Ostlund, Modern Quantum Chemistry: Introduction to Advanced Electronic Structure Theory, Dover, 1996.

[85] B. Thaller, The Dirac Equation, Springer, Berlin, 1991.

[86] D. ter Haar, The Old Quantum Theory, Pergamon Press, Oxford, 1967.

[87] L.H. Thomas, The motion of a spinning electron, *Nature* **117** (1926), 514.

[88] J. Vanderlinde, Classical Electromagnetic Theory, Fundamental Theories of Physics 145, Kluwer Academic Publishers, Dordrecht, 2004.

[89] B. L. van der Waerden, Group Theory and Quantum Mechanics, Springer, Berlin, 1974.

[90] B. L. van der Waerden (ed.), Sources in Quantum Mechanics, North-Holland, Amsgerdam, 1967.

[91] S. Weinberg, The Quantum Theory of Fields. Vol. 1. Foundations, Cambridge University Press, Cambridge, 2005.

[92] H. Weyl, The Theory of Groups and Quantum Mechanics, Dover, NY, 1949.

[93] G. K. Woodgate, Elementary Atomic Structure, Clarendon Press, Oxford, 2002.

[94] E. P. Wigner, Über die Operation der Zeitumkehr in der Quanten-mechanik, Nachr. Akad. Wiss. Göttingen Math.-Phys. Kl. (1932), 546–559.

[95] C.G. Wells, S.T.C. Siklos, The adiabatic invariance of the action variable in classical dynamics, *Eur. J. Phys* **28** (2007), 105–112.

[96] W. Yourgrau, S. Mandelstam, Variational Principles in Dynamics and Quantum Theory, Pitman, New York, 2007.

[97] E. Zeidler, Applied Functional Analysis. Main Principles and Their Applications, Applied Mathematical Sciences Vol. 109, Springer, Berlin, 1995.

Photoelectric effect

[98] V. Bach, F. Klopp, H. Zenk, Mathematical analysis of the photoelectric effect, *Adv. Theor. Math. Phys.* **5** (2001), no. 6, 969–999.

[99] O. Costin, R.D. Costin, J.L. Lebowitz, A. Rokhlenko, Evolution of a model quantum system under time periodic forcing: conditions for complete ionization, *Commun. Math. Phys.* *221* (2001), no. 1, 1–26.

[100] O. Costin, R.D. Costin, J.L. Lebowitz, Time asymptotics of the Schrödinger wave function in time-periodic potentials, *J. Stat. Phys.* **116** (2004), no. 1–4, 283–310.

[101] O. Costin, J.L. Lebowitz, C. Stucchio, Ionization in a 1-dimensional dipole model, *Rev. Math. Phys.* **20**, no. 7 (2008), 835–872.

[102] O. Costin, J.L. Lebowitz, C. Stucchio, S. Tanveer, Exact results for ionization of model atomic systems, *J. Math. Phys.* **51** (2010), no. 1, Paper No. 015211.

[103] M. Griesemer, H. Zenk, On the atomic photoeffect in nonrelativistic QED, arXiv:0910.1809.

[104] A. Sommerfeld, G. Schur, *Ann. d. Phys.* **4** (1930), 409.

[105] G. Szegő, Orthogonal Polynomials, Amer. Math. Soc., Providence, RI, 1975.

[106] G. Wentzel, *ZS. f. Phys.* **43** (1927), 1, 779.

[107] H. Zenk, Ionisation by quantized electromagnetic fields: The photoelectric effect, *Rev. Math. Phys.* **20** (2008), 367–406.

Diffraction of electrons

[108] R. Bach, D. Pope, S.-H. Liou, H. Batelaan, Controlled double-slit electron diffraction, *New J. Phys.* **15** (2013), 033018.

[109] http://iopscience.iop.org/1367-2630/15/3/033018/media/njp458349 suppdata.pdf

[110] L. Biberman, N. Sushkin, V. Fabrikant, Diffraction of successively travelling electrons, *Doklady AN SSSR* **66** (1949), no. 2, 185–186, 1949.

[111] R. G. Chambers, Shift of an electron interference pattern by enclosed magnetic flux, *Physical Review Letters* **5** (1960), 3–5.

[112] C. Davisson, L. Germer, The scattering of electrons by a single crystal of nickel, *Nature* 119 (1927), 558–560.

[113] D. M. Eidus, The principle of limit amplitude, *Russ. Math. Surv.* **24** (1969), no. 3, 97–167.

[114] D. M. Eidus, The limiting amplitude principle for the Schrödinger equation in domains with unbounded boundaries, *Asymptotic Anal.* **2** (1989), no. 2, 95–99.

[115] S. Frabboni, G. C. Gazzadi, G. Pozzi, Young's double-slit interference experiment with electrons, *Amer. J. Phys.* **75** (2007), Issue 11, 1053–1055.

[116] M. Jammer, The Conceptual Development of Quantum Mechanics, McGraw-Hill, New York, 1966.

[117] A. Tonomura, J. Endo, T. Matsuda, T. Kawasaki, H. Ezawa, Demonstration of single-electron buildup of an interference pattern, *Amer. J. Phys.* **57** (1989), no. 2, 117–120.

Omega-Hyperon

[118] V. E. Barnes *et al.*, Observation of a hyperon with strangeness minus three, *Phys. Rev. Lett.* **12** (1964), 204–206.

[119] M. Gell-Mann, Symmetries of baryons and mesons, *Phys. Rev. (2)* **125** (1962), 1067–1084.

[120] F. Halzen and A. Martin, Quarks and Leptons: an Introductory Course in Modern Particle Physics, John Wiley & Sons, New York, 1984.

[121] Y. Ne'eman, Unified interactions in the unitary gauge theory, *Nuclear Phys.* **30** (1962), 347–349.

[122] L.B. Okun, Leptons and Quarks, Elsevier, Amsterdam, 2013.

[123] J.C. Pati and A. Salam, Lepton number as the fourth "color", *Phys. Rev. D* **10** (1974), 275–289.

Laser radiation

[124] H. Haken, Laser Theory, Springer, Berlin, 1984.

[125] W. E. Lamb Jr., M. Sargent III, M. O. Scully, Laser Physics, Addison Wesley, Reading, 1978.

[126] H. Nussenzveig, Introduction to Quantum Optics, Gordon and Breach, London, 1973.

[127] M. O. Scully, M. S. Zubairy, Quantum Optics, Cambridge University Press, Cambridge, 1997.

[128] O. Svelto, Principles of Lasers, Springer, 2010.

Maxwell–Schrödinger equations

[129] I. Bejenaru, D. Tataru, Global wellposedness in the energy space for the Maxwell–Schrödinger system, *Commun. Math. Phys.* **288** (2009), 145–198.

[130] G. M. Coclite, V. Georgiev, Solitary waves for Maxwell–Schrödinger equations, Electronic Journal of Differential Equations 94 (2004), 1–31. arXiv:math/0303142 [math.AP]

[131] Y. Guo, K. Nakamitsu, W. Strauss, Global finite-energy solutions of the Maxwell–Schrödinger system, *Comm. Math. Phys.* **170** (1995), no. 1, 181–196.

[132] M. Nakamura, T. Wada, Global existence and uniqueness of solutions of the Maxwell–Schrödinger equations, *Comm. Math. Phys.* **276** (2007), 315–339.

Analysis and PDEs

[133] S. Agmon, Spectral properties of Schrödinger operators and scattering theory, *Ann. Scuola Norm. Sup. Pisa Cl. Sci. (4)* **2** (1975), 151–218.

[134] N. Boussaïd, A. Comech, Nonlinear Dirac equation. Spectral stability of solitary waves, Mathematical Surveys and Monographs, vol. 244. American Mathematical Society, 2019.

[135] T. Cazenave, Semilinear Schrödinger Equations, AMS, NY, 2003.

[136] T. Cazenave, A. Haraux, An Introduction to Semilinear Evolution Equations, Clarendon Press, Oxford, 1998.

[137] G. I. Gaudry, Quasimeasures and operators commuting with convolution, *Pacific J. Math.*, **18** (1966), 461–476.

[138] M. Grillakis, J. Shatah, W.A. Strauss, II. *J. Func. Anal.* **94** (1990), no. 2, 308–348.

[139] V. Guillemin, S. Sternberg, Geometric Asymptotics, American Mathematical Society, Providence, R.I., 1977.

[140] E. Hopf, Über die Anfangswertaufgabe für die hydrodynamischen Grundgleichungen, *Math. Nachr.* **4** (1951), 213–231.

[141] L. Hörmander, Distribution theory and Fourier analysis, The Analysis of Linear Partial Differential Operators. I, Springer, Berlin, 1990.

[142] W. Ignatowsky, Reflexion elektromagnetischer Wellen an einem Draft, *Ann. Phys.* **18** (1905), 495–522.

[143] A. Jensen, T. Kato, Spectral properties of Schrödinger operators and time-decay of the wave functions, *Duke Math. J.* **46** (1979), 583–611.

[144] K. Jörgens, Das Anfangswertproblem im Grossen für eine Klasse nichtlinearer Wellengleichungen, *Math. Z.* **77** (1961), 295–308.

[145] A.I. Komech, Attractors of nonlinear Hamilton PDEs, *Discrete and Continuous Dynamical Systems A* **36** (2016), no. 11, 6201–6256. arXiv:1409.2009.

[146] A.I. Komech, E. Kopylova, On eigenfunction expansion of solutions to the Hamiltonian equations, *J. Stat. Physics* **154** (2014), no. 1–2, 503–521.

[147] A.I. Komech, E. Kopylova, On the eigenfunction expansion for the Hamiltonian operators, *J. Spectr. Theory* **5** (2015), no. 2, 331–361.

[148] A.I. Komech, E. A. Kopylova, Dispersion Decay and Scattering Theory, Wiley, Hoboken, New Jersey, 2012.

[149] A.I. Komech, A.E. Merzon, Stationary Diffraction by Wedges. Method of Automorphic Functions on Complex Characteristics, Lecture Notes in Mathematics 2249, Springer, 2019.

[150] A.I. Komech, E. A. Kopylova, Attractors of Hamiltonian nonlinear partial differential equations, *Russ. Math. Surv.* **75** (2020), no. 1, 1–87.

[151] A.I. Komech, E. A. Kopylova, Attractors of Hamiltonian Nonlinear Partial Differential Equations, Cambridge University Press, Cambridge, 2021.

[152] O.A. Ladyženskaya, On the principle of limiting amplitude, *Uspekhi Mat. Nauk* **12** (1957), 161–164. [In Russian]

[153] P.D. Lax, Asymptotic solutions of oscillatory initial value problems, *Duke Math. J.* **24** (1957), 627–646.

[154] B.Y. Levin, Lectures on Entire Functions, vol. 150 of Translations of Mathematical Monographs, American Mathematical Society, Providence, RI, 1996, In collaboration with and with a preface by Yu. Lyubarskii, M. Sodin and V. Tkachenko, Translated from the Russian manuscript by Tkachenko.

[155] L. Lewin, Advanced Theory of Waveguides, Iliffe and Sons, Ltd., London, 1951.

[156] J.-L. Lions, Quelques Méthodes de Résolution des Problèmes aux Limites non Linéaires, Dunod, Gauthier-Villars, Paris, 1969.

[157] C.S. Morawetz, The limiting amplitude principle, *Comm. Pure Appl. Math.* **15** (1962), 349–361.

[158] E. Noether, Invariante Variationsprobleme, *Nachr. d. König. Gesellsch. d. Wiss. zu Göttingen, Math-phys. Klasse* (1918), 235–257. English translation in: M. A. Travel, *Transp. Theory Statist. Phys.* **1** (1971), no. 3, 186–207 (see also arXiv:physics/0503066 [physics.hist-ph]).

[159] M. Reed, B. Simon, Methods of Modern Mathematical Physics, Academic Press, NY, I (1980), II (1975), III (1979), IV (1978).

[160] W. Rudin, Functional Analysis, McGraw-Hill, New York, 1991.

[161] M. A. Shubin, Pseudodifferential Operators and Spectral Theory, Springer, Berlin, 2001.

[162] I.M. Sigal, Nonlinear wave and Schrödinger equations. I. Instability of periodic and quasiperiodic solutions, *Comm. Math. Phys.* **153** (1993), 297–320.

[163] E.C. Titchmarsh, The zeros of certain integral functions, *Proc. London Math. Soc.* **S2-25** (1926), no. 1, 283–302.

[164] K. Yajima, Resonances for the AC-Stark Effect, *Comm. Math. Phys.* **87** (1982), 331–352.

Global attractors of dissipative PDEs

[165] L. Landau, On the problem of turbulence, *C. R. (Doklady) Acad. Sci. URSS (N.S.)* **44** (1944), 311–314.

[166] A. V. Babin and M. I. Vishik, Attractors of Evolution Equations, North-Holland Publishing Co., Amsterdam, 1992.

[167] V.V. Chepyzhov and M.I. Vishik, Attractors for Equations of Mathematical Physics, American Mathematical Society, Providence, RI, 2002.

[168] C. Foias, O. Manley, R. Rosa, R. Temam, Navier-Stokes Equations and Turbulence, Cambridge University Press, Cambridge, 2001.

[169] J.K. Hale, Asymptotic Behavior of Dissipative Systems, American Mathematical Society, Providence, RI, 1988.

[170] A. Haraux, Systémes Dynamiques Dissipatifs et Applications, R.M.A. 17, Collection dirigée par Ph. Ciarlet et J.L. Lions, Masson, Paris, 1990.

[171] D. Henry, Geometric Theory of Semilinear Parabolic Equations, Springer-Verlag, Berlin-New York, 1981.

[172] R. Temam, Infinite-Dimensional Dynamical Systems in Mechanics and Physics, Springer, New York, 1997.

Local energy decay

[173] C.S. Morawetz, Time decay for the nonlinear Klein–Gordon equations, *Proc. Roy. Soc. Ser. A* **306** (1968), 291–296.

[174] C.S. Morawetz, W.A. Strauss, Decay and scattering of solutions of a nonlinear relativistic wave equation, *Comm. Pure Appl. Math.* **25** (1972), 1–31.

[175] I. Segal, Quantization and dispersion for nonlinear relativistic equations, pp. 79–108 in: Mathematical Theory of Elementary Particles (Proc. Conf., Dedham, Mass., 1965), M.I.T. Press, Cambridge, Mass., 1966.

[176] I. Segal, Dispersion for non-linear relativistic equations. II, *Ann. Sci. École Norm. Sup.* (4) **1** (1968), 459–497.

[177] W.A. Strauss, Decay and asymptotics for $\Box u = F(u)$, *J. Functional Analysis* **2** (1968), 409–457.

[178] W.A. Strauss, Nonlinear scattering theory at low energy, *J. Funct. Anal.* **41** (1981), 110–133.

[179] W.A. Strauss, Nonlinear scattering theory at low energy: sequel, *J. Funct. Anal.* **43** (1981), 281–293.

Global attraction to stationary states for Hamiltonian PDEs

[180] Lamb H, On a peculiarity of the wave-system due to the free vibrations of a nucleus in an extended medium, *Proc. London Math. Soc.* **32** (1900), 208–211.

[181] A. Komech, On the stabilization of interaction of a string with a nonlinear oscillator, *Moscow Univ. Math. Bull.* **46** (1991), no. 6, 34–39.

[182] A.I. Komech, On stabilization of string-nonlinear oscillator interaction, J. Math. Anal. Appl., 196 (1995), 384–409.

[183] A. Komech, On the stabilization of string-oscillator interaction, *Russian J. Math. Phys.* **3** (1995), 227–247.

[184] A. Komech, On transitions to stationary states in one-dimensional nonlinear wave equations, *Arch. Ration. Mech. Anal.* **149** (1999), 213–228.

[185] A. Komech, H. Spohn, M. Kunze, Long-time asymptotics for a classical particle interacting with a scalar wave field, *Comm. Partial Differential Equations* **22** (1997), 307–335.

[186] A. Komech, H. Spohn, Long-time asymptotics for the coupled Maxwell–Lorentz equations, *Comm. Partial Differential Equations* **25** (2000), 559–584.

[187] A. Komech, A. Merzon, Scattering in the nonlinear Lamb system, *Phys. Lett. A* **373** (2009), 1005–1010.

[188] A. Komech, A. Merzon, On asymptotic completeness for scattering in the nonlinear Lamb system, *J. Math. Phys.* **50** (2009), 023514.

[189] A. Komech, A. Merzon, On asymptotic completeness of scattering in the nonlinear Lamb system, II, *J. Math. Phys.* **54** (2013), 012702.

[190] E. Kopylova, On global attraction to stationary states for wave equation with concentrated nonlinearity, *J. Dynamics and Diff. Equations* **30** (2018), no. 1, 107–116.

[191] H. Spohn, Dynamics of Charged Particles and their Radiation Field, Cambridge University Press, Cambridge, 2004.

Global attraction to solitons for Hamiltonian PDEs

[192] W. Eckhaus, A. van Harten, The Inverse Scattering Transformation and the Theory of Solitons, North-Holland Publishing Co., Amsterdam–New York, 1981.

[193] A.I. Komech, H. Spohn, Soliton asymptotics for a classical particle interacting with a scalar wave field, *Nonlinear Anal.* **33** (1998), 13–24.

[194] V. Imaykin, A.I. Komech, N. Mauser, Soliton-type asymptotics for the coupled Maxwell–Lorentz equations, *Ann. Henri Poincaré* **P** (2004), 1117–1135.

[195] V. Imaykin, A.I. Komech, H. Spohn, Soliton-type asymptotics and scattering for a charge coupled to the Maxwell field, *Russ. J. Math. Phys.* **9** (2002), 428–436.

[196] V. Imaykin, A.I. Komech, H. Spohn, Scattering theory for a particle coupled to a scalar field, *Discrete Contin. Dyn. Syst.* **10** (2004), 387–396.

[197] A.I. Komech, N.J. Mauser, A.P. Vinnichenko, Attraction to solitons in relativistic nonlinear wave equations, *Russ. J. Math. Phys.* **11** (2004), 289–307.

[198] E. Kopylova, A. Komech, On asymptotic stability of moving kink for relativistic Ginzburg–Landau equation, *Comm. Math. Phys.* **302** (2011), 225–252.

[199] E. Kopylova, A. Komech, On asymptotic stability of kink for relativistic Ginzburg–Landau equations, *Arch. Ration. Mech. Anal.* **202** (2011), 213–245

[200] G.L. Lamb Jr., Elements of Soliton Theory, John Wiley & Sons, Inc., New York, 1980.

Adiabatic effective dynamics of solitons of Hamiltonian PDEs

[201] V. Bach, T. Chen, J. Faupin, J. Fröhlich, I. M. Sigal, Effective dynamics of an electron coupled to an external potential in nonrelativistic QED, *Ann. Henri Poincaré*, **14** (2013), 1573–1597.

[202] S. Demoulini, D. Stuart, Adiabatic limit and the slow motion of vortices in a Chern–Simons–Schrödinger system, *Comm. Math. Phys.*, **290** (2009), 597–632.

[203] J. Fröhlich, T.-P. Tsai, H.-T. Yau, On the point-particle (Newtonian) limit of the nonlinear Hartree equation, *Comm. Math. Phys.*, **225** (2002), 223–274.

[204] J. Fröhlich, S. Gustafson, B. L. G. Jonsson, I. M. Sigal, Solitary wave dynamics in an external potential, *Comm. Math. Phys.*, **250** (2004), 613–642.

[205] A. Komech, M. Kunze, H. Spohn, Effective dynamics for a mechanical particle coupled to a wave field, *Comm. Math. Phys.*, **203** (1999), 1–19.

[206] M. Kunze, H. Spohn, Adiabatic limit for the Maxwell–Lorentz equations, *Ann. Henri Poincaré*, **1** (2000), 625–653.

[207] E. Long, D. Stuart, Effective dynamics for solitons in the nonlinear Klein–Gordon–Maxwell system and the Lorentz force law, *Rev. Math. Phys.*, **21** (2009), 459–510.

[208] D. Stuart, Existence and Newtonian limit of nonlinear bound states in the Einstein–Dirac system, *J. Math. Phys.*, **51** (2010), 032501, 13.

Global attraction to stationary orbits for Hamiltonian PDEs

[209] A. Comech, On global attraction to solitary waves. Klein–Gordon equation with mean field interaction at several points, *J. Differential Equations* **252** (2012), 5390–5413.

[210] A. Comech, Weak attractor of the Klein–Gordon field in discrete space-time interacting with a nonlinear oscillator, *Discrete Contin. Dyn. Syst.* **33** (2013), 2711–2755.

[211] A. Comech, Solutions with compact time spectrum to nonlinear Klein–Gordon and Schroedinger equations and the Titchmarsh theorem for partial convolution, *Arnold Math. J.* **5** (2019), 315–338.

[212] A.I. Komech, On attractor of a singular nonlinear U(1)-invariant Klein–Gordon equation, in *Progress in analysis, Vol. I, II (Berlin, 2001)*, World Sci. Publ., River Edge, NJ, 2003, 599–611.

[213] A.I. Komech, A.A. Komech, On the global attraction to solitary waves for the Klein–Gordon equation coupled to a nonlinear oscillator, *C. R. Math. Acad. Sci. Paris* **343** (2006), 111–114.

[214] A.I. Komech, A.A. Komech, Global attractor for a nonlinear oscillator coupled to the Klein–Gordon field, *Arch. Ration. Mech. Anal.* **185** (2007), 105–142.

[215] A.I. Komech, A.A. Komech, On global attraction to solitary waves for the Klein–Gordon field coupled to several nonlinear oscillator, *J. Math. Pures Appl. (9)* **93** (2010), 91–111.

[216] A.I. Komech, A.A. Komech, Global attraction to solitary waves for Klein–Gordon equation with mean field interaction, *Ann. Inst. H. Poincaré Anal. Non Linéaire* **26** (2009), 855–868.

[217] A.I. Komech, A.A. Komech, Global attraction to solitary waves for a nonlinear Dirac equation with mean field interaction, *SIAM J. Math. Anal.* **42** (2010), 2944–2964.

[218] E. Kopylova, Global attraction to solitary waves for Klein–Gordon equation with concentrated nonlinearity, *nonlinearity* **30** (2017), no. 11, 4191–4207.

[219] E. Kopylova, A.I. Komech, On global attractor of 3D Klein–Gordon equation with several concentrated nonlinearities, *Dynamics of PDE*, **16** (2019), no. 2, 105–124.

[220] E. Kopylova, A.I. Komech, Global attractor for 1D Dirac field coupled to nonlinear oscillator, *Comm. Math. Phys.* **375** (2020), no. 1, 573–603. Open access. http://link.springer.com/article/10.1007/s00220-019-03456-x

[221] A.A. Komech, A.I. Komech, A variant of the Titchmarsh convolution theorem for distributions on the circle, *Functional Analysis and Its Applications* **47** (2013), no. 1, 21–26.

[222] V. Imaikin, A.I. Komech, H. Spohn, Rotating charge coupled to the Maxwell field: scattering theory and adiabatic limit, *Monatsh. Math.* **142** (2004), 143–156.

Index